5th International Conference on Mechanics and Mechatronics (ICMMR 2018)

IOP Conference Series: Materials Science and Engineering
Volume 417

Tokyo, Japan
19 – 21 July 2018

ISBN: 978-1-5108-7364-3
ISSN: 1757-8981

Printed from e-media with permission by:

Curran Associates, Inc.
57 Morehouse Lane
Red Hook, NY 12571

CURRAN ASSOCIATES INC.
proceedings.com

Some format issues inherent in the e-media version may also appear in this print version.

Printed by Curran Associates, Inc. (2018)

Institute of Physics
Dirac House, Temple Back
Bristol BS1 6BE UK

Phone: 44 1 17 929 7481
Fax: 44 1 17 920 0979

techtracking@iop.org

Additional copies of this publication are available from:

Curran Associates, Inc.
57 Morehouse Lane
Red Hook, NY 12571 USA
Phone: 845-758-0400
Fax: 845-758-2633
Email: curran@proceedings.com
Web: www.proceedings.com

Table of contents

Volume 417

5th International Conference on Mechanics and Mechatronics Research (ICMMR 2018)

19–21 July 2018, Tokyo, Japan

Accepted papers received: 15 August 2018
Published online: 19 October 2018

Preface

Papers

Materials Science and Engineering

Intelligent robot design and implementation

IOP Conf. Series: Materials Science and Engineering **417** (2018) 011001 doi:10.1088/1757-899X/417/1/011001

PREFACE

The 5th International Conference on Mechanics and Mechatronics Research (ICMMR 2018) was held at Morito Memorial Hall of Tokyo University of Science, Japan during July 19-21, 2018.

The conference has received 80 papers, 42 papers are accepted for publication. The participants are from Japan, USA, France, China, Korea, Indonesia, Ecuador, Iran, Pakistan, Kazakhstan, Malaysia, etc. The conference program included keynote, oral and poster presentations from scholars working in the areas to establish platforms for collaboration. This conference covered recent trends and progresses made in the field of mechanics and mechatronics research.

The objective of the conference is to bring academia, industry together to exchange and share their experiences and research results, and discuss the practical challenges encountered and the potential solutions.

International scientific activities, like conference, symposium and workshop are big scientific platforms for knowledge exchange among the scientists from all over the world. International scientific activities are the best opportunities for the students, researchers and engineers to interact with the experts and specialists to get their advice or consultation on technical matters. The theme of this conference was the key issues associated with science and technology in this rapidly evolving field of research and to promote contact between basic researches and technological needs for real advanced mechanics and mechatronics.

We express our deepest gratitude to all members of the scientific committee, invited speakers, reviewers, sponsors, participants in the conference and members of the ICMMR organizing committee for their valuable contribution to the successful organization of ICMMR 2018.

Conference Committee Chair
Prof. Shih-Chieh Lin
National Tsing-Hua University, Taiwan
July, 2018

Committee

Conference Committee Chairs

Prof. Kazuo Umemura, Tokyo University of Science, Japan
Prof. Shih-Chieh Lin, National Tsing-Hua University, Taiwan
Prof. Sabri Cetin, University of Illinois at Chicago, USA

Program Committee Chairs

Prof. Arunachala Kannan, Arizona State University, USA
Assoc. Prof. Guocheng Lv, China University of Geosciences, China

International Technical Committees

Prof. Jian-Gang Guo, Tianjin University, China
Prof. Livija Cveticanin, Faculty of Technical Sciences, Serbia
Prof. Zhuo Wang, Harbin Engineering University, China
Prof. Belkacem OULD BOUAMAMA, Lille University, France
Prof. Jung-Min Yang, Kyungpook National University, South Korea
Prof. Dr.Evgenia Kirillova, RheinMain University of Applied Sciences, Germany
Prof. S.M Sapuan, Universiti Putra malaysia, Malaysia
Prof. Ch. Sanjay, GITAM University Hyderabad, India
Assoc.Prof.Shu-San Gan, Petra Christian University, Indonesia
Assoc. Prof. Li-Jun Zhou, Tianjin University of Technology and Education, China
Assoc. Prof. Kuo-Chih Chuang, Zhejiang University, China
Assoc. Prof. Willyanto Anggono, Petra Christian University, Indonesia
Assoc. Prof. Hamdan Haji Ya, Universiti Teknologi PETRONAS (UTP), Malaysia
Assoc. Prof. Kaykhosrow Khojier, Islamic Azad University, Iran
Assoc. Prof. Hilton Ahmad, Tun Hussein Onn University of Malaysia
Assoc. Prof. Rajkumar Durairaj, Universiti Tunku Abdul Rahman
Assoc. Prof. Megat Ahmad Kamal Megat Hanafiah, Universitit Teknologi MARA Cawangan Pahang
Assoc. Prof. Saravanan Pichiah, Indian Institute of Technology, India
Asst. Prof. Chih-Ling Huang, Kaohsiung Medical University, Taiwan
Dr. Sammy V. Militante, University of Antique, Philippines
Dr. S Krishnakumar, Defence Research and Development Organisation, India
Dr. Grzegorz Ilewicz, University of Bielsko-Biała, POLAND
Dr. Zheng Wei, Beijing University of Chemical Technology, China
Dr. Reza Rashidi, State University of New York, Alfred State College, USA
Dr.Sumit Desai, Veermata Jijabai Technological Institute, India
Maria Shevtsova, RheinMain University of Applied Sciences, Germany
Uğur Arabacı, Gazi University, Ankara, Turkey
Ervina Efzan Mhd Noor, Multimedia University, Malaysia
Atilla Altintas, Chalmers University of Technology, Sweden
Khadijah Khalid, Universitit Teknologi MARA Cawangan Pahang, Malaysia

Peer review statement

All papers published in this volume of *IOP Conference Series: Materials Science and Engineering* have been peer reviewed through processes administered by the proceedings Editors. Reviews were conducted by expert referees to the professional and scientific standards expected of a proceedings journal published by IOP Publishing.

IOP Conf. Series: Materials Science and Engineering **417** (2018) 012001 doi:10.1088/1757-899X/417/1/012001

Comparison Sound Absorption Methods of Multilayer Micro-perforated Panels

Shen Min and Bai Cong

School of Mechanical Engineering and Automation, Wuhan Textile University, 1Fangzhi Road, Wuhan, Hubei Province, 430073, China

min_shen18@163.com

Abstract. Micro-perforated panel (MPP) absorbers as the next-generation sound absorbers are getting more widely use in aviation, automobiles, construction and other fields. It's a complex problem to predict the acoustic performance using surface impedance methods. The acoustic absorption performance of micro-perforated panels were compared in terms of Transfer matrix method and Equivalent electric circuit approach (EECA). The calculated results show that transfer matrix method is more convenient and effective than the equivalent circuit approach especially to multiply layer micro-perforated absorbers.

1. Introduction

Micro-perforated panels (MPP) has been widely used in harsh and corrosive environment. The diameter of MPP is sub-millimeter providing enough acoustic resistance and low acoustic [1-2]. Ma[3] has developed a theoretical model for circular cross-sectional MPP structures. A single-layer mechanical impedances structure combining a MPP exhibits a sound peak at the low frequency. However, the absorption frequency band is narrow for a single layer MPP.

Most researchers [4-6] focused on estimate the absorption performances of MPP absorbers by equivalent electro-acoustic approach (EECA). By means of the circuit analysis, the resultant surface acoustic impedance is derived, if there is multiple layer perforated panel system, the circuit analysis becomes very complicated. The transfer matrix method [7-8] has become very popular for multi-layer absorber's analysis. This paper predicted the sound absorption performance of by transfer matrix method comparison with EECA. The results show that the transfer matrix method is more simple and effective than EECA for multiple layer micro-perforated absorbers.

2. Transfer matrix method

Let us consider a rigid single layer micro-perforated panel backing with an air-space which makes a simple resonator as shown in Figure 1. For the case of a perforated panel with grazing flow, there is a impedance model could be expressed in terms of the porosity, thickness, hole diameter and frequency, well established by Rao and Munjal[9] as follows

$$\zeta = \left[7.337\times10^{-3} + j\delta_x 2.2245\times10^{-5}(1+51t)(1+204d)f\right]/\phi \tag{1}$$

where ϕ is porosity, t panel thickness in mm, d hole diameter in mm, f frequency in Hz, and M Mach number.

According to the Chang et al. [10] prediction the absorption coefficients compared with experimental results for correction the factor. Therefore, impedance of the perforated panel could be expressed as:

$$\zeta = \left[7.337\times10^{-3} + j\delta_x 2.2245\times10^{-5}(1+51t)(1+204d)f\right]/\phi \tag{2}$$

For micro-perforated absorbers, it is not necessary to provide extra fibrous or porous materials.

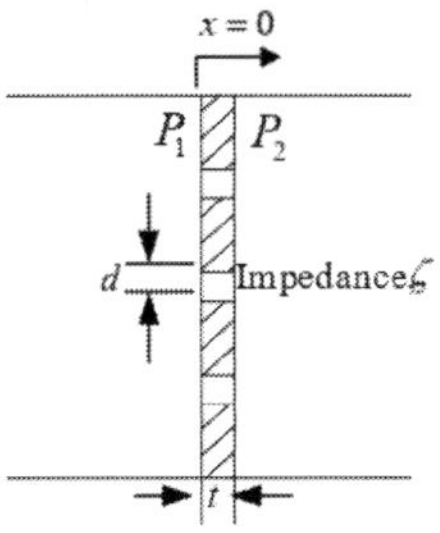

Figure 1. Perforated panel and

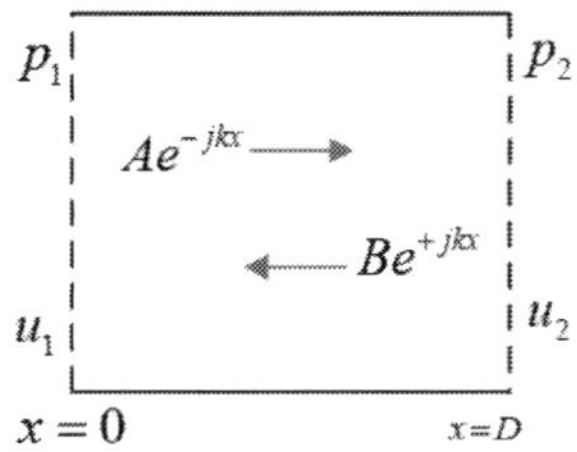

Figure 2. Four-pole parameters of an acoustic cavity

Figure 2. shows an acoustic cavity with incidence pressures p_1 and particle vibration velocity u_1 and reflection pressure p_2 , air particle vibration velocity u_2 . The depth of the cavity is D. The relation for the pressure and particle velocity at the incidence and reflection of element is expressed as the unit transfer matrix as follows:

$$\begin{bmatrix} p_1 \\ u_1 \end{bmatrix} = \begin{bmatrix} E_{11} & E_{12} \\ E_{21} & E_{22} \end{bmatrix} \begin{bmatrix} p_2 \\ u_2 \end{bmatrix} \tag{3}$$

where E_{11} , E_{12} , E_{21} and E_{22} are the four-pole parameters.

In the case of a perforated panel, the unit transfer matrix can be expressed by

$$[P] = \begin{bmatrix} P_{11} & P_{12} \\ P_{21} & P_{22} \end{bmatrix} = \begin{bmatrix} 1 & \rho c\zeta \\ 0 & 1 \end{bmatrix} \tag{4}$$

where ρ is the density of air, c the speed of sound in air and ζ is the normalized acoustic impedance of the panel defined by

$$\rho c\zeta = \Delta p / u \tag{5}$$

we can obtain the matrix elements as follow:

$$\begin{bmatrix} p_1 \\ u_1 \end{bmatrix} = \begin{bmatrix} \cos kD & j\rho c\sin kD \\ j\rho c\sin kD & \cos kD \end{bmatrix} \begin{bmatrix} p_2 \\ u_2 \end{bmatrix} \tag{6}$$

For an airspace of depth D , the unit transfer matrix can be expressed:

$$[S] = \begin{bmatrix} S_{11} & S_{12} \\ S_{21} & S_{22} \end{bmatrix} = \begin{bmatrix} \cos kD & (j/\rho c)\sin kD \\ (j/\rho c)\sin kD & \cos kD \end{bmatrix} \tag{7}$$

where $j = \sqrt{-1}$ and k is the wave number.

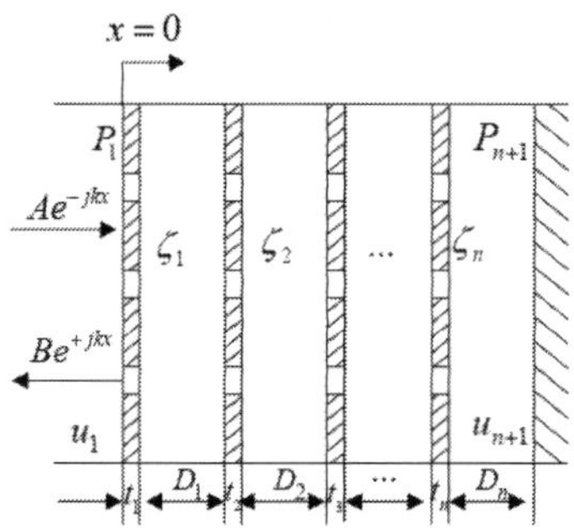

Figure 3. Configuration of multi-layer micro-perforated panel

To broaden the range frequency of absorption, it is effective to use more layers of micro-perforated panels. Figure 3. shows the multiple layer micro-perforated panels backing with associated cavities.

The overall transfer matrix $[T]$ for a multi-layer perforated panel system can be obtained by multiplying all the unit transfer matrices as follows:

$$[T]=\begin{bmatrix} T_{11} & T_{12} \\ T_{21} & T_{22} \end{bmatrix}=[P]_1[S]_1\cdots[P]_n[S]_n \tag{8}$$

Then the state variables on the surface of the left and panel, number 1, can be expressed in terms of the overall transfer matrix and the variables at the right end number $n+1$, as follows:

$$\begin{bmatrix} p_1 \\ u_1 \end{bmatrix}=\begin{bmatrix} \mathrm{T}_{11} & \mathrm{T}_{12} \\ \mathrm{T}_{21} & \mathrm{T}_{22} \end{bmatrix}\begin{bmatrix} p_{n+1} \\ u_{n+1} \end{bmatrix} \tag{9}$$

Since the particle velocity $u_{n+1}=0$ on a rigid plate, the acoustic pressure reflection coefficient R can be expressed by the transfer matrix elements as

$$R=\left(T_{11}-\rho cT_{21}\right)/\left(T_{11}+\rho cT_{21}\right) \tag{10}$$

For normal incidence, the sound absorption coefficient is defined by

$$\alpha=1-|R|^2 \tag{11}$$

3. Equivalent Electric Circuit Approach

In acoustic system analysis, the equivalent electric circuit approach (ECA) is one of the most popular approaches. The ECA method is very useful to analyse the acoustic propagation through a series parallel assembly acoustic perforation elements. Ma [3] found a simple approximate formula for the acoustic impedance of the micro-perforated panel Z_{panel}, which is:

$$R=32\eta t\left(\sqrt{1+x^2/32}+\sqrt{2}xd/8t\right)/\phi d^2 \tag{12}$$

$$M=t\rho\left(1+1/\sqrt{9+x^2/2}+0.85d/t\right)/p \tag{13}$$

$$x=\left(d\sqrt{\rho\omega/\eta}\right)/2 \tag{14}$$

$$Z_{panel}=R+j\omega M \tag{15}$$

where η is dynamic viscosity of air($=17.9\mu Pa\ s$), ρ is air density, c the sound speed in air and ω the angular frequency of the sound, ϕ the perforation ratio, d the hole diameter, t the panel thickness. R and M are acoustic resistance and acoustic mass, respectively.
The impedance of airspace is

$$Z_D=-j\rho c\cot(\omega D/c) \tag{16}$$

where D is the thickness of the airspace. The configuration of single layer micro-perforated panel backing with airspace can be described by the equivalent electric circuit shown in Figure 4.

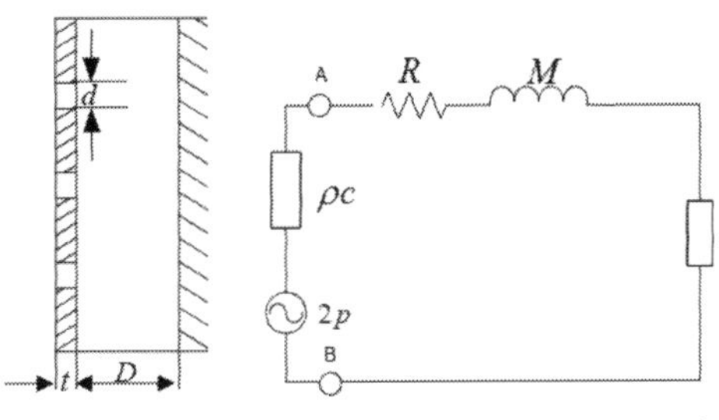

Figure 4. Perforated panel structure and equivalent

From the equivalent circuit, the resonance system contains the mass-resistance element of the panel in series with the cavity reactance, the resultant impedance Z of single layer absorber is:

$$Z=Z_{panel}+Z_D \tag{17}$$

IOP Conf. Series: Materials Science and Engineering **417** (2018) 012001 doi:10.1088/1757-899X/417/1/012001

Figure 5 shows the structure of double micro-perforated panel absorber and equivalent electrical circuit and the subscripts 1 and 2 indicate the two resonator elements.

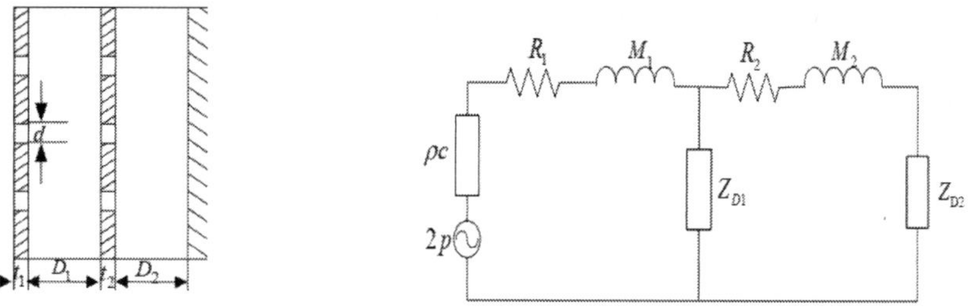

(a)Double layer perforated panel structure (b) Equivalent electric circuit

Figure 5. Configuration of double perforated panel and equivalent electric circuit

According to the equivalent circuit in Fig.5, the resultant acoustic impedance Z of the parallel circuit is

$$Z = R_1 + j\omega M_1 + \frac{(R_2 + j\omega M_2 + Z_{D2})Z_{D1}}{R_2 + j\omega M_2 + Z_{D2} + Z_{D1}} \tag{18}$$

where Z_{D1} and Z_{D2} are the acoustic impedances of the airspaces, respectively.

By substituting Eq.(18) to Eq.(12), the sound absorption coefficient of double layer MPM structure is obtained. All of the equations are solved by Matlab software since program becomes simply with complex number computation function.

4. Results and discussions

For comparing the effect of two methods, the normal incidence absorption coefficients are calculated using the transfer matrix(TFM) method and the equivalent electro-acoustic(ECA) method, respectively. In this part simulation, the parameter are as listed in Tabel.1

Table.1 Geometric parameter of the perforated panel

Parameter	Panel I	Panel II
Porosity ϕ	1%~3%	2%
Thickness h	2~6 mm	3mm
hole diameter d	0.8mm	0.8mm
Airspace thickness D	7mm	7mm

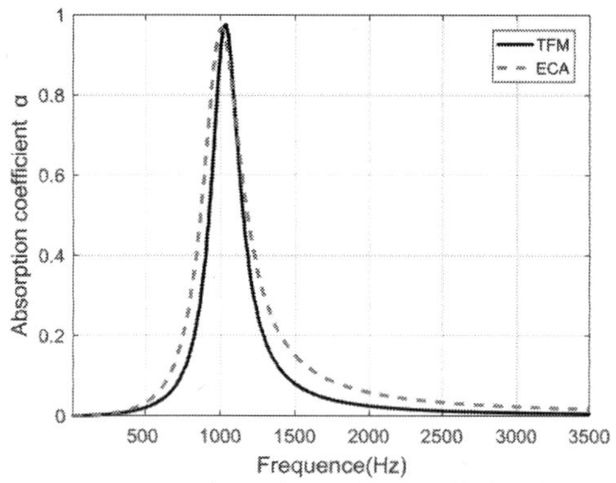

Figure 6. Comparison absorption coefficient by TFM method with ECA approach

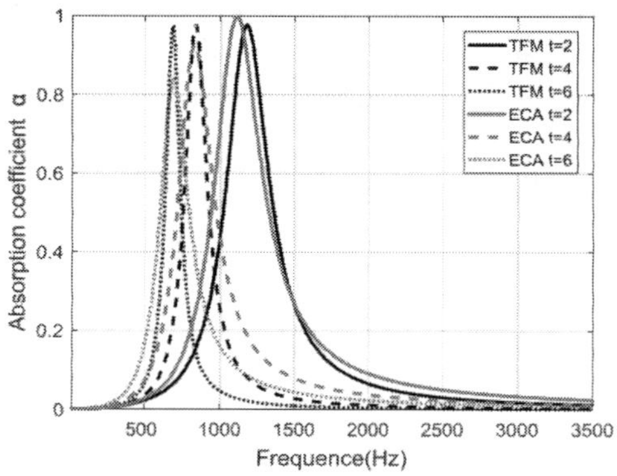

Figure 7. Variation of the panel thickness of the single layer MPP absorber

Figure 6. reveals that the results solved by transfer matrix method also agree well with those computed by electro-acoustic analogy approach. Its absorption performance is limited to the resonance frequency region. |Some more predictions are used to demonstrate the effect of important parameters.

Figure 7. shows the effects of the panel thickness on sound absorption coefficients of single layer micro-perforated panel. In this case, except for panel thickness t, all the parameters are the same as in the first case in previous case. It is seen from Fig.7 that reducing thickness of perforated panel increase the higher acoustic resonance frequency.

Figure 8. compares the sound absorption performance of single layer MPP by changing the porosity. In this case, except for panel porosity ϕ , all the parameters are the same as in the first case. It is seen from Fig.8 that the absorption coefficients calculated by transfer matrix method agree with the values by EECA method except the porosity over the 1%. The lower porosity of the micro-perforated panel has a lower acoustic resonance frequency and higher acoustic absorption performance. This is because the acoustic impedance of the panel increases with inversion proportion to the porosity, decrease in porosity results in increase of impedance of the system.

In order to make an efficient sound absorbing performance, double layer micro-perforated absorber is used, where each compartment is composed of one layer micro-perforated panel and airspace. Herein, the normal incidence absorption coefficients of the double layer micro-perforated panel absorber are discussed.

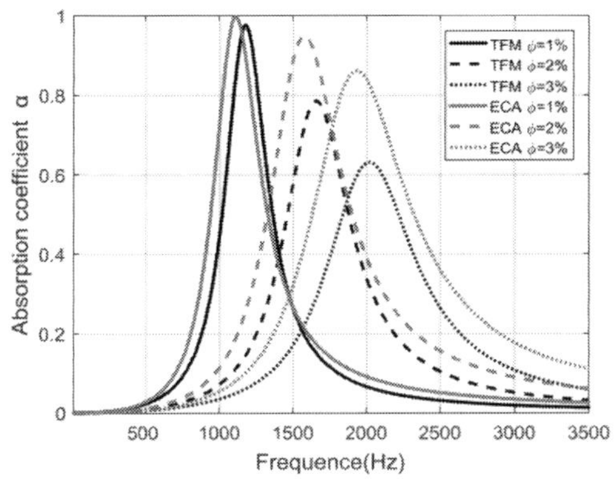

Figure 8. Variation of the porosity of the single layer MPP absorber

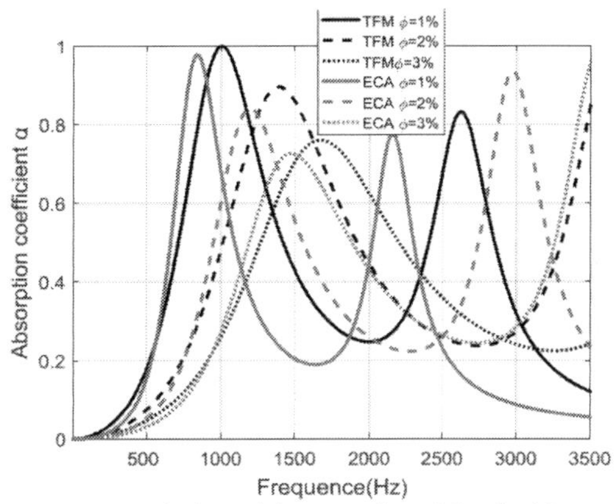

Figure 9. Variation of the porosity of the double layer MPP absorber

Figure 9. reveals that the acoustic resonance frequencies of the double layer absorbers are more broadband than those of the single layer absorbers. The highest acoustic resonance frequencies of those double layer absorbers are mostly higher than single layer compartments. The reason is that when the incident frequency is equal to the acoustic resonance frequency of the double layer structure, the air medium existing in the absorber will vibrate with the incident sound and the incident sound energy will be largely absorbed. They are suitable for any reverberation or noise control applications.

5. Conclusions

In this paper, the applicability and accuracy of the acoustic absorption performance analysis by transfer matrix method have been successfully demonstrated. As expected, the micro-perforated constructions can supply highly efficient sound absorbing performance without any porous materials. Besides, several the influence factors on sound absorption performance are also discussed include the thickness, holes diameter, porosity of the perforated panel and length of airspace. This provides a reliable reference for the design structure of the multi-layer micro-perforated panel absorbers.

Acknowledgments

This work was financially supported by the Natural Science Foundation of Hubei Province of China (No.2014CFB766), China National Science Foundation (No.51505344), Scientific Research Foundation for Returned Scholors Ministry of Education of China and Hubei Provincial Education department outstanding young and middle-aged scientific and technological innovation team project (No.T201607).

References

[1] Liu Z Q, Zhan J X, Mohammad F, John L D, Acoustic properties of multilayer sound absorbers with a 3D printed micro-perforated panel, Applied Acoustics, 2017, 121:25-72.

[2] Ning J F, Ren S W, Zhao GP, Acoustic properties of micro-perforated panel absorber having arbitrary cross-sectional perforations, Applied Acoustics, 2016, 11:135-142.

[3] Ma D Y, General theory and design of microperforated –panel absorbers, China J Acoust 1997, 3:193-202.

[4] Pieren R, Kurt H, Predicting sound absorption coefficients of lightweight multilayer curtains using equivalent circuit method, Applied Acoustics, 2015, 92: 27-41.

[5] Shen M, Nagamura , Nakagawa N, Okamura M, Design of compact micro-perforated membrane absorbers for polycarbonate pane in automobile, Applied Acoustics, 2013,74(4):622-627.

[6] Zhao X D, Yu Y J, Wu Y J, Improving low-frequency sound absorption of micro-perforated panel absorbers by using mechanical impedance plate combined with Helmholtz resonators, Applied Acoustics, 2016, 114:92-98.

[7] Kevin V, Raymond P, and Elkoun S, Transfer matrix method applied to the parallel assembly of sound absorbing materials, J.Acoust.Soc.Am.2013, 134(6):4648-4658.

[8] Zhao X D, Wang X, Yu Y J, Enhancing low-frequency sound absorption of micro-perforated panel absorbers by combining parallel mechanical impedance, Applied Acoustics, 2018,130:300-304.

[9] Rao K N, Munjal M L, Experimental evaluation of impedance of perorated with grazing flow, Journal of Sound and Vibration, Vol.108, No.2,(1986),283-295.

[10] Chang D Q, Lu F Jin W N and Liu B L, Low frequency sound absorptive properties of double-layer perforated plate under grazing flow, Applied Acoustics, 130:115-123.

Reflection Coefficients of Plane Waves at a Seawater-magnetorheological Fluid Interface

Min Shen[1] **and Bai Cong**[2]

School of Mechanical Engineering and Automation, Wuhan Textile University, 1Fangzhi Road, Wuhan, Hubei Province, 430073, China.

min_shen18@163.com

Abstract. The Magnetorheological fluid (MRF) is a class of smart material with tunable acoustic properties and could be used as active sound coating material on underwater structure. This paper presents an acoustic model to investigate the reflection of a plane wave incidence at seawater-MRF interface. The acoustic model is developed by modelling the MRF as Biot's type porous material. In this paper, the effect of three major parameters of MRF material on the reflection coefficients are introduced. It is concluded that reflection coefficients can achieve low values by increasing shear modulus, particle volume fraction and decrease permeability.

1. Introduction

Magnetrohelogical fluids (MRF) consist of micrometric magnetic particles suspended in a viscous or viscoelastic fluid[1]. The MR fluids exhibit variable acoustic and rheology properties by applying an external magnetic fields. Recently, MRF materials are proposed to as active sound barriers or acoustical metamaterial [2-3]. Submarine are often covered with sound absorbing or perforated materials called anechoic coatings. Hence, the MRF material can be used as active sonar anechoic material in submarines.

The MRF behaves as a Newtonian fluid in the absence of magnetic field and the particles are scattered randomly in the liquid carrier. The fluid shows a non-Newtonian behaviour in the presence of an external magnetic field and the dispersed magnetic particles make chain-like structures that align in the direction of the applied field. The transition between these two phases is highly reversible, which provides a unique feature of magnetic field controllability of the acoustic properties of the MRF. Hence, the structure of the MRF is considered as a skeletal frame filled with a base fluid, the acoustic mechanics of MRF is complicate because MRF is coupled multiphase medium: the solid phase of ferromagnetic grain and the fluid phase of the pore fluid.

A series of acoustic experiments have been conducted for comparing sound speed in MRF. Lo´pez et al.[4] have studied the effect of particle volume fraction and magnetic field intensity on the velocity and attenuation in MRF using ultrasonic techniques. Bramantya et al.[5] have experimentally studied the sound speed in some commercial MRF as function of the anisotropy, intensity and uniformity of the applied field. Shen et al.[6] have theoretically analysed the dependence of magnetic fields strength on sound velocity and attenuation in MRF based on Biot's model.

In order to reduce the reflection of underwater structure, it is necessary to study the reflectivity of MRF material. However, very limited number of researches has so far been studied underwater acoustic properties of MRF as active underwater acoustic coating. Hence, this paper introduced the reflection coefficients of plane wave incidence at the interface of seawater and MRF material.

2. Acoustic model of plane wave incident at a seawater-MRF interface

The particles in MRF suspension are interconnected and small cavities generated in the presence of a magnetic field. Hence, the MRF is considered as a porous material and composed of the skeletal frame filled with silicon oil fluid. It suggests an attempt of modelling its mechanical behaviour by the Biot[7-8] theory for porous elastic media saturated by a compressional fluid.

According to Biot's theory[7-8], the wave propagating in the porous material has three independent components: a shear wave and two dilatational waves. The derivation of sound speed propagation in porous material can be found in several papers [9-10]. Firstly, the fast V_{p1} and slow dilatational wave velocities V_{p2} and shear wave velocities V_s have been calculated based on Biot's theory.

The wave number of fast dilatational wave k_1 and the slow dilatational wave k_2 can be written by

$$k_1 = \omega / V_{p1} \quad \text{and} \quad k_2 = \omega / V_{p2} \tag{1}$$

The wave number of shear wave k_t

$$k_t = \omega / V_s \tag{2}$$

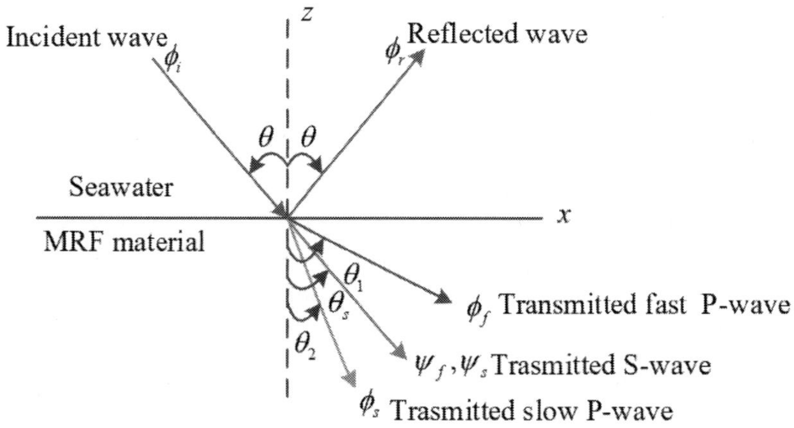

Figure 1. Reflection and refraction of plane Wave at seawater-MRF interface

We consider a plane wave incident at an angle θ to the steady, laminar and incompressible flows of MRF and seawater at $z = 0$, where the externally imposed magnetic field may be spatially non-uniform as shown in Figure 1. There exit incident (ϕ_i) and reflected waves (ϕ_r) in the seawater, and three refracted waves that the fast P-wave (ϕ_f) and the slow P-wave (ϕ_s) of dilatational waves, and the shear wave (ψ_s)in the MRF material.

In a two-component material, such as seawater-MRF material, two vectors, u and U represents the displacements of the frame and fluid, respectively.

$$u = \nabla \phi_s + rot\psi_s \tag{3}$$

$$U = \nabla \phi_f + rot\psi_f \tag{4}$$

The fluid displacement relative to the frame is introduced as:

$$w = -\beta(U - u) \tag{5}$$

where β is porosity.

The incident (ϕ_i) and reflected waves (ϕ_r) in the seawater will have the displacement potentials

$$\phi_i = A_i \exp[i(\omega t - k_w \cos\theta . z - k_w \sin\theta . x)] \tag{6}$$

$$\phi_r = A_r \exp[i(\omega t - k_w \cos\theta . z - k_w \sin\theta . x)] \tag{7}$$

where $k_w = \omega / c_w$, c_w is the sound velocity in seawater. A_i is the amplitude of incident wave, A_r is the amplitude of reflected wave and three refracted wave potentials

In the MRF material, the scalar and vector potentials defined in Eqs.(3) and (4) are

$$\phi_s = A_{s1} \exp[i(\omega t - k_{1z}.z - k_{1x}.x)] + A_{s2} \exp[i(\omega t - k_{2z}.z - k_{2x}.x)] \quad (8)$$

$$\phi_f = A_{f1} \exp[i(\omega t - k_{1z}.z - k_{1x}.x)] + A_{f2} \exp[i(\omega t - k_{2z}.z - k_{2x}.x)] \quad (9)$$

$$\psi_s = A_{st} \exp[i(\omega t - k_{tz}.z - k_{tx}.x)] \quad (10)$$

$$\psi_f = A_{ft} \exp[i(\omega t - k_{tz}.z - k_{tx}.x)] \quad (11)$$

where $k_{nx} = k_n \sin\theta$ ($n = 1,2,t$), k_{1z} and k_{2z} are z direction component of the fast and slow longitudinal wavenumbers, k_{1x} and k_{2x} are x direction component of the fast and slow longitudinal wavenumbers, k_{tx} are x direction component of the shear wavenumbers. A_{s1} is the refraction coefficient of fast dilatational wave, A_{s2} is the refraction coefficient of slow dilatational wave and A_{st} is the shear wave refraction coefficient in porous solid respectively. A_{f1} is the fast dilatational wave, A_{f2} is the slow dilatational wave and A_{ft} is the shear wave reflection coefficient in porous fluid medium, respectively.

The following four boundary conditions are required at a seawater-MRF material interface [11]

(1)For continuity of fluid movement

$$\frac{\partial\phi_i}{\partial z} + \frac{\partial\phi_r}{\partial z} = \frac{\partial\phi_s}{\partial z} + \frac{\partial\psi_s}{\partial x} - \frac{\partial\phi_f}{\partial z} - \frac{\partial\phi_f}{\partial x} \quad (12)$$

(2)For equilibrium of normal traction

$$H(\frac{\partial^2\phi_s}{\partial x^2} + \frac{\partial^2\phi_s}{\partial z^2}) - 2\mu(\frac{\partial^2\phi_s}{\partial x^2} + \frac{\partial^2\phi_s}{\partial x\partial z}) - c_i(\frac{\partial^2\phi_f}{\partial x^2} + \frac{\partial^2\phi_f}{\partial z^2}) = \rho_f(\frac{\partial^2\phi_i}{\partial t^2} + \frac{\partial^2\phi_r}{\partial t^2}) \quad (13)$$

(3)For equilibrium of fluid pressure

$$M(\frac{\partial^2\phi_f}{\partial x^2} + \frac{\partial^2\phi_s}{\partial z^2}) - c_i(\frac{\partial^2\phi_f}{\partial x^2} + \frac{\partial^2\phi_f}{\partial z^2}) = -\rho_f(\frac{\partial^2\phi_i}{\partial t^2} + \frac{\partial^2\phi_r}{\partial t^2}) \quad (14)$$

where $M = k_r^2 / (D_1 - k_b)$

(4)For equilibrium of tangential traction

$$2\mu\frac{\partial^2\phi_s}{\partial x\partial z} - \mu(\frac{\partial^2\psi_f}{\partial z^2} - \frac{\partial^2\psi_f}{\partial x^2}) = 0 \quad (15)$$

When considered four boundary conditions at the interface $z = 0$, the following four linear equations can be obtained,

$$\begin{bmatrix} a_{11} & a_{12} & a_{13} & a_{14} \\ a_{21} & a_{22} & a_{23} & a_{24} \\ a_{31} & a_{32} & a_{33} & a_{34} \\ a_{41} & a_{42} & a_{43} & a_{44} \end{bmatrix} \begin{bmatrix} A_r \\ A_1 \\ A_2 \\ A_t \end{bmatrix} = \begin{bmatrix} 1 \\ -\rho_f \\ \rho_f \\ 0 \end{bmatrix} \quad (16)$$

A_r is the complex amplitude of seawater and MRF material. Therefore, the reflection coefficient $R = |A_r / A_i|$ can be obtained by solving Eq.(16), once A_i is specified. The solution of reflection coefficients can be obtained straight by applying Grammer's rule. The components of Eq.(16) are listed in Appendix A.

3. Results and Discussion

In this section, the reflection coefficients at seawater-MRF interface are derived. The MRF is named MRHCCS4-A manufactured by the UK LIQUID REASEARCH company. The parameters of the model for the MRF are listed in Table 1. All parameters of MRF in braces are either directly supplied by the manufacturer or have been derived by Carmignani et al.[12]. For the MRF description, the fluid bulk modulus can be assumed elastic, because in our case the base fluid is low-viscosity silicone oil.

Table 1. The Parameters of MRHCCS4-A by UK LIQUID REASEARCH Corporation

Parameters	Symbol(Unit)	MRF
Solid density	$\rho_r(\mathrm{kg}/m^3)$	2980
Fluid density	$\rho_f(\mathrm{kg}/m^3)$	820
Grain bulk modulus	$K_r'(N/m^2)$	1.26×10^{10}
Fluid bulk modulus	$K_f(N/m^2)$	2.6×10^{8}
Grain diameter	$d_m(\mu\mathrm{m})$	$3\sim10$
Fluid viscosity	$\eta(Pa/s)$	0.013
Frame shear modulus	$G_b'(N/m^2)$	$0\sim7\times10^{5}$
Shear specific loss	δ_{Gb}	$0\sim0.5$
Frame bulk modulus	$K_b'(N/m^2)$	$0\sim4\times10^{8}$
Volumetric specific loss	δ_{Kb}	$0\sim0.5$
Particle volume fraction	ϕ	$0.01\sim0.35$
Structure factor	α	1.4
Permeability	$\kappa(m^2)$	4×10^{-12}

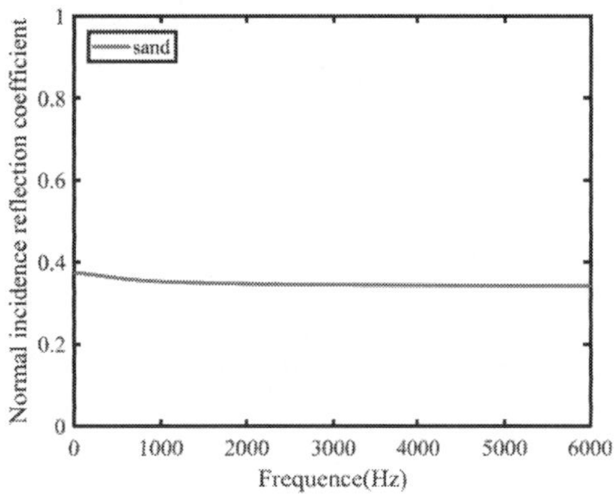

Figure 2. Normal reflection coefficient as a function of frequency for the sand

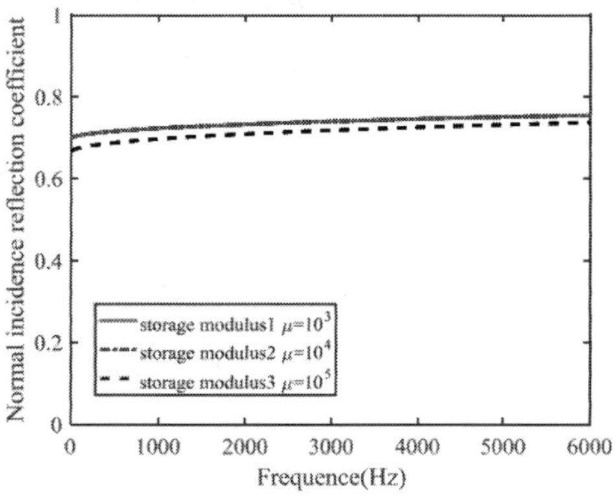

Figure 3. Normal reflection coefficient as a function of frequency for three different storage modulus μ

To veryfine the correctiness of this model, the reflection coefficients at a seawater-sediment interface for a typical kinds of marine sediment sand is shown in Figure 2. The physical parameters of sand were taken from Kimura[11]. In this calculation, sound velocity in seawater is assumed to be 1500m/s. In order to analyse the effects of different frequency, we suppose that the plane wave is normal incidence.

Figure 3. shows the magnitude of the reflection coefficient as the frequency for three different values of the storage modulus $\mu=10^3$, $\mu=10^4$ and $\mu=10^5$. The intensity of the magnetic field is 100mT, 200mT and 500mT, respectively. It is seen from the Figure 3. that the reflection coefficients are obviously improved when the shear modulus increase from 10kPa to 100kPa. The reason is that an external magnetic field induces chain-like particulate structures in the direction of the field. The new strcture alters the shear resistance and fluid viscosity.The fluid viscosity and shear modulus increase with the increasing of the strength of magnetic field. The more sound energy is dissipated by viscoelastic properties and fluid viscosity. Hence, the reflection coefficient is greatly reduced by applying a small strength of magnetic field.

Figure 4. shows the magnitude of the reflection coefficient as the frequency for three different values of the particle volume fraction ϕ=0.1 , ϕ=0.2 and ϕ=0.3, respectively. The particle volume fraction is a parameter related to the suspension itself and independent of the external field strength. It is seen from Figure 4. that the reflection coefficients decrease with the increasing of the volume fraction. Because the iron particle is more connected formed a frame with high volume fraction(ϕ=0.3), which is weak-link regime. The contribution of the solid part is more important than fluid when the particle volume fraction is increased.

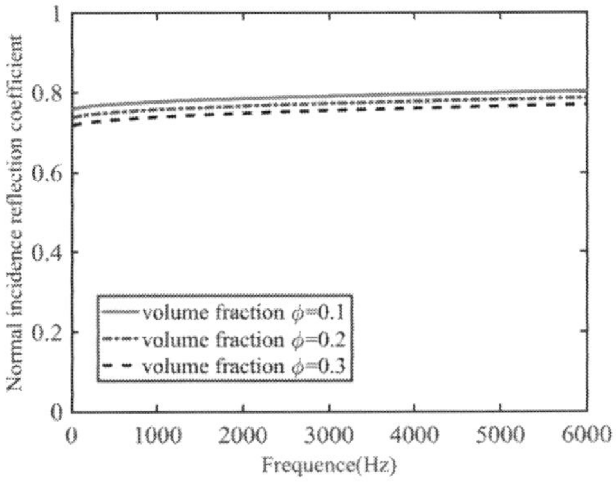

Figure 4. Reflection coefficient as a function of frequency for three different particle volume fraction ϕ.

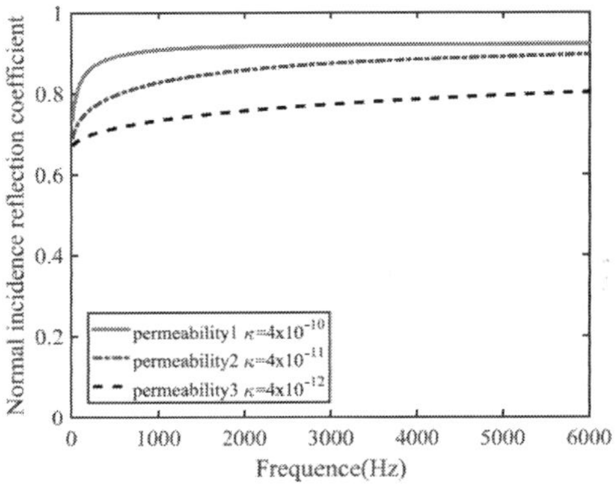

Figure 5. Reflection coefficient as a function of Frequency for three different permeability κ.

Figure 5. shows the magnitude of the reflection coefficient as a function of the frequency for three different values of permeability κ=4×10^{-10} , $\kappa = 4\times10^{-11}$ and $\kappa = 4\times10^{-12}$. It is seen from Figure 5. that reflection coefficients are obviously decrease with the permeability decrease over a wide range of frequency. Permeability as a function of porosity, fluid viscosity and pore distribution. The permeability also is affected by imperfect pore structures and non-uniformity in particle geometry. It is a complicated parameter influenced the acoustic performance of MRF.

4. Conclusions

The MRF could be used as active sound coating material on underwater structure.In order to reduce the reflection of underwater structure, this paper presents an acoustic model to investigate the reflection of a plane wave incidence at seawater-MRF interface. We highlighted the effect of the major parameter on reflection coefficient in the presence of an external magnetic field.

(1)We find that applying appropriate magnetic density could improve the reflective coefficient of MRF and there is no obvious change when the strength of magnetic field is below 200mT.

(2) The reflection coefficients decrease with the increasing of particle fraction volume from 0.1 up to 0.3. The contribution of the solid part is more important than fluid if the value of particle volume fraction is above 0.3.

(3) The reflection coefficients are obviously decrease with the permeability decrease over a wide range of frequency.

Acknowledgments
This work was financially supported by the Natural Science Foundation of Hubei Province of China (No.2014CFB766), China National Science Foundation (No.51505344), Scientific Research Foundation for Returned Scholars, Ministry of Education of China.

References

[1] Ghaffari A, Hashemabadi S H, and Ashtiani M, A review on the simulation and modeling of magnetorheological fluids, Journal of Intelligent Material Systems and Structures, 26(2015)881-904.
[2] Szary M, The analytical model of rheological fluid for vibration and noise control, Proceedings of Meetings on Acoustics, Montreal, Canada, 19(2013)1-8.
[3] Donskoy D M and Malinovsky V S, Broadband acoustic metamaterials with electro-magnetically controlled properties, Proceedings of Meetings on Acoustics, Montreal, Canada, 19(2013)1-8.
[4] Lo′pez J R, Castro P, Elvira L and Espinosa F M D, Study of the effect of particle volume fraction on the microstructure of magnetorhelogical fluids using ultrasound: Transition between the strong-link to the weak-link regimes, Ultrasonics, 61(2015)10-14.
[5] Bramantya M A, Motozawa M and Sawada T, Ultrasonic propagation velocity in magnetic and magnetorheological fluids due to an external magnetic field, Journal of Physics:Condensed Matter, 22(2010)1-5.
[6] Shen M and Huang Q B, Acoustic velocity and attenuation coefficient of magnetorhelogical fluids under electromagnetic fields, Applied Acoustics, 107(2016)27-33.
[7] Biot M A, Theory of propagation of elastic waves in a fluid-saturated porous solid, Part I:Low frequency range, Journal of the Acoustical Society of America, 28(1956)168-178.
[8] Biot M A, Theory of propagation of elastic waves in a fluid-saturated porous solid, Part II:Higher frequency range, Journal of the Acoustical Society of America, 28(1956)179-191.
[9] Stoll R D and Bryan G M, Reflection of acoustic waves at a water-sediment interface, Journal of the Acoustical Society of America, 70(1981) 149-156.
[10] Wu K, Xue Q and Adler L, Reflection and transmission of elastic waves from a fluid-saturated porous solid boundary, Journal of the acoustical society of America, 1990,87:2349-2358.
[11] Kimura M, Reflection of plane acoustic waves at a Seawater-marine sediment interface, Journal Applied Physics, 35(1996) 2948-2951.
[12] Carmignani C, Forte P and Rustighi E, Design of a novel magneto-rheological squeeze-film damper, Smart Materials and Structures, 15(2006)164-170.

Appendix A

$a_{11}=1$; $a_{12}=(1-r_1)c_w\cos\theta_1/(c_1\cos\theta)$;

$a_{13}=(1-r_2)c_w\cos\theta_2/(c_2\cos\theta)$; $a_{14}=(1-r_3)c_w\sin\theta_t/(c_t\cos\theta)$;

$a_{21}=\rho_f$; $a_{22}=(r_1C-H)/c_1^2+2\mu\sin^2\theta/c_w^2$;

$a_{23}=(r_2C-H)/c_2^2+2\mu\sin^2\theta/c_w^2$ $a_{24}=-2\mu\sin\theta\cos\theta_t/c_wc_t$;

$a_{31}=-\rho_f$; $a_{32}=(C-Mr_1)/c_1^2$; $a_{33}=(C-Mr_2)/c_2^2$; $a_{34}=0$;

$a_{43}=2\sin\theta\cos\theta_2/c_wc_2$ $a_{41}=0$; $a_{42}=2\sin\theta\cos\theta_1/c_wc_1$; $a_{44}=\sin^2\theta/c_w^2-\cos^2\theta_t/c_t^2$;

Research on the Mechanical Properties of Hard Rock

Longyun ZHANG[1], Shucai LI[1],Shangyang YANG[2*], Qiangyong ZHANG[1]

[1]Geotechnical & Structural Engineering Research Center, Shandong University, Jinan, 250061, China
[2]Department of Physics and Mathematics, Shandong Jiaotong University, Jinan 250023, China

77959140@qq.com

Abstract: Large hydropower stations in the western region of China play very important roles in the development of lacal economy, and most of them locate in complex geological conditions, the deeper the rock is buried, the more obvious the geothermal phenomenon is, and the geothermal phenomenon does great damage for the underground construction. In order to study the mechanical properties of hard rock, some tests on granite have been accomplished. Through the uniaxial compression test, the triaxial compression test and the triaxial compression temperature creep test, some mechanical properties are obtained: (1) The peak strength of the granite is 92.9MPa in the uniaxial compression test, 291.0MPa in the triaxial compression test, 260MPa in the triaxial compression test when T=25℃, 220MPa in the triaxial compression test when T=50℃, 200MPa in the triaxial compression test when T=70℃. (2) Compared with the triaxial compression test, when the specimen is broken, the increased percentage of the axial strain of the temperature creep test is respectively 37% whenT=25℃,53.3% whenT=50℃,80.6% whenT=70℃. And the value of the lateral strain is respectively 20% whenT=25℃,59.5% whenT=50℃,101% whenT=70℃. (3) A threshold exists during the creep of the gneissic granite, which is affected by temperature. The temperature increases the creep deformation of the rock.

1. Introduction

Large hydropower stations in the western region of China play very important roles in the development of lacal economy, and most of them locate in complex geological conditions, where the magmatism is frequent and intrusion is developed. The surrounding rock of the diversion tunnels tends to be unstable under the magmatism, and the high temperature steam erupts from the rock fractures, which will further destabilize the rock mass, such as deformation increase and strength reduction.This phenomenon is known as "geothermal phenomenon". Generally speaking, the deeper the rock is buried, the more obvious the geothermal phenomenon is. Moreover, the geothermal phenomenon does great damage for the underground construction, and many scholors have dedicated to the study of failure mechanism caused by high temperature.Shao et al. [1]studied the effect of temperature on the mechanical behaviour of Strathbogie granite under unconfined stress conditions. The results showed that the increasing temperature reduces the stress thresholds for crack initiation and crack damage and extends the duration of stable crack propagation. Liu et al. [2] conducted the dynamic compressive tests of sandstone at seven temperatures in the range of room temperature to 1000 C and five impact velocities in the range of 11.0-15.0 m/s by using the SHPB test system improved with high temperature device, and the results showed that the temperatures the strain rates affectting dynamic compressive strength and peak strain most are 800℃ and 1000℃ respectively, while at 1000℃and room temperature the strain rates affect dynamic compressive strength and peak strain weakest.Yang et al.[3] carried out uniaxial compression tests to evaluate the effect of high temperature treatments (200, 300, 400, 500, 600, 700 and 800◦C) on the crack damage, strength and deformation failure behavior of a granite, and the results showed that the cracking process of granite depended on the heat treatment temperature. Ding et al. [4] studied the AE signal nonlinear characteristics of thermally damaged sandstone samples under uniaxial compression, and the results showed that after treatment at different temperatures the AE signals were positively correlated with the stress. Chen et al.[5] used uniaxial compression and three-point bending to measure the macroscopic mechanical properties of granite,which was exposed to high temperatures before measuring, and theresults indicated that the uniaxial compressive strength (UCS), Young's modulus and fracture toughness of the granite specimens decrease with an increase in heat treatment temperature up to 800 °C, above which there is no obvious change.Muhammad et al. [6] studied the effect of elevated temperatures on the mechanical properties of limestone, quartzite and granite concrete, which were subjected to temperatures ranging from 25 to 650 °C for a duration of 2 h, and the results indicated that the mechanical properties of concrete are largely affected from elevated temperatures and the type of coarse aggregate used.

So, how the deformation and strength will change on earth as the granite specimen creep under high temperature, high axial stress and high confining pressure. According to practical condition of engineering, high temperature creep tests on granite are done to research the characteristics of creep deformation and creep strength under the complex conditions of different high temperatures and different high stresses. So as to find out the failure mechanism.

2. High temperature creep test

2.1. Test Condition

Take an important large hydropower station with geothermal phenomenon for instance, the intrusion is composed of a gneissic granite, with hard quality and good integrity. The average teperature of the surrounding rock in diversion tunnel is over 75℃, and the maximum temperature even up to 110℃. Many tunnel sections arise high temperature steam, with gas temperature between 160℃ and 170℃.

The specimens used in this test are all taken from the diversion tunnel with embedded depth of more than 1000m, and they are designed as cylinder with a height of 100mm and a diameter of 50mm. The rock automatic three-axis servo rheometer self-developed completely by Shandong university will help to do the high temperature creep test, which is composed of main engine, pressure chamber, loading device, computer control system and other components. The maximum test force along axial direction is 1000KN with an accuracy of ±1%, while the maximum confining pressure is 50MPa with an accuracy of ±0.5%. The warming system comprises two parts, a pressure chamber and a hesting ring, and the highest permission temperature is 120℃ an accuracy of ±0.1℃.

2.2. Test method

The experiment was divided into three parts. Firstly, the uniaxial compression test was carried out; Secondly, the triaxial compression test was carried out, and the confining pressure was designed as 20MPa according to the in-situ stress;Finally, the triaxial compression temperature creep test was carried out, and the confining pressure was also designed as 20MPa.

The purpose of the test is to find out how the temperature to influence the creep deformation of the granite. And then the test methods are introduced respectively.

From the uniaxial compression test, the results show that the peak strength of the granite is 92.9MPa, the residual strength is 4.1MPa. Moreover, we can obtain the elasticity modulus is 52.3GPa , deformation modulus is 38.2GPa and the Poisson's ratio is 0.24. When the specimen is broken, the axial strain is about $1.18\times10^{-2}\varepsilon$, the lateral strain is about $-3.11\times10^{-2}\varepsilon$.

In the triaxial compression test, the stress path is designed as follows: First of all, loading the confining pressure to the desired value of 20MPa. And then, loading the axial stress continuously until the specimen is broken.

From the triaxial compression test, the results show that the peak strength of the granite is 291.0MPa, the residual strength is 125.6MPa, internal friction angle is 56°, and the cohesion is 14.25MPa under the confining pressure of 20MPa. Obviously, the triaxial compression stress - strain relation curve is divided into three parts-stable stage, unstable stage and broken stage. In the stable stage, stress - strain relation curve is similar to a straight line, and we can obtain the elasticity modulus is 55.8GPa ,the deformation modulus is 51.3GPa and the Poisson's ratio is 0.26. In the unstable stage, the stress is changing, and it arrives the peak strength,and then, the stress of the specimen reducts sharply .In the broken stage, the bearing capacity of the specimen reduces to the minimum, the specimen is broken, and the stress arrives its residual strength. When the specimen is broken, the axial strain is about $1.65\times10^{-2}\varepsilon$, the lateral strain is about $-2.05\times10^{-2}\varepsilon$. Here, ε represents the strain of diverse direction.

3. Test results

In this work, the stress path is designed as loading the axial stress step by step, and keeping an constant confining pressure in the meanwhile. The testing scheme is designed as three types of temperature(25℃、50℃、70℃) with 20MPa as the confining pressure.

Test procedure is as follows:

Firstly, using the heat shrink sleeve to seal the specimen and the setting blocks on both ends, and then installing the fiber axial and lateral displacement sensors in order to monitor the deformation of the specimen.

Secondly, installing the packaged specimen and the temperature sensor into the pressure chamber, and then putting the heating ring on the outside of the pressure chamber.

Thirdly, Opening computer control system, loading an set temperature, at the same time, loading stress at a rate of 50N/s up to the scheduled value of confining pressure. Waiting until the deformation tends to be constant, loading the axial stress at the same rate. Starting the computer control system to record the test data when the deformation trends to be stable.

Finally, loading the next stress level until the former creep deformation being stable, and the interval of records are different at different stages, generally, every 10 to 20 minutes at the preliminary stage, every 1 to 5 minutes when the deformation is unstable, and every 1 to 10 seconds when near the destruction. Ending the test till the specimen is broken, and saving the data.

Data processing is a key link before test analyzing. Using computer software and other data processing software to deal with the huge amount of data, and then obtaining the different axial creep deformation curves under different conditions of confining pressure (written as σ_3 in figure) and temperature, as shown in Figure 1.

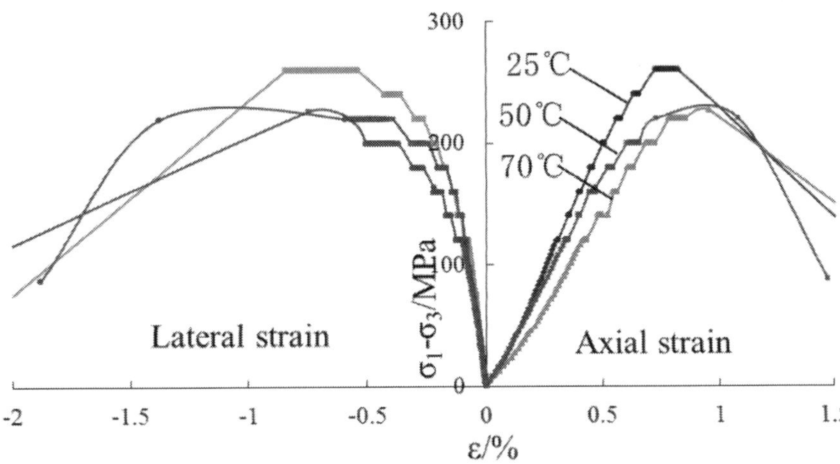

Figure 1. The axial creep deformation curves of gneissic granite specimen under different conditions of confining pressure and temperature(σ_3=20Mpa)

The increased percentage of the temperature creep test compared with the triaxial compression test is shown in Table 1.

Table 1. The increased percentage of the temperature creep test compared with the triaxial compression test

	Axial strain	Increased percentage	Lateral strain	Increased percentage
Triaxial compression test, T=25℃	$1.65\times10^{-2}\varepsilon$	/	$-2.05\times10^{-2}\varepsilon$	/
The temperature creep test, T=25℃	$2.26\times10^{-2}\varepsilon$	37.0%	$-2.46\times10^{-2}\varepsilon$	20%
The temperature creep test, T=50℃	$2.53\times10^{-2}\varepsilon$	53.3%	$-3.27\times10^{-2}\varepsilon$	59.5%
The temperature creep test, T=70℃	$2.98\times10^{-2}\varepsilon$	80.6%	$-4.12\times10^{-2}\varepsilon$	101%

All of the above shows that high temperature and high stress level both have significant influence on the creep properties of the rock, the higher the temperature and stress level, the more drastic the rock breaking, and the greater the harm.

4. Conclusions

In this article, some tests on granite have been accomplished, through the analysis of the deformation, main research conclusions are as below:

(1) A threshold exists during the creep of the gneissic granite, which is affected by temperature. When at the same confining pressure condition, the higher the temperature, the smaller the threshold.

(2) The temperature increases the creep deformation of the rock, the higher the temperature, the greater the creep deformation, and the faster the deformation development.

(3) The peak strength decreases with the increase of temperature at the same confining pressure condition.

(4)The axial strain and the lateral strain of the temperature creep test are much larger than that of the triaxial compression test.

(5) The test results can help offer significant experimental supports in terms of the long-term stability, design and construction of the similar kind of engineering.

Acknowledgments

The work was supported by China Postdoctoral Science Foundation funded project (No.2016M600537). We would also like to express our sincere gratitude to the editor and the reviewers for their valuable contributions to this paper.

References

[1] Shao, S.S., Ranjith, P.G., Wasantha, P.L.P., Chen, B.K., 2015. Experimental and numerical studies on the mechanical behaviour of Australian Strathbogie granite at high temperatures: An application to geothermal energy.Geothermics 54, 96-108

[2] Liu,s., Xu, J.Y.,2015. Effect of strain rate on the dynamic compressive mechanical behaviors of rock material subjected to high temperatures. Mechanics of Materials 82, 28–38

[3] Yang, S.Q., Ranjith, P.G., Jing, H.W., Tian, W.L., Ju, Y., 2016. An experimental investigation on thermal damage and failure mechanical behavior of granite after exposure to different high temperature treatments. Geothermics 65, 180-197

[4] Ding, Q.L., Ju,F., Mao, X.B., Ma, D., Yu, B.Y., Song, S.B.,2016. Experimental Investigation of the Mechanical Behavior in Unloading Conditions of Sandstone After High-Temperature Treatment. Rock Mechanics and Rock Engineering 49, 2641–2653

[5] Chen,Y.L., Wang,S.R., Ni,J., Rafig Azzam, Tomás Manuel Fernández-steeger, 2017. An experimental study of the mechanical properties of granite after high temperature exposure based on mineral characteristics. Engineering Geology 220, 234–242

[6] Muhammad Tufail, Khan Shahzada, Bora Gencturk, Wei, J.Q.,2017. Effect of Elevated Temperature on Mechanical Properties of Limestone, Quartzite and Granite Concrete. International Journal of Concrete Structures and Materials, 11 (1), 17-28

Development of a Support Bump for Increasing the Energy Absorption of a Jounce Bump

C H Lee[1], M J Han[1], T W Park[1], S J Lee[2] and J S Park[2]

[1] Department of Mechanical Engineering, Ajou University, Gyeonggi 16499, Korea
[2] DONGYANGP&C, 149 Wondang-ro, Chilgok-ri, Wongok-myeon, Anseong-si, Gyeonggi, 1755, Korea

park@ajou.ac.kr

Abstract. In this study, we developed a support bump to increase the energy absorption of the jounce bump. The support bump is made of TPE material, and the jounce bump is made of TPU material. The support and jounce bump each verified the simulation model using the load deflection test. The developed support and jounce bump were satisfied with the durability test. The support bump was assembled with the jounce bump and confirmed that the energy absorption amount was increased compared with the single jounce bump. Finally, we verified the validity of the simulation by comparing the load deflection data under the same conditions as the test using the verified the support and jounce bump simulation model.

1. Introduction

The jounce bump is a shock absorber located at the top of the shock absorber. In the past, the jounce bump was mostly made of rubber, which simply acted as a stopper to protect the shock absorber, so it was a safety assisting device to protect the strut complete from being stopped at the height required for compression. However, nowadays, it is expected not only to function as a safety device but also to improve ride comfort and driving performance. The jounce bump disperses the impact energy generated when passing through an irregular road or passing through a bump, thereby reducing impact and reducing noise. Therefore, the development of jounce bump with increased energy absorption has been carried out to improve ride comfort and steering stability of the vehicle. Schudt J A *et al* developed a dual rate jounce bump with increased energy absorption and were first applied to vehicles in 2009. [1] And Wang Y *et al* developed a jounce bump which made of polyurethane with skin and transition layer and confirmed that the energy absorption was increased.[2] And a negative poisson's ratio structure was introduced and applied on the jounce bumper.[3]

In this study, we developed a component called a support bump to increase the energy absorption of the jounce bump. This support bump was made by benchmarking the rebound bump. [4] The support bump is made of a thermoplastic elastomeric (TPE) material and is assembled with jounce bump which is made of thermoplastic polyurethane (TPU) to increase the energy absorption. The load-deflection test was carried out for the verification of the support and the jounce bump simulation model. Next, by comparing the energy absorption amount of the support-jounce bump assembly and the single jounce bump, the effect of increasing the energy absorption amount of the support bump was confirmed. The analysis was carried out under the same conditions as the test conditions, using the previously proven support and the jounce bump simulation model. Test and simulation results were compared and validated for testing and analysis. Finally, it was confirmed that the support and jounce bump satisfies the durability test, respectively, and the permanent compression strain was confirmed by measuring the height before and after the test.

Figure 1. Rebound bump

2. Support-jounce bump model

2.1. Material property test

TPU material for the support bump and TPU material for the jounce bump were selected. The TPE material has high durability against repeated fatigue and large deformation cycles and is selected as the material of the support bump. The TPU material has higher strength, durability and energy absorption than rubber, and is selected as the material of the jounce bump. For the simulation model, specimens were fabricated and material properties were tested to confirm the physical properties of the TPE and TPU materials.

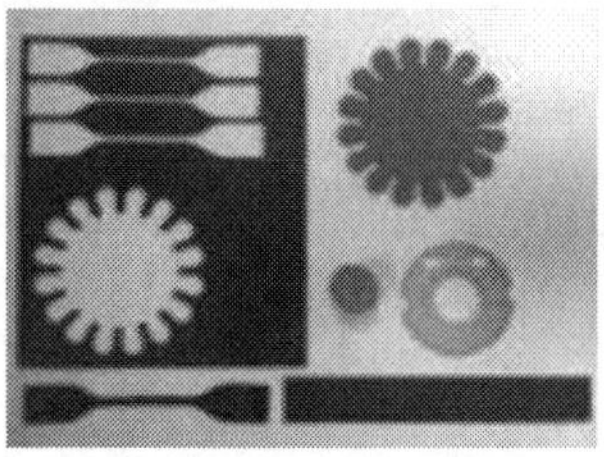

Figure 2. Specimens fabrication

Figure 3. Physical material test

2.2. Support bump model

The support bump model is shown in figure 4. The support bump has been developed with a limitation in increasing the energy absorption amount by a single jounce bump. The support bump was made to benchmark the rebound bumps to have the same performance as the rebound bumps. Figure 5 shows the load-deflection test result of the developed support bump and simulation model. From the test results, it was confirmed that the results agree well with no large error. Abaqus software was used in this study.

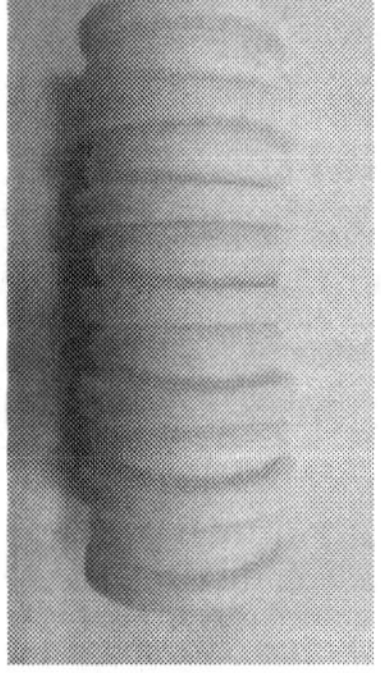

Figure 4. Support bump

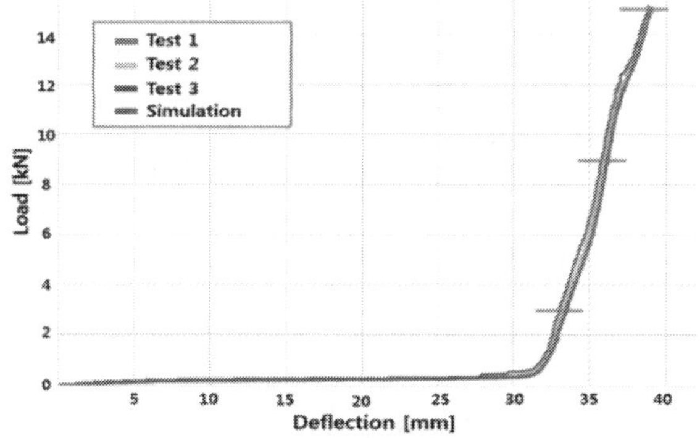

Figure 5. Load-deflection results of support bump test & simulation

2.3. Jounce bump model

Jounce bump is made of TPU material. The production model is shown in figure 6. Load-deflection test was conducted to verify the simulation model of the jounce bump. From the results, the jounce bump simulation model is verified.

Figure 6. Jounce bump

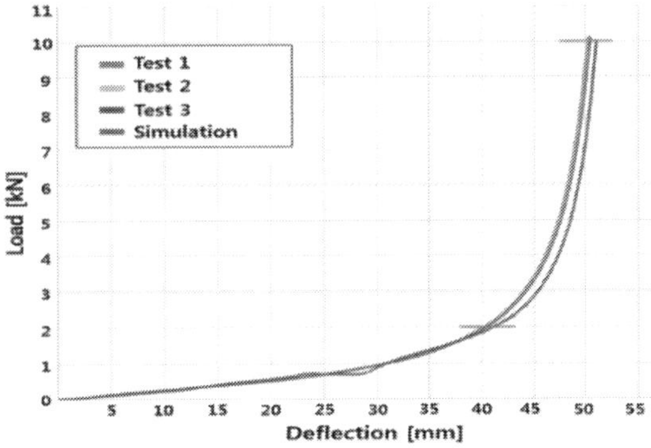

Figure 7. Load-deflection results of jounce bump test & simulation

3. Support, jounce bump durability test

3.1. Support bump durability test

The support bump is made of TPE material. TPE material has compressive strain due to its characteristics. Therefore, before and after the durability test, the post compression length was measured to confirm the permanent compressive strain. The durability test result was satisfied and the permanent compression strain, which was originally aimed, was satisfied within 10%. The test conditions are shown in table 1. And the length before and after the test is shown in figure 8.

$$\Delta H\ (\%) = \frac{|(H-H_0)\times 100|}{H_0} = \frac{|(60.02-65.88)\times 100|}{65.88} = 8.89 \tag{1}$$

Table 1. Support bump durability test conditions

Contents	Condition
Temperature [℃]	20±5
Load [kgf]	225±125
Frequency [Hz]	3.3
Cycle	100,000

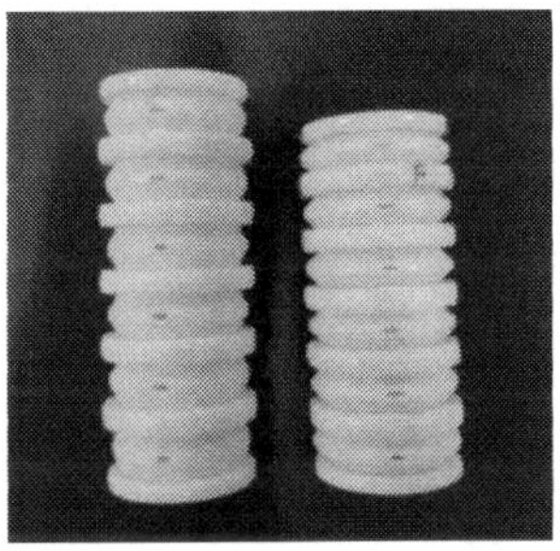

(a) Before (b) After

Figure 8. Support bump after durability test

3.2. Jounce bump durability test

The jounce bump was fabricated TPU material and has higher strength and durability than the rubber. The permanent compressive strain was confirmed before and after the durability test, and the results were satisfied within 10%. The durability test conditions of the jounce bump are shown in table 2. As a result of the durability test, it was confirmed that the test conditions were satisfied.

$$\Delta H\ (\%) = \frac{|(H-H_0)\times 100|}{H_0} = \frac{|(68.97-72)\times 100|}{72} = 4.21 \qquad (2)$$

Table 2. Jounce bump durability test conditions

Contents	Condition
Temperature [℃]	20±5
Stroke [mm]	0 ~ 39.3
Frequency [Hz]	1.5
Cycle	300,000

※ The load response at the time of excitation should be more than 4.9kN.

Figure 9. Durability test equipment

Figure 10. Jounce bump durability test

4. Support-jounce bump assembly

To verify the increase in energy absorption of the support-jounce bump assembly, it was compared with the load-deflection test of the single jounce bump. The energy absorption amount corresponds to the area in the load-deflection. As a result of the test, it can be confirmed that the energy absorption amount of the support-jounce bump assembly is increased by about 8%.

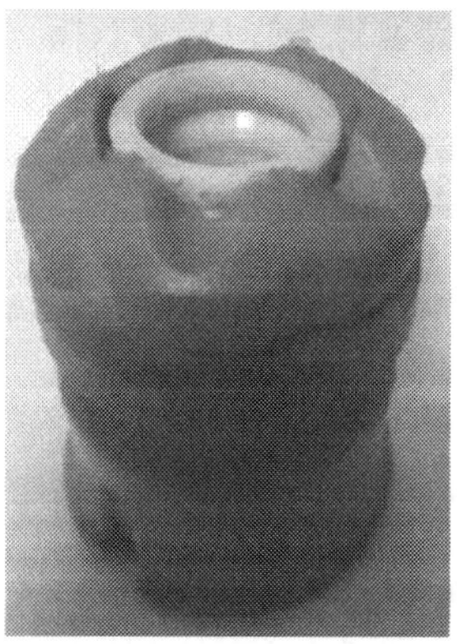

Figure 11. Support-jounce bump assembly

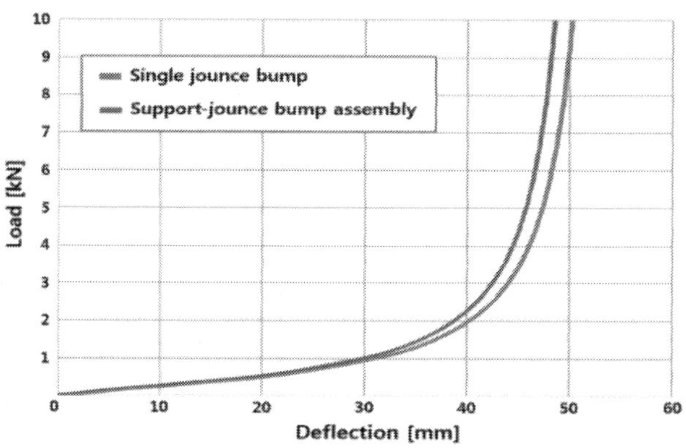

Figure 12. Energy absorption comparison between a single jounce bump & a support-jounce bump assembly

Next, the previously verified support and jounce bump model were analyzed in the same test conditions. The amount of deformation was confirmed by controlling the load of the upper plate while the support-jounce bump was fixed to the lower plate. The validity of the test and simulation results of the support-jounce bump assembly was verified by comparing the test and simulation results.

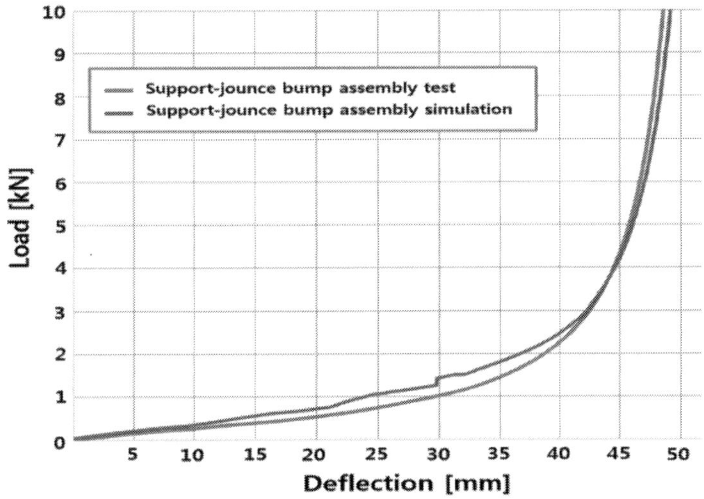

Figure 13. Energy absorption comparison between support-jounce bump assembly test & simulation

5. Conclusion

The support bump was developed to increase the energy absorption of the jounce bump. There is a limitation in increasing the amount of energy absorption by a single jounce bump, and it is disadvantageous in that tuning is difficult depending on the purpose of use. In this study, it was confirmed that the support bump was assembled with the jounce bump and increased by about 8% compared with the single jounce bump energy absorption. When the developed support bump is assembled to the jounce bump, the energy absorption amount of the support bump can be adjusted according to the purpose of the user while increasing the energy absorption amount. Also, in this study, we verified the support and the jounce bump simulation model using the load-deflection test. The validated simulation model was verified by comparing the actual test with the load-deflection test with the support-jounce bump assembly.

The simulation models that have been verified in this study can perform shape optimization to increase energy absorption. Further, it is possible to develop support bumps of various performances according to the purpose of the user.

Acknowledgments

This research is supported by the global expertise development project (10076515) supported by the Ministry of Commerce, Industry and Energy and the Korea Industrial Technology Evaluation and Management Institute (KEIT).

References

[1] Schudt J A, Tsai M, Patil R and Geisler R 2011 *Dual Rate Jounce Bumper Design* SAE Technical Papers pp 2011-01-0791
[2] Wang Y, Ma Z and Wang L 2016 *A finite element stratification method for a polyurethane jounce bumper* Automobile Engineering vol 230 no7 pp 983-992
[3] Wang Y, Wang L, Ma Z.D and Wang T 2016 *A negative Poisson's ratio suspension jounce bumper* Material&Design vol 103 pp 90-99
[4] Han C S 2016 *Development of Rebound Bumper Using Thermoplastic Elastomer in Suspension System of Vehicle* SAE Technical Papers vol 2016 no 01 pp 2016-01-0518

IOP Conf. Series: Materials Science and Engineering **417** (2018) 012005 doi:10.1088/1757-899X/417/1/012005

Optimization for Welding Parameters of Magnetic Media Box based on Orthogonal Experiment

Zhiyong Wang[1a], Xuan Sun[1b*] and Dehui Sun[2]

1 School of Mechanical Engineering, University of Jinan, Jinan, China
2 Hanlin school, Shenxian, Shandong, China

[a] 971917165@qq.com,[b]me_sunx@ujn.edu.cn,*Corresponding Author

Abstract. In The magnetic media box of the magnetic separator is used as the research object, because at present the argon arc welding speed and the welding quality of the magnetic media box are relatively low, so the orthogonal test is used to optimize the welding technology of the magnetic media box. According to the characteristics of magnetic media box welding technology, the influence of welding current, arc voltage and welding speed on the tensile strength of magnetic media box is investigated. By experimenting and recording the experimental results, then using the range analysis method to analyze the experimental results to determine the optimal combination of welding process parameters. The reliability of the test results is proved by experiments. Finding the law of welding process parameters affecting the tensile strength of magnetic media boxes. It provides reference for TIG.

1. Introduction

The reasonable setting of welding parameters of TIG has great influence on the welding quality of magnetic media box.In previous productions, the choice of welding process parameters based on experience or inquiries on welding manuals was often too conservative due to safety considerations. The choice of optimum welding process parameters was not achieved, and the welding quality of magnetic media boxes was also reduced, and the production efficiency could not be guaranteed.The optimization of welding process parameters is a key technology to ensure the quality of magnetic media boxes, improve processing efficiency, and reduce production costs.

In this paper, the orthogonal test is used to optimize the parameters, and the calculation is small. At the same time, it can meet the quality requirements of the optimized processing and it is easy to promote in actual production[1].

2. Selection of experimental indicators and factor levels

In this paper, the parameters of argon arc welding are optimized, and the main requirement of magnetic media box is the connection strength of magnetic rod and magnetic plate.Because when the magnetic separator is working, the magnetic media box is used to sift iron powder from iron ore, and the selected iron powder will be washed out by high-pressure water, thus the screening of iron ore will be realized.In the long-term operation, due to the friction of iron powder, the impact and corrosion of high-pressure water will cause the magnet bar to fall off from the magnetic plate, thereby affecting the beneficiation quality and efficiency.At present, the method of detecting the quality of the weld is to analyze the microstructure, phase composition and mechanical properties of the weld by means of metallographic microscope and X-ray diffractometer.The mechanical properties mainly consider the tensile strength and hardness of the weld.The greatest influence on the overall welding quality of the magnetic media box in the detected microstructure characteristics and mechanical properties of the weld is the tensile strength of the weld.Therefore, ensuring the tensile strength at the joint between the magnetic plate and the magnetic bar of the magnetic media box is a key factor to ensure the quality of the magnetic media box.By analyzing the mechanical and technical requirements of the magnetic media box, it must be ensured that the tensile strength of the weld seam of the media box is greater than 240 MPa[2].

The structure of the magnetic media box is shown in Figure 1. Both the magnetic plate and the magnetic bar are nickel-iron alloy.The purpose of this machining is to weld both ends of the magnet to the magnetic plate. The thickness of the magnetic plate is 3 mm and the magnetic bar is 3 mm in diameter.The test of tensile strength is shown in Figure 2.

Figure 1. Magnetic media box

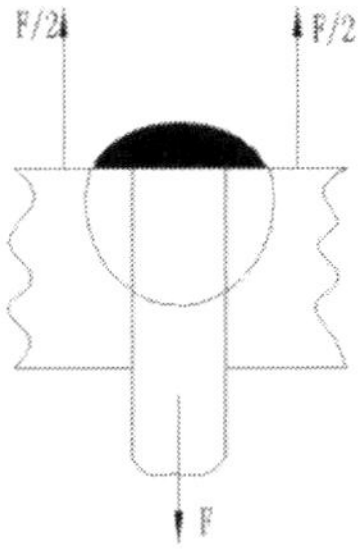

Figure 2. Tensile strength test

The welding method of magnetic media box adopts argon arc welding, welding process parameters mainly include welding current, welding speed, arc voltage, argon flow, tungsten electrode diameter, welding gun angle, etc[3].There are many process parameters affecting the welding quality of the media box, and many process parameters have little effect on the welding quality. After reviewing the relevant literature and field tests, this paper selects three influencing factors of welding current, arc voltage, and welding speed[4], and examines its influence on the tensile strength of magnetic media boxes.Consider the actual welding situation, select five levels for each factor to study. The specific level values are shown in Table 1.

Table 1. Factor Level Table

Factors	Unit	Levels				
		1	2	3	4	5
Welding current（A）	A	160	180	200	220	240
Welding speed（B）	mm/s	14	16	18	20	22
arc voltage（C）	V	12	14	16	18	20

Table 2. experimental scheme

Column number \ Test number	1	2	3	4	5	6	7	8	9	10	11	12	13	14	15	16	17	18	19	20	21	22	23	24	25
A	1	1	1	1	1	2	2	2	2	2	3	3	3	3	3	4	4	4	4	4	5	5	5	5	5
B	1	2	3	4	5	1	2	3	4	5	1	2	3	4	5	1	2	3	4	5	1	2	3	4	5
C	1	2	3	4	5	2	3	4	5	1	3	4	5	1	2	4	5	1	3	2	5	1	2	3	4

3. Select orthogonal table

The orthogonal table is a pre-prepared form by mathematicians, which has the outstanding advantages of uniform dispersion and uniformity.The general orthogonal table format is $L_N(q^s)$.In this paper, the factors of A, B and C are selected, and each factor has five levels, so the degree of freedom is $f_A=f_B=f_C=5-1=4$.Therefore, the orthogonal test of this paper is arranged 25 times. According to the above discussion, $L_{25}(5^6)$orthogonal table is selected in this paper[5].

Table 3. Factor Level Table

Experiment times	Welding current (A)	Welding speed (B)	arc voltage (C)	tensile strength (MPa)
1	160	14	12	185
2	160	16	14	220
3	160	18	16	238
4	160	20	18	231
5	160	22	20	203
6	180	14	14	193
7	180	16	16	216
8	180	18	18	223
9	180	20	20	211
10	180	22	12	183
11	200	14	16	210
12	200	16	18	201
13	200	18	20	247
14	200	20	12	218
15	200	22	14	207
16	220	14	18	205
17	220	16	20	217
18	220	18	12	223
19	220	20	16	212
20	220	22	14	207
21	240	14	20	181
22	240	16	12	192
23	240	18	14	214
24	240	20	16	209
25	240	22	18	207

Table 4. Range Analysis

Test number \ Column number	1	2	3	tensile strength (MPa)
1	1	1	1	190
2	1	2	2	200
3	1	3	3	210
4	1	4	4	210
5	1	5	5	185
6	2	1	2	190
7	2	2	3	210
8	2	3	4	213
9	2	4	5	211
10	2	5	1	183
11	3	1	3	225
12	3	2	4	211
13	3	3	5	247
14	3	4	1	228
15	3	5	2	217
16	4	1	4	210
17	4	2	5	217
18	4	3	1	213
19	4	4	3	212
20	4	5	2	207
21	5	1	5	190
22	5	2	1	192
23	5	3	2	204
24	5	4	3	209
25	5	5	4	207
K1	995	1005	1006	
K2	1007	1030	1018	
K3	1128	1087	1066	
K4	1059	1070	1051	
K5	1002	999	1050	
k1	199	201	201.2	
k2	201.4	206	203.6	
k3	225.6	217.4	213.2	
k4	211.8	214	210.2	
k5	200.4	199.8	210	
Range	26.6	17.6	12	
Excellent solution	A3	B3	C3	

4. Design header of orthogonal test and make experiment plan

The final design of the orthogonal table is shown in Table 2.

According to the 25 sets of data proposed in the above table[6], this paper uses a 400A welding power source to weld 25 magnetic media boxes. Measure the magnetic media box tensile strength. The measurement results are recorded in Table 3, and the resulting orthogonal test data is shown in Table 3.

5. Range analysis of Multifactor Orthogonal Experiments

The range analysis method is the most commonly used method for analyzing the results of orthogonal experiments.The excellent level and excellent combination of experimental factors can be obtained through the range analysis of tensile strength. Also, it was found that each factor influences the strength of tensile strength.The range analysis table is calculated as shown in Table 4.

According to the analysis of table 4, the welding current size has the greatest influence on the tensile strength of the magnetic media box, the welding speed is the second and the arc pressure is the smallest.The higher the tensile strength, the higher the quality of the magnetic media box, so the optimal level of the process parameters is welding current 200A, welding speed 18mm/s, and arc voltage16V.

6. Verify the results of orthogonal experiments

According to the experimental results, new welding process parameters were used to weld 5 magnetic media boxes. Then the tensile strength of the magnetic media box is calculated, as shown in table 5.

Table 5. Magnetic Media Box Experiment Results

Product	1	2	3	4	5
tensile strength (MPa)	263	257	271	264	266

From the above table, it can be seen that the welding quality of magnetic media boxes has been significantly improved. To achieve the purpose of this experiment.

7. Conclusion

This article orthogonal test is a three-factor five levels. The results show that when the welding current is 200A, the welding speed is 18mm/s, and the arc voltage is 16V, the magnetic media box has the highest tensile strength and satisfies the technical requirements.The optimization objective of this paper is the tensile strength of magnetic dielectric box, and its optimization experiment can be popularized, and its research idea also provides reference for relevant researches.

References

[1] Analysis of influencing factors of mining subsidence in Jinjie coal mine based on orthogonal test [J]. Wang Kai ye, Ceng Xiangzhu, Luan Chun Xue. Coal technology. 2015 (02):56-60

[2] Zhao Yuanbo. Study on welding technology of Tungsten TIG Welding of TC2 Thin Plate[J].Rural Science Experiment,2017,(11):110-114

[3] YSuper Audio DC Pulse TIG Welding of TC4 Titanium Alloy[J].Xu Haiying,Qi Bojin,Zhang Wei,Zhao Haitao.Journal of Aeronautical Materials.2011(01):48-52

[4] Stainless Steel Conveyor Belt and Bellows TIG Welding[J].CHEN Su-Bin, GAO Hui-Yang.Welding Technology.2016(S1):12-17

[5] Zhan Liang, Li Xia, Sun Libin, etal. Design optimization of process parameters of crankshaft die forging based on orthogonal experiment [J]. Forging & Stamping Technology, 2014, 39(7):10-13.

[6] Optimization of crankshaft hot forging process parameters based on orthogonal test[J].Zhan Liang,Li Xia,Sun Ribin,Wang Yuxiao,Li Lei.Forging and Stamping Technology.2014,(07):25-30

An Efficient Numerical Model for Squeeze Film Air Damping of Microplates in the Free Molecular Regime

Cunhao Lu[1], **Pu Li**[1a]

[1]School of Mechanical Engineering, Southeast University, Nanjing, 210096, China

[a] Corresponding author: seulp@seu.edu.cn

Abstract: The squeeze film air damping of microplates in the free molecular regime has been received particular attention for many years. Based on Energy Transfer Model (ETM) proposed by Bao et al and Monte Carlo (MC) model proposed by Hutcherson and Ye, this paper proposes an efficient numerical model, compared with experiment data and present models, the present model costs less computational work than the previous MC model and is more accurate than ETM.

1. Introduction

Squeeze film air damping has been identified as an important mechanism of energy dissipation in microplates, even encapsulated in rarefied air [1]. The accurate prediction of squeeze film air damping is important for the design of microplates [2-5].

Based the free molecular approach, by considering the squeeze film air damping effect, Bao et al [6] proposed a closed form Energy Transfer Model (ETM) in the free molecular regime, which based on some assumptions for simplicity. However, the deviation with experiments remains large. Hutcherson and Ye [7] developed a Monte Carlo (MC) model by releasing some assumptions in the ETM. The MC model is verified to be more accurate than the ETM, however, the process of the MC model costs a lot of computational work and is difficult to handle. Based on the ETM and MC model, this paper gives an efficient numerical model for squeeze film air damping in the free molecular regime, which shows relatively high accuracy and efficiency.

This paper will introduce the ETM and MC model in Section 2. In Section 3, the new numerical model is given. The comparison with experiment data is shown in Section 4. At last, a conclusion of this paper is given in Section 5.

2. Previous models

2.1 Process of ETM [4]

As the vibrating plate experiences a time dependent sinusoidal motion parallel to the substrate, the displacement of the plate is $z=A\sin\omega t$, A and ω are the amplitude and vibration frequency of the plate. Bao et al [6] made several assumptions:

(1) Collisions of molecule and wall are fully elastic and specula.

(2) Velocity of molecules going through the gap is constant.

(3) During the travel of a molecule staying in the gap, the velocity of the plate is assumed to be constant.

(4) The gap length is assumed to be constant during the travel of the molecule.

The number of molecules entering the gap per unit time is $N=1/4\cdot n\bar{v}Ld$, where n is the concentration of molecules, $L=2(a+b)$ the peripheral length of the plate, d the gap length. $\bar{v}=\sqrt{8kT/(\pi m)}$ the average velocity of the molecules, k the Boltzmann constant, T and m the air temperature and the molecule mass, The total energy loss of the plate in one cycle is:

$$\Delta E=\frac{1}{4}n\bar{v}L\frac{1}{\omega}\int_0^{2\pi}\frac{\overline{l^2}\,\overline{v_{z0}^2}}{2d\cdot\overline{v_{xy0}^2}}A^2\omega^2\cos^2\omega t\cdot d(\omega t)=\frac{\pi}{16}\overline{l^2}A^2\omega p\bar{v}\frac{L}{d} \quad (1)$$

l^2, v_{xy0}^2 and v_{z0}^2 are approximated as the average traveling distance $\overline{l^2}$, $\overline{l^2}\approx\bar{l}^2=2ab/\pi$, the molecule velocity $\overline{v_{xy0}^2}$ and $\overline{v_{z0}^2}$, meanwhile $\overline{v_{xy0}^2}=2\overline{v_{z0}^2}$, $\overline{v_{xy0}^2}+\overline{v_{z0}^2}\approx\bar{v}^2$, p the pressure, ρ the gas density and ρ_P the plate density. The quality factor of ETM is:

$$Q_{Bao}=(2\pi)^{\frac{3}{2}}\rho_P H\omega(\frac{d}{L})\sqrt{\frac{RT}{M_m}}\frac{1}{p} \quad (2)$$

2.2 Process of MC model[7]

In the MC model, the assumptions (2), (3), (4) are released. The vibration period of the microplate is divided into M time divisions with a length of Δt. At the ith time division, the number of molecules entering the gap is $1/4\cdot n\bar{v}L(d-z_i)\Delta t$, $z_i=A\sin(\omega i\Delta t)$. The energy change Δe_i of a representative molecule entering at the ith time division is calculated by tracking the motion of this molecule and recording velocity changes after each collision with the plate. The initial position of the molecule is assigned randomly on the basis of a uniform distribution, and the initial velocity is assigned as the average velocity $\bar{v}=\sqrt{8kT/(\pi m)}$. By summing up over M divisions, the quality factor of the MC model is obtained:

$$Q_{HY}=\pi abH\rho_p A^2\omega^2/\Delta E \text{ and } \Delta E=\sum_{i=1}^{M}\frac{1}{4}n\bar{v}L(d-z_i)\Delta e_i\Delta t \quad (3)$$

3. Numerical model

As can be seen, the MC model discretizes the vibration period of the plate, the initial velocity and initial position of molecules entering the gap to release the assumptions of the ETM. However, these discretizations cost much computational work. Based on the ETM and MC model, this paper proposes a new numerical model, which just releases the assumption (3). The process of the model is shown as follows.

Like the MC model, the vibration period of the plated is divided into M time divisions with a length of Δt. When the molecule going through the gap at the ith time division, the velocity of the plate is considered to be constant $z'_i=A\omega\cos(\omega i\Delta t)$. Due to the fully elastic and specula collisions of molecule and wall, the time of molecule staying in the gap is calculated as $\bar{l}/\overline{v_{xy0}}$, $\overline{v_{xy0}}=\sqrt{6}/3\cdot\bar{v}$. At each collision of molecule and the plate, the molecule has a speed increment $2z'_i$. For the first collision of molecule and the plate, the time cost is $2d/\overline{v_{z0}}$, $\overline{v_{z0}}=\sqrt{3}/3\cdot\bar{v}$. For the second collision, the time cost is $2d/(\overline{v_{z0}}+2z')$. If the total times of collisions of a single molecule and the plate at the ith time division is K_i, the time cost is $\frac{2d}{\overline{v_{z0}}}+\frac{2d}{\overline{v_{z0}}+2z'_i}+\frac{2d}{\overline{v_{z0}}+2\cdot 2z'_i}\cdots\frac{2d}{\overline{v_{z0}}+K_i\cdot 2z'_i}$, which should be less than or equal to $\bar{l}/\overline{v_{xy0}}$:

$$\frac{2d}{\overline{v_{z0}}}+\frac{2d}{\overline{v_{z0}}+2z'_i}+\frac{2d}{\overline{v_{z0}}+2\cdot 2z'_i}\cdots\frac{2d}{\overline{v_{z0}}+K_i\cdot 2z'_i}\le\frac{\bar{l}}{\overline{v_{xy0}}} \quad (4)$$

In order to calculated the value of K_i, we start K_i from 1, if equation (4) is true, $K_i= K_i+1$ until equation (4) is false, which means the time cost of （K_i+1） collisions would be larger than $\bar{l}/\overline{v_{xy0}}$. Then we can get the value of K_i at the ith time division. By calculating all values of K_i at all time divisions, the quality factor of the new numerical model is obtained:

$$Q_{New} = \pi abH\rho_p A^2\omega^2 / \Delta E_{New} \quad \text{and} \quad \Delta E_{New} = \sum_{i=1}^{M}[\frac{1}{4}n\bar{v}Ld\cdot\frac{1}{2}m(4K_i z_i'\overline{v_{z0}} + 4(K_i z_i')^2)\Delta t] \qquad (5)$$

4. Experiments

To validate the new numerical model, Zook's experiment data is used [1]. The parameters of Zook's experiment are show in table 1. Firstly, the quality factor is calculated at different time dicretizations of the vibration period and shown in figure 1. As can be seen, the simulation results have converged to within 0.01% with less than 1000 time divisions, and the computational time is less than 0.001s, which shows high efficiency. As a consequence, the time division M is set as 1000, which can meet the requirement of convergence.

Figure 2 gives the comparison of quality factors. It is obvious that at Kn>10, the results of HY's MC model and the new model are both more accurate than those of Bao's ETM. The cure of the new model are very close to those of the MC model at the whole range of Kn. Table 2 lists the detailed values of MC and the new models at Kn>10.

Table 1. Parameters of Zook's experiment [1]

Parameters	
Plate density ρ_P (kg/m^3)	2330
Length a (μm)	200
Width b (μm)	40
Thickness H (μm)	1.8
Gap length d (μm)	1.1
Frequency f (kHz)	550
Amplitude A (μm)	<0.1
Temperature T (K)	293

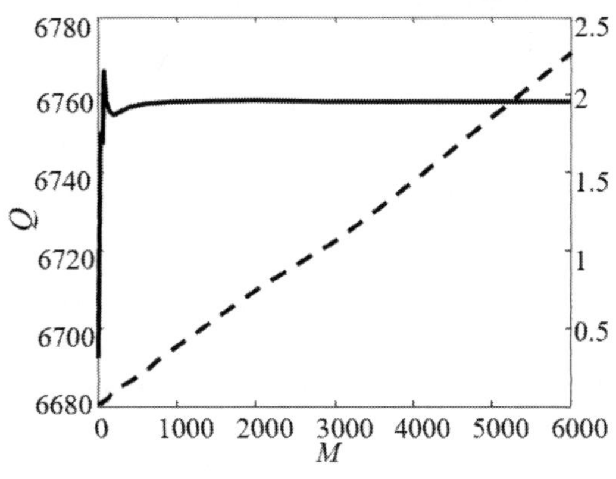

Figure 1. Convergence of simulation results at Kn=561

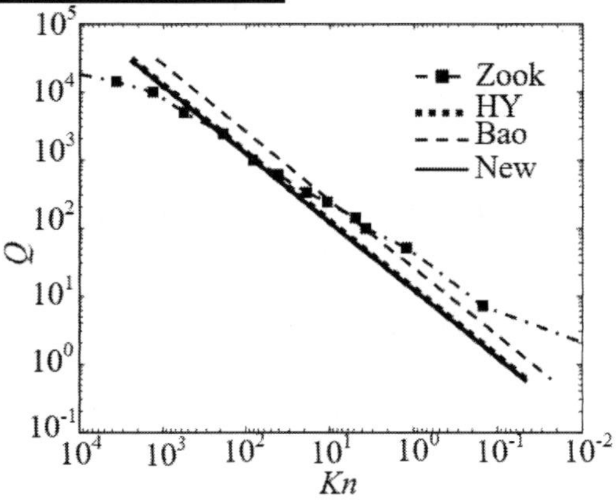

Figure 2. Comparison of quality factors

Table 2. Quality factors at Kn>10

Kn	Q_{Exp}	Q_{HY}	Q	$(Q_{New}-Q_{Exp})/Q_{Exp}$%	$(Q_{New}-Q_{HY})/Q_{HY}$%
1320	10000	16724	15901	59.01	4.92
561	5000	7107	6758	35.16	4.91
189	2400	2369	2252	-6.14	4.91
83	1000	1053	1000	-0.01	5.05
41	600	526	494	-17.68	6.10
19	330	236	229	-30.64	3.02
10	240	132	120	-49.81	8.74

It can be seen in table 2, the results of the new model are slightly less than those of the MC model and are slightly closer to the results of experiment. The largest deviation of the new model and the MC model is less than 10%. HY [7] pointed that the main factor that influences the accuracy of the ETM is the assumption (2), which may take 90% compared with the MC model. The assumptions (3), (4) may take 10%. The new numerical model has just released the assumption (2) and the deviation is less than 10%, which verifies the assumption (2) is the main factor that influences the accuracy pointed by HY.

It may be confusing that the results of the new model are slightly closer to experiment data than those of the MC model. It should be noted that, the assumption (3) considers the travelling time of each molecules entering the gap is smaller than the vibration period, which means during the vibration period, all molecules entering the gap can go through the gap. Actually, some of the molecules cannot go through the gap within the vibration period. As a result, the assumption (3) would slightly overestimate the energy change of molecules in the gap and underestimate the quality factor, which can explain that the new model is "more accurate" than the MC model.

5. Conclusion

For predicting squeeze film air damping in the free molecular regime, Bao et al proposed an analytical ETM [6]. However, the deviation with experiment remains large. HY developed a MC model more accurate than ETM by releasing some assumptions in ETM [7]. However, the MC model costs a lot computational work. Based on ETM and MC model, this paper proposes an efficient numerical model, which just releases the assumption (2)—the main factor that influences the accuracy. Compared with experiment, the new model is more accurate than the ETM, the deviation of the new model and the MC model is less than 10%, while the simulation process is simple and the computational time is reduced to 0.01s.

In general, owing to relatively high accuracy and efficiency, the new numerical model can have a practical value in design applications of MEMS devices.

Acknowledgments

This work was supported by the National Natural Science Foundation of China (Grant No 51375091).

References

[1] Zook J, Burns D and Guckel H 1992 *Sensors Actuators* A **35** 51-9
[2] Veijola T, Kuisma H, Lahdenpera J and Ryhanen T 1995 *Sensors Actuators* A **48** 239-48
[3] Pandey A and Pratap R 2008 *J. Micromech. Microeng.* **18**: 105003
[4] Qiu H, Schwarz P, Feili D, Wu X and Seidel H 2015 *J. Micromech. Microeng.* **25** 055016
[5] Ishfaque A and Kim B 2016 *J. Sound Vib.* **375** 422-435
[6] Bao M, Yang H, Yin H and Sun Y 2002 *J. Micromech. Microeng.* **12**:341-346
[7] Hutcherson S and Ye W 2004 *J. Micromech. Microeng.* **14** 1726-33

Combustion Characteristics Behavior of *Pterocarpus indicus* Leaves Waste Briquette at Various Particle Size and Pressure

Willyanto Anggono[1], Fandi D. Suprianto[2], Gabriel Jeremy Gotama[3], Sutrisno[4] and Jovian Evander[5]

[1,2,3,4,5]Mechanical Engineering Department, Petra Christian University, Surabaya, Indonesia
[1,2,3,4,5]Centre for Sustainable Energy Studies, Petra Christian University, Surabaya, Indonesia

E-mail: willy@petra.ac.id

Abstract. As one of the countries with a high variety of plant species, Indonesia becomes a potential source for biomass fuel. One plant that commonly found in the street of Surabaya, Indonesia is *Pterocarpus indicus*. The plant exists in large number and produced litters from its leaves waste. The litters negatively affect Surabaya city cleanliness. The previous study suggested a solution for this problem is by converting these litters into biomass fuel in the form of briquette. The solution may help this problem by reducing the wastes in Surabaya street and avoiding unnecessary handling of these wastes by burning or burying. In order to maximize the potential of *Pterocarpus indicus* leaves waste briquette, two primary parameters of briquette production were investigated in this study. These parameters are particle size of briquette and pressure used for compacting the briquette. This study used four burning characteristics (flame temperature, ignition time, burning time and combustion rate) to determine the best particle size and pressure to obtain the highest quality of briquette. The result of the study indicates *Pterocarpus indicus* produced with particle size of 60 mesh and compacting pressure of 2 MPa yield the best quality of briquette.

1. Introduction

One major cause in the decrease of fossil fuel reservoir comes from the growth of human population on earth [1,2]. The decrease of fossil fuel reservoir brings unfavorable condition in fuel market by putting the price of energy resources higher. A new sustainable and renewable energy resources are in demand with the decrease of fossil fuel reservoir. By utilizing new resources of renewable energy, the fluctuation of energy resources' market price will stabilize [3,4].

Biomass is one of many renewable energy sources that has been studied thoroughly. Many sources of biomass include industrial waste, algae, aquatic plants, wood, wastes of wood, urban waste (solid), agriculture crops, agriculture wastes, and animal wastes. The major advantage of biomass is its availability in nature which makes them easily acquired without requiring an exorbitant cost in the process [5]. Another advantage of biomass in term of usage as fuel comes from the capability of its sources (plants) to reduce CO_2. The ability of plants to reduce CO_2 may cause biomass fuel to leave no additional carbon footprint when utilized and thus be considered as CO_2 neutral fuel [5].

There are two processes used to utilize biomass as fuel. These two processes are bio-chemical process and thermo-chemical process. The bio-chemical process works by decomposing biomass material. The thermo-chemical process works by creating high energy products and using biomass as the main material. The thermo-chemical process is divided further into five categories, combustion, gasification, liquefaction, pyrolysis, and hydrogenation [1].

The most regular process for utilizing biomass is by directly ignite biomass and converting them into chemical energy. The chemical energy obtained is used by transforming them into heat, electricity, machine-driving energy and man more. This process is known as direct combustion [6]. Although convenient, directly combusting biomass without pre-processing might not always work. Biomass with water content higher than 50% cannot be burned and it is common to acquire biomass with this level of water content [1]. Pre-processing is needed to reduce the water content of biomass to a suitable level. Another weakness by directly combusting biomass is the transportation issue. Biomass requires more space during transportation because of its low bulk density. In a condition where a large amount of biomass need to be transported, the transportation requires high expenditure [7].

These weaknesses in direct combustion can be subdued by processing the biomass into briquette [7,8]. Briquette is a process to transform biomass into a product with higher bulk density and energy density [8]. Higher heating value, better energy properties, lower moisture content and higher density are some advantages of briquette compared to its raw material [9]. Steps of creating biomass are reducing the water content of biomass, shred the biomass into the desired size and compacting and forming briquette by applying pressure. The energy required to create briquettes is lower than the energy obtained from utilizing them [10]. The effect of briquette in the social environment are also positive. Briquette is found to reduce energy shortages, improving the ecological condition, and growing people's earning [11].

The wastes of plants have been one of many primary sources of biomass [12]. Countries with a tropical climate and myriads of floras varieties have a high level of biomass resources. Developing countries are also known to generate a high number of biomass [5]. Indonesia fits these characteristics and it has many plants wastes ready to be utilized as biomass fuel. One plants that are usually found in Indonesia is *Pterocarpus indicus*. *Pterocarpus indicus* is mainly found in the street of Surabaya (a major city in Indonesia) as greening plants. *Pterocarpus indicus* is a type of plant that generates a lot of wastes from its leaves. The wastes of these plant dropped in Surabaya street and become litters. The appearance of *Pterocarpus indicus* tree and its wastes are presented in figure 1. In taking care of these litters, people tried to burn them. The burning of these wastes results in additional GHG (greenhouse gases) in Surabaya air and exhausting viable source of energy. To avoid these kinds of careless treatments of *Pterocarpus indicus* leaves waste, the leaves waste can be processed into briquette fuel.

Figure 1. *Pterocarpus indicus* tree and leaves waste in Surabaya

Processing *Pterocarpus indicus* leaves waste by converting them into briquette may reduce the leaves waste presence in Surabaya street without adding GHG in the environment. Briquette will help the energy potential of *Pterocarpus indicus* leaves waste to be utilized and the high bulk density of briquette helps in distributing the fuel. The previous study has discovered the potential of *Pterocarpus indicus* leaves waste as briquette [4]. To further this solution by increasing the quality of briquette produced, the optimal briquette processing parameters need to be discovered. Parameters in manufacturing briquette such as particle size and pressure affect the quality of briquette [2,13]. This study was conducted to discover these parameters level to obtain the optimum size for *Pterocarpus indicus* leaves waste briquette material, and the optimum level of pressure used in compacting *Pterocarpus indicus* leaves waste briquette.

2. Experimental method

The steps of creating briquette begin by obtaining *Pterocarpus indicus* leaves waste. Obtaining *Pterocarpus indicus* leaves waste can be done by simply collecting the fallen leaves beneath *Pterocarpus indicus* plant. After acquiring the leaves waste, the leaves waste was dried in the sun for three days to reduce the water content of leaves waste. Afterward, the dried leaves waste was shredded into various sizes for the purpose of this study. The sized leaves waste was then mixed with tapioca flour mixture to create briquette materials. Subsequently, the blend of these materials was put under pressure to shape the briquette. After acquiring the briquettes, they are tested as part of obtaining data for the study. Final result of *Pterocarpus indicus* leaves waste briquette product is given in figure 2.

Figure 2. Briquette from *Pterocarpus indicus* leaves waste

In studying the effect of particle size of *Pterocarpus indicus* leaves waste briquette, three different sizes were used, 20 mesh or 800 μm, 40 mesh or 425 μm, and 60 mesh or 250 μm. These particle sizes were investigated to understand their effect on *Pterocarpus indicus* leaves waste briquette's strength and combustion characteristics. For studying the effect of compacting pressure in briquette, two levels of pressure were used, 1 MPa and 2 MPa. Pressures were applied by utilizing the hydraulic machine to a mold. The mold has 25 mm diameter and 50 mm length [4].

Briquettes created from two various pressures and three particle sizes were tested to investigate the optimum pressure and particle size for producing *Pterocarpus indicus* leaves waste briquette. The effect of this parameter on combustion characteristics was examined. The combustion characteristics investigated were flame temperature, ignition time, burning time, and combustion rate. Temperature evaluated in the briquette's surface when combustion occurred is the flame temperature. the time measured for briquette to continuously burn is known as ignition time. While the time evaluated for briquette to turn into ashes is known as burning time. The rapidity of briquette's change of mass is known as combustion rate.

3. Results and discussion

A previous study on *Pterocarpus indicus* leaves waste briquette discovered the optimum proportion of *Pterocarpus indicus* leaves waste and tapioca mixture. This study also conducted proximate and ultimate analyses on *Pterocarpus indicus* leaves waste briquette and determine its calorific value. The study suggested that 90% *Pterocarpus indicus* leaves waste and 10% tapioca mixture ratio is the optimum ratio for highest calorific value. The calorific value of *Pterocarpus indicus* leaves waste briquette was found to reach 4648 kcal/kg, which is higher compared to the other type of biomass briquettes such as sawdust, rice straw and sugar cane [4].

The previous study has also conducted proximate and ultimate analyses of *Pterocarpus indicus* leaves waste briquette. The proximate analysis results are 5.2 % water content, 6.2 %wt ash content, 73.3% volatile matter, 15.3 %wt fixed carbon, and 0.25 %wt total sulfur. The ultimate analysis results are 49.12 %wt carbon. 5.67 %wt hydrogen, 0.19 %wt nitrogen, 0.25 %wt sulphur, and 33.4 %wt oxygen. These two analyses results indicated that *Pterocarpus indicus* leaves waste briquette is a viable renewable energy source [4].

The effect of pressure and particle size to the flame temperature is shown in figure 3(a). The higher is the pressure and particle size, the higher the flame temperature reached by briquette. The highest flame temperature is obtained with a value of 613 °C from briquette composed of 60 mesh particle size and compacted with the hydraulic pressure of 2 MPa.

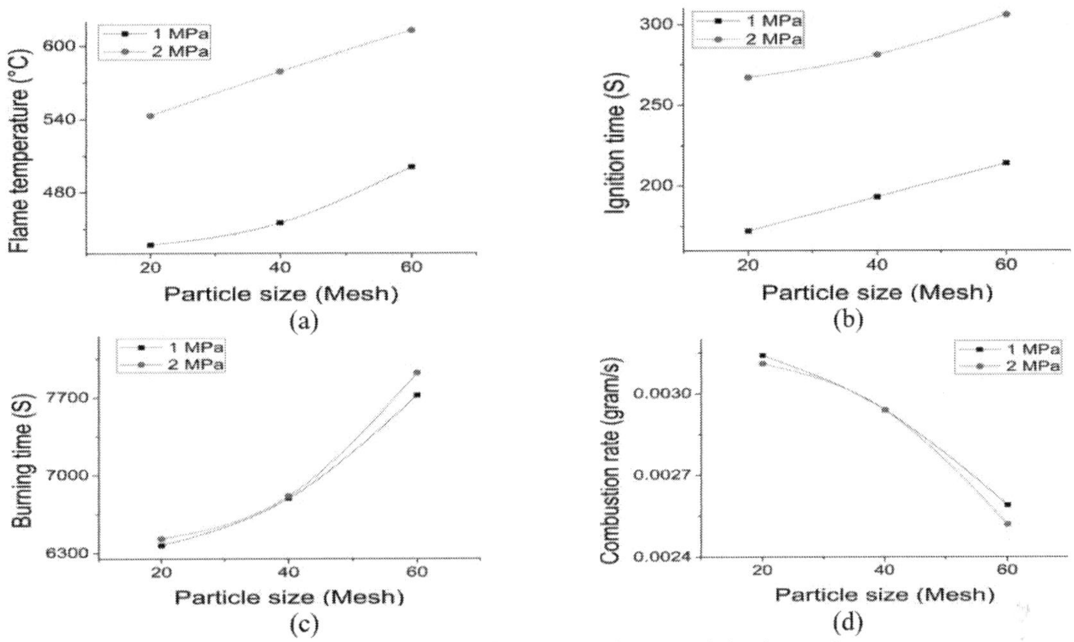

Figure 3. Combustion characteristics at various particle size and pressure
(a) Flame Temperature Characteristics, (b) Ignition Time Characteristics, (c) Burning Time Characteristics, (d) Combustion rate Characteristics

The effect of pressure and particle size to the ignition time value is shown in figure 3(b). The higher pressure and finer particle size lead to longer ignition time of *Pterocarpus indicus* leaves waste briquette. The longest duration of ignition time is obtained with a value of 306 seconds from *Pterocarpus indicus* leaves waste briquette composed of 60 mesh particle size and compacted with the hydraulic pressure of 2 MPa.

The effect of pressure and particle size to the burning time value is shown in figure 3(c). The higher pressure and finer particle size lead to longer burning time of *Pterocarpus indicus* leaves waste briquette, similar to ignition time. The longest duration of burning time is obtained with a value of 7823 seconds from *Pterocarpus indicus* leaves waste briquette composed of 60 mesh particle size and compacted with the hydraulic pressure of 2 MPa.

The effect of pressure and particle size on the combustion rate is shown in figure 3(d). The higher pressure and finer particle size lead to lower combustion rate of *Pterocarpus indicus* leaves waste briquette. The lowest combustion rate is obtained with a value of 0.00252 gram/seconds from briquette composed of 60 mesh particle size and compacted with the hydraulic pressure of 2 MPa.

4. Summary

This study shows the interaction between compacting pressure and particle size to the quality of briquette produced. As the particle size becomes smaller and compacting pressure rises, the flame temperature, ignition time, and burning rate increase. In contrast, the combustion rate decreases with finer level of particle size and higher pressure. It was discovered that the optimum pressure for compacting *Pterocarpus indicus* leaves waste briquette is 2 MPa. The optimum particle size for shredding *Pterocarpus indicus* leaves waste was found to be 60 mesh (800 μm). *Pterocarpus indicus* leaves waste briquette with these levels of parameters has a flame temperature of 613 °C, ignition time of 306 seconds, burning time of 7923 seconds and combustion rate of 0.00252 gram/second.

Although 2 MPa and 60 mesh are the best parameters found in this study, these levels of parameters might not be the most optimum for the other type of leaves waste briquette. Future research may incorporate broader range of compacting pressure and particle size to provide a more thorough study in hope for supporting the notion to utilize leaves waste as means of alternative renewable energy and promote cleaner environment.

Acknowledgments
Thanks to Kementerian Riset Teknologi dan Pendidikan Tinggi Republik Indonesia (Hibah Penelitian Produk Terapan 2016-2017) and Universitas Kristen Petra Indonesia for their support during this research.

References
[1] Goyal H B, Seal D, and Saxena RC 2008 *Ren. & Sust. Ene. Rev.* **12** 504–17
[2] Sutrisno, Anggono W, Suprianto F D, Kasrun A W and Siahaan I H 2017 *ARPN J. Eng. and App. Sci.* **12** 931 – 6
[3] Sarkar A and Praveen G 2016 *Biofuels and Bioenergy Int. Conf. (India)* (Switzerland: Springer International Publishing AG) p 117
[4] Anggono W, Sutrisno, Suprianto F D and Evander J 2017 IOP Conf. Ser.: Mater. Sci. Eng. Vol. **241** 012043
[5] Sugumaran P and Seshadri S 2010 *Biomass Charcoal Briquetting Technology for Alternative Energy Based Income Generation in Rural Areas* (India: Shri AMM Murugappa Chettiar Research Centre)
[6] Thipkhunthod P, Meeyoo V, Rangsunvigit P, Kitiyanan B, Siemanond K and Rirksomboon T 2005 *Fuel* **84** 849–57
[7] Sengar S H, Mohod A G, Khandetod Y P, Patil S S and Chendake A D 2012 *Int. J. Ene. Eng.* **2** 28–34
[8] Mwampamba T H, Owen M and Pigaht M 2013 *Ene. Sust. Dev.* **17** 158–170
[9] Stolarski M J, Szczukowski S, Tworkowski J, Krzyżaniak M, Gulczyński P and Melczek m 2013 *Renew. Ene.* **57** 20–6
[10] Sakkampang C and Wongwuttanasatian T 2014 *Fuel* **115** 186–9
[11] Hu J, Lei T, Wang Z, Yan X, Shi X, Li Z, He X and Zhang Q 2014 Energy **64** 557–66
[12] Anggono W, Suprianto F D, Sutrisno, Gotama G J, Evander J, Kasrun A W 2018 ARPN J. Eng. and App. Sci. **13** 1080-4
[13] Anggono W, Suprianto F D, Sutrisno and Kasrun A W 2016 Int. J. Ind. Res. and App. Eng. **1** 11 –4

IOP Conf. Series: Materials Science and Engineering **417** (2018) 012008 doi:10.1088/1757-899X/417/1/012008

Research on the Deicing Robot and Motion Characteristic of the Four Divisions High-Voltage Line

Bo Zhang[1], Qiao-chu Wang[2*], and Tao Wang [3]

[1]College of Mechanical and Electrical Engineering, Harbin Engineering University, Harbin 150001 China
[2]Autonomous University of Madrid, Madrid, 28049, Spain
[3]School of Mechanical Engineering, Hebei University of Technology, Tianjin, 300401 China

*zhangbo_heu@hrbeu.edu.cn

Abstract. This paper is aimed at the problem of less dehumidification robot for four-split high-voltage line and with some obstacle-removing ability and high efficiency de-icing, developed a four-split high-voltage de-icing robot with barrier capability, Based on the combination of theoretical calculation and virtual prototype simulation, this paper focuses on the key technology of robot, the overall structure design and virtual prototype modeling, and the rigid-flexible coupling of the robot under flexible high voltage line. The Lagrangian method is used to establish the rigid and flexible coupling dynamics model of the robot. The ADAMS dynamic simulation software is used to simulate the obstacle process of the robot to test the rigid and flexible coupling stability of the robot. Through the simulation results, it can be seen that the vibration of the robot in the process of obstacle crossing, by the high-voltage flexible, with the high-voltage lines along the vibration, the vibration in the direction of the space to meet the stability requirements.

1. Introduction

At present, China's long-distance power transmission capacity and operating voltage level continues to increase, to high pressure, high pressure, and even special high pressure direction. High-voltage transmission lines exposed to contact with the outside world and most of the harsh areas in the environment, the biggest challenge is the high ice [1]. With the continuous development of robot technology, the use of robots on the transmission line de-icing has become possible. Compared with the traditional de-icing method, the use of robots for high-voltage lines for de-icing with no casualties, no need to transfer load and power outages [2]. Therefore, the development of a four-split high-voltage line and has a high degree of autonomy navigation and barrier capability of the de-icing robot has important significance.

Many scholars at home and abroad have attached great importance to the technology related to the grid, and have made many researches on the maintenance and deicing of the high voltage transmission line [3]. 2004 to 2010, Shenyang Institute of Automation, Chinese Academy of Sciences (SIACAS) developed three generations of high-voltage line inspection robot [4-7], Respectively AApe-A、AApe-B 、 AApe-C. 2006 to 2010 Canada Quebec Hydroelectric Research Institute Serge Montambault, Nicolas Pouliot, who developed the high-voltage line inspection robot LineScout mobile platform. 2008 Tokyo Technical University of Japan [8] developed a four-split high-voltage line inspection robot Expliner. 2010 Wuhan University Professor Wu Gongping research team developed a 220kV high-voltage transmission line of de-icing robot [9]. 2013 Shandong Electric Power Research Institute (SEPRI) [10, 11] has improved the LineROVer de-icing robot to optimize its performance. 2006 University of Illinois at Chicago, USA Sugiyama [12] and so on the absolute node coordinate method. The flexible body is discredited by a uniform mass finite element method. 2008 Iranian Sheriff University of Technology Zohoor H and Khorsandijou S M [13] the nonlinear dynamic model is established by Hamilton principle. The floating coordinate system is chosen as the tangent coordinate system to study the flexible structure. 2011 Harbin Institute of Technology Yang Zhengxian, Kong Xianren and so on[14]Based on the basic theory of continuum mechanics, an efficient and high-precision approximation rigid-flexible coupling dynamic model for the design of the actual control system is established for the flexible body. Peking University, Tianjin University, Northwestern Polytechnical University, Huazhong University of Science and Technology, and many other famous universities in China, in the field of rigid-flexible coupling dynamics of their respective research, and achieved some research results.

2. Design of de-icing robot

Deicing robot debris barrier - walking mechanism is a barrier mechanism and the organic combination of walking, when the obstacle is not encountered, the robot moves along the high voltage line, encounter obstacles through the obstacle barrier through the obstacles. The obstacle mechanism of the robot is a humanoid robot designed according to the principle of bionics, and the walking mechanism is installed at the end of the robot arm.

When the robot is on-line for de-icing, the front drive arm and the rear drive arm are suspended from the two high-voltage lines on the underside of the four-split high-voltage line, the drive wheels at the end of the intermediate auxiliary arm are located below the two high-voltage lines on the underside of the four-split high-voltage line, and the front end of the arm, the rear end of the robot arm drive wheel with each other to match the high-voltage line, the robot walking process, to prevent the robot fall. It's work as showed in Figure 1,where the part one is percussion deicing, part two front arm, part three auxiliary arm, part four back-end robotic arm, part five Center of gravity adjustment.

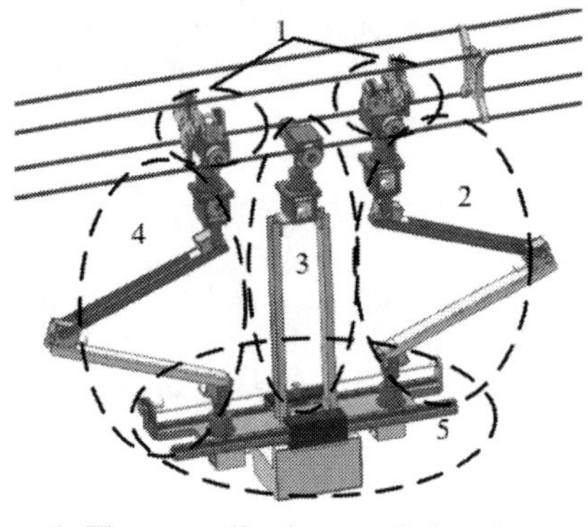

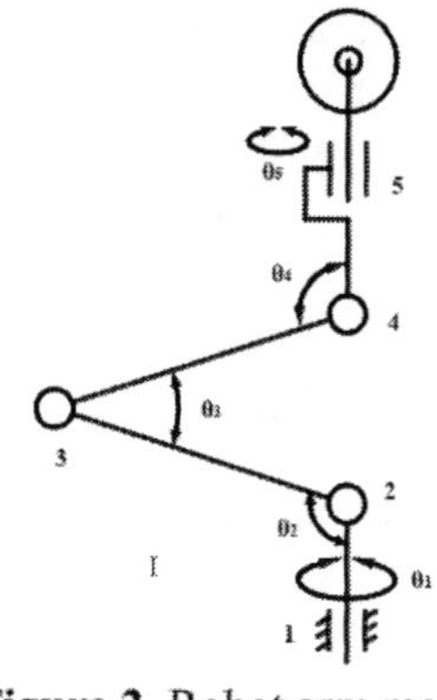

Figure 1. The overall scheme of virtual prototype of deicing robot **Figure 2.** Robot arm motion diagram

The key technology of the de-icing robot is the obstacle function of the robot, and the robot movement diagram is shown in Figure 2.In summary, the robot arm I consists of a robot support platform (base), shoulder joint, arm, elbow, arm, wrist and drive wheel. There are five degrees of freedom in the arm I, followed by the rotation and pitch of the shoulder joint, the pitch between the arms and the joints, the pitch between the arms and the arms, the pitch and rotation between the wrist and the arm. In order to meet the robot in the obstacle position on the requirements of the position, the arm of the joints of the pitch angle and rotation angle should meet a certain range of motion. The motion parameters are shown in Table 1.

Table 1. The range of movement of each joint

Joint	Pitch/rotation angle	Movement range
1	θ_1	-90°~90°
2	θ_2	30°~180°
3	θ_3	0°~180°
4	θ_4	90°~270°
5	θ_5	-90°~90°

3. Simulation method

3.1. The rigid and flexible coupling dynamics model of the robot is established

Consider the elasticity of the flexible high-voltage line caused by the elastic force of virtual work, acting on the flexible high-voltage line of all the power of virtual power:

$$\delta W = \delta U + \delta W_e = -q^T K \delta q + F_e^T q = F^T \delta q \tag{1}$$

Where: $F = -qK + F_e$, F_e is generalized force vector associated with generalized force coordinates q.

According to the Lagrangian function and the generalized force vector, the Lagrangian kinetic equation of the flexible high voltage line can be obtained [15]:

$$\frac{d}{dt}\left(\frac{\partial L}{\partial \dot{q}}\right)^T - \left(\frac{\partial L}{\partial q}\right)^T = F \tag{2}$$

The constraint equation for the complete constraint is:

$$C(q,t) = 0 \tag{3}$$

In a socioeconomic constraint, the constraint relation is velocity dependent and not integral, Select the appropriate Lagrange multiplier vector $\lambda = \left[\lambda_1, \lambda_2, \ldots\ldots, \lambda_{n_c}\right]^T$, make

$$\lambda^T C\left(\dot{q}, q, t\right) = 0 \tag{4}$$

Among them: n_c -The Number of Mechanical System Constraint Equations. Equation (4), (3) and (2) are combined to obtain the rigid and flexible coupling dynamic equations of the robot with contact constraints.

$$\begin{cases} \dfrac{d}{dt}\left(\dfrac{\partial L\left(\dot{q}, q, t\right)}{\partial \dot{q}}\right)^T - \left(\dfrac{\partial L\left(\dot{q}, q, t\right)}{\partial q}\right)^T + C^T\left(\dot{q}, q, t\right)\lambda = F \\ C(q,t) = 0 \end{cases} \tag{5}$$

3.2. Stability analysis of rigid - flexible coupling based on ADAMS

Robot in flexible four-split high-voltage line walking de-icing process, the de-icing device in the percussion process, percussion shock on the high-pressure line shaking less impact. n the process of obstacle crossing, the front end of the manipulator needs to extend forward and the motion amplitude is large, which will affect the deformation and shaking of the flexible four-split high-voltage line, thus affecting the spatial position of the robot's center of gravity.

The influence of the obstacle process and the flexible high voltage line on the spatial position of the robot's center of gravity is analyzed by the rigid-flexible coupling simulation of the robot by ADAMS. The flexible four-split high-voltage line with the actual span is simplified, and the flexible four-split high-voltage line model with a 10m pitch is established. The influence of the dancing of the flexible high-voltage line on the stability of the robot is simulated, Robot and flexible four-split high-voltage rigid-flexible coupling model showed in Figure 3.

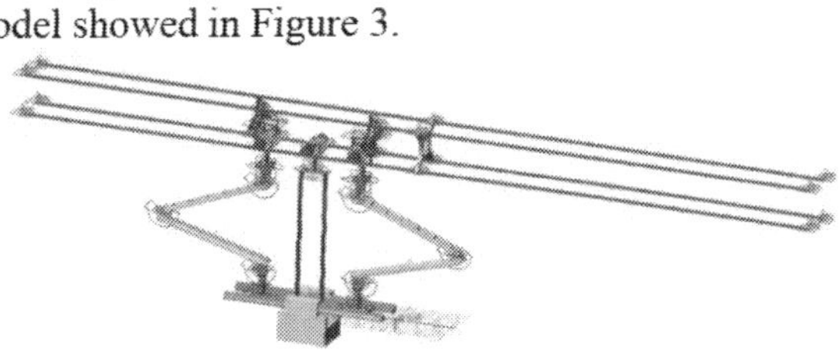

Figure 3. Rigid and Flexible Coupling Model of Robot and Flexible High Voltage Line

When the robot hangs on the flexible four-split high-voltage line, the de-icing robot's rear-end arm drive wheel, the middle arm drive wheel and the flexible high-voltage line contact, the robot's spatial position will change with the flexible high-voltage line dancing. Robot and flexible four-split high-voltage line rigid-flexible coupling obstacle process is shown in Figure 4 (a) ~ (d) .

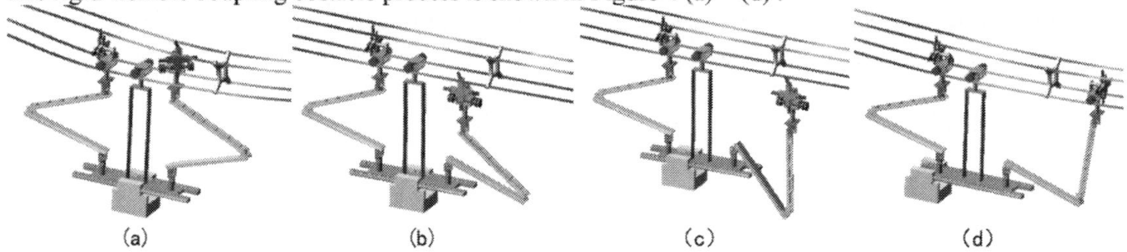

(a) (b) (c) (d)

Figure 4. Robot Rigid and Flexible Coupling

4. Simulation Results and analysis

Taking the overall center of gravity of the robot as the reference point, the spatial position change of the robot is analyzed by the change rule of the reference point spatial position. The vibration of the reference point lateral vibration, the vertical direction vibration and the horizontal direction (the movement direction of the robot along the high voltage line) are as showed in Figs. 5 (a) ~ (b).

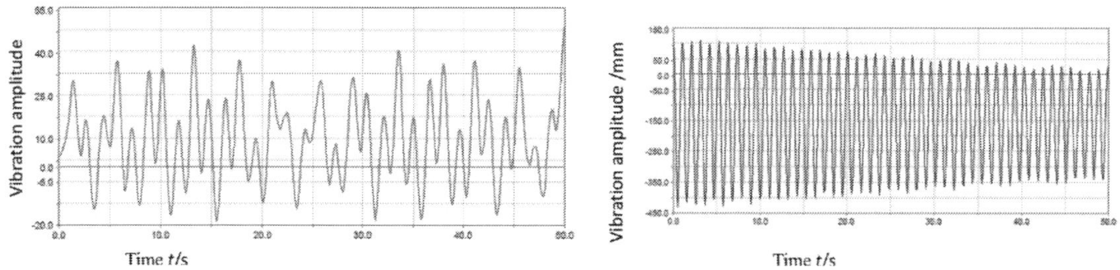

(a) (b)

Figure 5. The vibration amplitude curve of each direction of the center of gravity of the robot

The center of the middle arm drive wheel is in the same vertical plane as the center of gravity of the robot, that is, the inclination angle of the robot is zero. The center of the middle arm drive wheel is in the same vertical plane as the center of gravity of the robot. High-pressure line dancing usually caused by the robot as a whole with the high-voltage line with the side of the swing, fluctuation within the vertical ups and downs, generally does not affect the lateral side of the robot; the lateral inclination of the robot is mainly caused by the wind load acting on the windward surface of the robot. As showed in Figure 6.

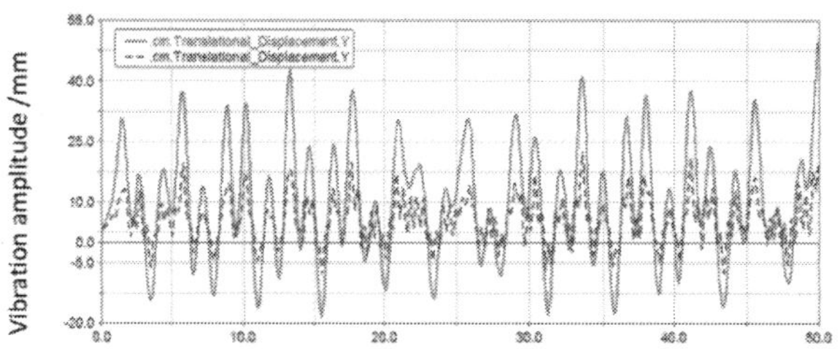

Figure 6. The lateral vibration curve of the center of the robot and the center of the middle arm drive wheel

Flexible four-split high-voltage line in the natural wind under the action of the dancing phenomenon, the impact of high-voltage line dancing on the robot from Figure 5, the direction of the center of the robot vibration curve analysis shows that. It can be seen from Figure 5 (a) that the whole center of gravity of the robot vibrates in the lateral direction, that is, the robot oscillates laterally, and the vibration amplitude is about 60mm.

From the analysis in Figure 5 (b), it can be seen that the speed of the center of gravity in the vertical plane is 350mm ~ 600mm under the influence of the flexible high voltage line, and the vibration process Although the vibration amplitude is large, but no mutation, the robot with the high voltage line vibration, the robot will not fall off.

It can be seen from Fig. 6 that the distance between the center of gravity of the center of the robot and the center of the drive arm is the distance between the center of gravity of the robot and the center of the center arm drive wheel, that is, the length of the middle arm $l_1 = 1860\text{mm}$ Lateral Inclination:

$$\theta_1 = \arcsin\frac{\Delta x}{l_1} = 1.54^\circ \tag{6}$$

By means of ADAMS, the lateral inclination angle $\theta_1 = 1.54\,^\circ$ of the robot is consistent with the inclination angle $\theta < 5^\circ$ of the robot under the action of wind load, and it is proved that the rigid-flexible coupling dynamics of the robot by ADAMS The accuracy of the analysis results, and the rigid and flexible coupling stability of the robot to meet the work requirements. This paper does not analyze the vibration under interference of wind because the de-icing robot is operating without wind.

5. Conclusion

The robot's soft motion analysis under the condition of flexible high voltage line, the robot's own movement and the wind load will cause the robot and the high voltage line to shake, need the robot has certain stability, to ensure the robot normal work. Firstly, the flexible model of high voltage is analyzed, and the rigid and flexible coupling equations of the robot under flexible high voltage line are established, which lays the theoretical foundation for the rigid and flexible coupling analysis of the robot. Based on ADAMS, the stability of the rigid and flexible coupling of the robot is analyzed. Finally, the stability of the robot is verified by the stability of the robot.

Acknowledge:
The National Natural Science Foundation of China project "Research of analysis and experiments of gripping and bearing mechanism for large-scale holding and lifting tools on ocean" (NSFC Grant No. 51479043).

References

[1] Chuncheng L,Hongnan L, Jiao L 2012 *JOURNAL OF NATURAL DISASTERS* **21** pp155-162
[2] Shuning W,Yaonan W, Feng Y 2012 *China Mechanical Engineering* **23 pp** 771-776P.
[3] Yunchu Z, Zize L,Min Tan 2004 *ROBOT* **26** pp 467-473
[4] Wang H G, Zhang F, Jiang Y, et al 2010 *Intelligent Robots and Systems (IROS IEEE)* pp 5007-5112
[5] Wang L D, Fang L J, Wang H G, 2006 *SICE-ICASE International Joint Conference IEEE* pp5400-5405
[6] Xinglong Z,Jiping Z,Hongguang W, et al 2009 *JOURNAL OF MECHANICAL ENGINEERING,* **45** pp 119-125
[7] Wang H G,Jiang Y,Liu A H, et al 2010 Applied Robotics for the Power Industry (CARPI) pp 1-7
[8] Debenest P, Guarnieri M, Takita K, et al 2008 *IEEE International Conference on IEEE* pp 3978-3984
[9] Wu G P, Xiao H, Xiao X H et al 2010 *Applied Robotics for the Power Industry (CARPI) IEEE* pp 1-6
[10] Zhang F, Cao L, Guo R, et al 2013 *10th IEEE International Conference on Control and Automation (ICCA)* pp 1415-1418
[11] Feng Z,Rui G, Jinlong Z,Lei C 2011 *Mechanical Science and Technology for Aerospace Engineering* **30** pp1917-1921
[12] Sugiyama H, Gerstmayr J, Shabana A . 2006 *Journal of Sound and Vibration* **298** pp1129-1149
[13] Zohoor H, Khorsandijou S M. 2008 *Applied Mathematical Modelling,* **32** pp2117-2132
[14] Zhengxian Y, Xianren K, Jun L et al 2011 *SPACECRAFT ENVIRONMENT ENGINEERING,* **28** pp141-146
[15] Shiji S,Chengxu H 2000 *Rigid - flexible Coupling Dynamic Analysis and Simulation of Mechanical System* (beijing:China Communications Press) pp157-158

Research on Impact and Steering Mechanism of the Puncturing Mud Robot

Zhuo Wang[1], Qiao-chu Wang[2*], and Tao Wang [3]

[1]College of Mechanical and Electrical Engineering, Harbin Engineering University, Harbin 150001 China
[2]Autonomous University of Madrid, Madrid, 28049, Spain
[3]School of Mechanical Engineering, Hebei University of Technology, Tianjin, 300401 China

zhangbo_heu@hrbeu.edu.cn

Abstract. The purpose of this paper was to propose an automatic device based on a trenchless underground perforation and a path planning method to learn. Based on conservation of mechanical energy, the authors propose a pneumatic impact mechanism. Through the establishment of dynamic analysis of mathematical Model, the authors point out the relationship between the displacement of the piston segments, mass and output parameters. Through the development of its test system, the parameters and path correlation curves are acquired to point out the mutual connection and restrictive relationship. Several data and experimental analyses are described, showing that puncturing mud robot can realize steering and its impact reaches the level of trenchless laying pipes technology at home and abroad. With the proposed approach， it is possible for puncturing mud robot to squeeze the surrounding soil layer to form a circular hole and lay the trenchless pipeline to improve perforation performance.

1. Introduction

There are mainly five kinds of methods on trenchless pipe laying in China, including pipe jacking, ramming method, impact spear, vibration squeeze soil, horizontal directional drilling technology. [1-2] Research of the Puncturing mud robot drew lessons from the working principles of impact spear, which paths can be controlled being researched by the Research Institutes at home and abroad [3-4].The robot ideal state can overcome the resistance in the soil to walk independently. Its structure is consist of tapered bit, knuckle mechanic, impact mechanism and other units. It can realize the PE or PVC pipe, cable and other small diameter pipelines underground trenchless laying. It can finish perforated underground operation with distance about 30~60m. The curve perforation is different from the impact spear which the path can be controlled. Puncturing mud robot can rotates with its rotating mechanism through transferring torque to influence mechanism. [5-7]This is the same principle as the study by Yang et al the ontology appearance of Puncturing mud robot looks like two long rods with ribs [8]. Xv et al. proposed that SDH-70 Pneumatic Impact Trajectory Controlled Puncher can change the working state of the bistable head through making flexible rod relative bistable head rotating 180 degrees [9]. The research of Wang et al. [10] revealed the change rule of the KCM-130 controlled impact spear's drilling trajectory, and designed the structure of impact spear.

At present, foreign puncturing mud robots (underground perforating equipment) are manipulated on land by manual operation, and can't be operated automatically. The impact spear can have independent motion characteristics. After it is installed steering mechanism, it can be controlled the trajectory of impact motion curve. However, the article of the robot research and experimentation, the special rotation mechanism of the robot body can achieve the rotation driven. The generated torque is transferred to the robot body, and the autonomous movement of the robot soil is realized.

2. Impact force test method and system composition

According to the test principle, the test system is composed of impacting spear, force sensor, transmitter, power supply and oscilloscope, single chip sampling system and computer system, as showed in Figure 1(a-d).

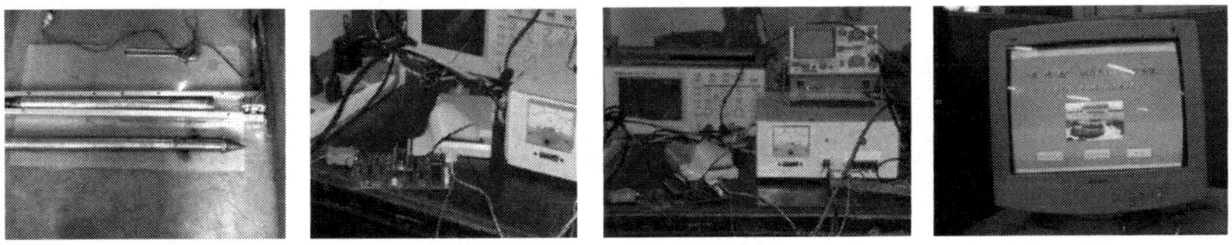

(a) (b) (c) (d)

Figure 1. Test system components

Notes: (a) Impact force transduce; (b) single chip sampling system; (c) power supply and oscilloscope; (d) computer system

1)Transmitter BK-2F force sensor measured voltage signal is 0-22mv, if the voltage signal is directly used as a microcontroller system A / D input, it will causes the large measurement error. So it must be amplified with a transmitter line to 0 -5v of the voltage signal.

2)Sampling system analog voltage signal output from the pressure sensor is amplified by the transmitter and sent to the A / D converter. The conversion results by AT89C51 via a serial port to a computer in interrupt mode.

3) The computer system mainly includes microprocessors, keyboard, printers, CRT display terminal, I / O interface and a wealth of system software and application software. Through using the computer system and Visual C ++ language programming, it completes serial data and data processing to accept the work.

3. Design method and analysis of impact and steering mechanism

3.1. Mathematical model establishment of pneumatic impact mechanism

The pneumatic impact mechanism of Puncturing mud robot is shown in Figure 2.

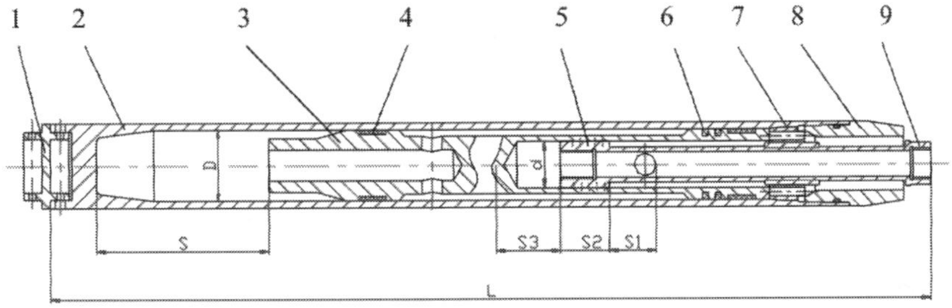

Figure 2. Structure diagram of impacting machine

1-short connection of the front, 2-cylinder, 3-Impact piston, 4-support ring, 5-valve rod core, 6-piston ring, 7-valve rod block, 8-cylinder rear-end plate, 9-intake pipe joints

Under a certain pressure, piston moves backwards and forwards in the cylinder, it can be divided into piston power stroke and return stroke. There are three kinds of motions in the power stroke, those are uniformly accelerating motion, variable accelerating motion and slowdown impact which are shown as piston displacement S_1, S_2 and S_3 accordingly. Return stroke can regard as the inverse process of the power stroke, that is, returning uniformly accelerating motion S_3, variable accelerating motion S_2, slowdown impact S_1. Then, after piston starting, piston will impact the cylinder front end cover and rear end cover respectively with certain velocity, which forms a kind of self-induced vibration which has certain velocity and frequency. Then, we will build its dynamic process mathematics model by taking the case of the impact process, its mathematical model is given by the following equations [11-12]:

$$W = {m_p v^2}/{2} \tag{1}$$

$$a_1 = \frac{dv_1}{dt} = (\frac{\pi}{4}(d^2 \cdot P_0 - D^2 \cdot P_1) - f_\mu)/m \tag{2}$$

$$t_1 = \sqrt{2S_1/a_1} \tag{3}$$

$$v_1 = \frac{ds_1}{dt} = \sqrt{2 \cdot S_1 \cdot a_1} \tag{4}$$

$$P_s = P_1(\frac{V_1}{V_1 - A_1 \cdot s})^k \tag{5}$$

$$a_s = {(P_0 \cdot A_0 - P_s \cdot A_1 - f_\mu)}/{m} \tag{6}$$

$$A_0 = \frac{\pi}{4} \cdot d^2 \quad A_1 = \frac{\pi}{4} \cdot D^2 \tag{7}$$

$$a_3 = \frac{\pi}{4}(D^2 - d^2) \cdot P_0 - f_\mu \tag{8}$$

$$v_3 = v_2 - a_3 \cdot t_3 \tag{9}$$

$$S_3 = v_2 \cdot t - 0.5 \cdot a_3 \cdot t_3^2 \tag{10}$$

$$v' = \frac{(m_p - m_w \cdot e)}{m_p + m_w} v_3 \tag{11}$$

$$u = \frac{(1+e)m_p}{(m_p + m_w)} v_3 \tag{12}$$

Where: W is the impact function, Nm; m_p is the quality of impact piston, kg; v is the piston impact velocity, m/s; P_1 is the pressure before the air chamber, MPa; f_μ is the friction, N; S_1, is the piston uniform acceleration, S_2 is the variable acceleration, S_3 is the deceleration impact displacement, m; a_1, and a_3 are the piston acceleration in displacement S_1,S_3, m/s^2; t_1,t_3 are the time of displacement S_1,S_3, s; After displacement S_1, S_2,S_3,the velocity are v_1,v_2,v_3, m/s; d, D are the cross-sectional diameter of after or before the gas chamber, m; A_0,A_1 are the Sectional area of after or before the gas chamber, m^2; P_s is the a gas chamber pressure when the displacement of the piston is s, MPa;V_1 is the initial volume of the former chamber,m^3; s is the piston displacement, m; a_s is the piston acceleration in displacement s, m/s^2; k is the adiabatic coefficient, k=1.41; v' is the speed after the piston hit, m/s; u is the impact mechanism forward speed, m/s; m_w is the total mass of impact mechanism apart from the quality of piston, kg; e is the coefficient of restitution, for steel to steel impact, e=0.56.

The formulas (2) - (4) are the uniform acceleration process, the equations (5) - (7) are the accelerating processes, and the equations (8) - (10) are the deceleration impact processes. We can get a differential equation (13) from Equation (5) - (7), such as:

$$\frac{d^2 s}{dt^2} = a_s = (\frac{\pi}{4}(P_0 \cdot d^2 - P_1(\frac{V_1}{V_1 - \frac{\pi}{4} \cdot D^2 \cdot s})^k \cdot D^2) - f)/m \tag{13}$$

Initial conditions: t_0=0, s_0=0, v_0=v_1; Ultimate displacement: s=S_2. It is difficult that an algebraic solution is worked out from this differential equation. So, the fourth-order Runge-Kutta method was used for the numerical solution. Equation (13) can be turned into a first order differential equation respectively with mathematical transformation as followed.

$$s' = V$$

$$V' = s'' = (\frac{\pi}{4}(P_0 \cdot d^2 - P_1(\frac{V_1}{V_1 - \frac{\pi}{4} \cdot D^2 \cdot s})^k \cdot D^2) - f) / m \tag{14}$$

Where:

$$s'' = a_s = \frac{d^2 s}{dt^2}$$

$$s' = V = \frac{ds}{dt}$$

First-order differential equation $y' = f(x, y)$ of the Runge-Kutta method was involved in, then:

$$y_{n+1} = y_n + \Delta y_n = y_n + \frac{1}{6}(K_1 + 2K_2 + 2K_3 + K_4) \tag{15}$$

$$\begin{cases} K_1 = f(x_n, y_n) \cdot h \\ K_2 = f(x_n + h/2, y_n + K_1/2) \cdot h \\ K_3 = f(x_n + h/2, y_n + K_2/2) \cdot h \\ K_4 = f(x_n + h, y_n + K_3) \cdot h \end{cases}$$

Where: K_1, K_2, K_3, K_4 is the coefficient; x_n, y_n is the first n-point independent variables and variables; h is the step size.After the establishment, the mathematical model can simulate the dynamic process of pneumatic impact mechanism.

3.2. The effect of main parameters of impact mechanism on performance

Output performance of impact mechanisms (impact energy, impact frequency etc.) are affected by operating pressure P_0, the quality of impact piston m_p, displacement parameters of the acceleration S_1, $S2$, $S3$ etc. Then the effect will be analyzed by changing the parameters, while the other parameters cannot be changed. According to the impact mechanism of the external diameter of φ60mm, the impact energy $E \geq 20$ J design conditions to initially determine the main parameters: impact piston quality m_p=5 kg, cylinder bore is φ50 mm, piston stroke s=133 mm, operating pressure P_0=0.6 MPa.

As depicted in the Figure 3, impact energy and impact velocity are increased with the increasing of S_1 and S_2. Impact frequency and reverse impact velocity are decreased with the increasing of S_1 and S_2.

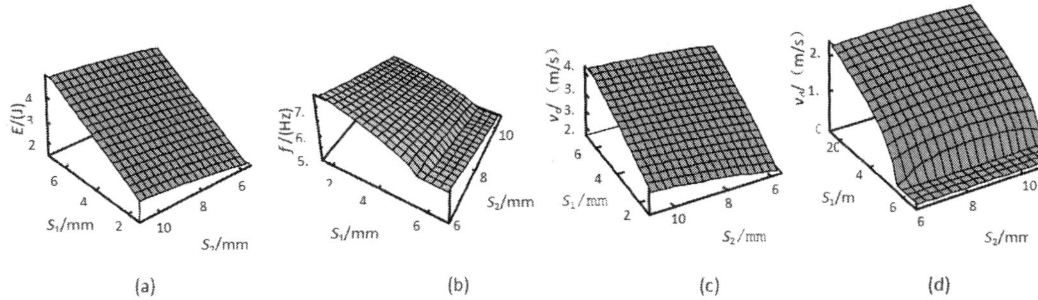

Figure 3. The effect of S_1 **and** S_2 on E, F, V_c and V_d

Analysed from Figrue 3 (a), S_1 decides the amount of reverse impact velocity, while a greater impact on reverse impact velocity from S_2 is shown when S_1 is larger, and there is space in the increasing of velocity. So, to ensure certain reverse impact velocity, S_1 will be less than a certain amount, about $S_1 \leq 58$mm as showed in the Figure 3. The effect on the returning impact velocity is the maximum when $S_2 \approx 90$mm; When impact energy and impact velocity are the maximum as shown in the Figrue3(b and c), it can conclude from Figrue3(d) that returning impact velocity will be zero, that is, piston can't reverse impact. So, the choice of S_1 and S_2 should make the impact energy and impact velocity maximum. Other effects on the performance like composition change of S_2 and S_3, S_1 and S_3 are similar to the above.

3.3. Design of impact steering mechanism

The steering mechanism of the robot body adopts the structure of the double-piston cylinder design. The reciprocating motion of the piston drives the tooth clutch to be engaged and separated, so that the head rotates and realizes the curve and linear motion. The mechanism diagram is shown in Figure4.

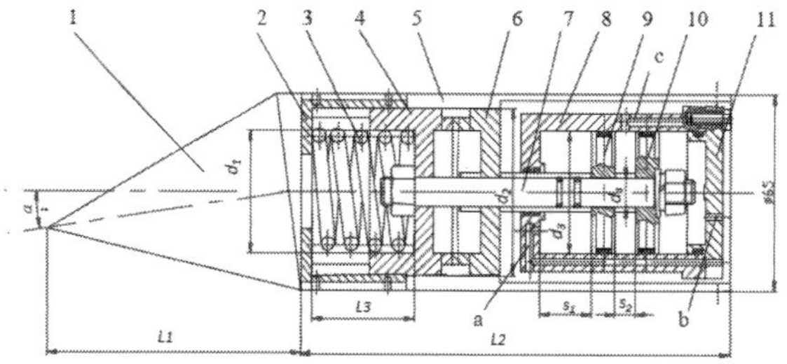

Figure 4. Structure diagram of turning machine

1. Tapered bit 2. Rotating sleeve 3. Compressed spring 4. Front clutch 5. Clutch bases 6. After the clutch 7. Piston 8. Cylinder 9. Front piston 10. After the piston 11. Cylinder back cover

When the trajectory of the curved drilling hole needs to be adjusted, the attitude of the robot is feed backed by the testing part, and the steering mechanism is rotated to a fixed angle according to the posture, so that the inclination angle of the conical drill bit is adjusted to the direction of the adjustment path. Only the impact does not rotate, then drill out of the path along the direction of the deviation angle of the drill, that is, to achieve the curve drilling.

In the drilling process, according to the feedback of the robot's attitude, at any time to adjust the drilling direction, curve prototype can be achieved in the soil track control drilling operation, therefore, it can be called real "puncturing mud dragon" Robot prototype.

4. Experiment and analysis

4.1. Impact mechanism performance experiments

Computer system was used for dealing with the test data. The data processing system is mainly composed of an interactive interface, data curve display, the relevant parameter setting, impact energy and frequency calculation. The test data are shown in Table 1 and Figure 5. Table 1 showed the optimize design result. It reached the specifications of the same kind products and means that it can finish perforation operation

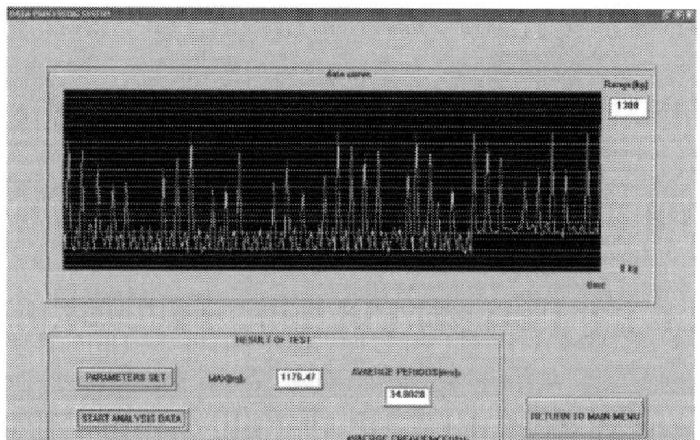

Figure 5. The measured curve of impact force

Table1 Testing of main performance parameters of puncturing mud robot

Gas pressure (MPa)	0.4	0.5	0.6	0.65	0.80
Impact frequency (Hz)	4.49	5.20	5.98	6.04	7.08
Impact force (N)	7480	9351	11221	11525	13113
Impact energy (N·m)	14.9	22.4	32.2	34.1	50.9

4.2. Rotational experiment of steering mechanism of curve prototype

The robot works as shown in Figure 6(a). Before the experiment, we will plan its trajectory. When the angle of the drill bit of the prototype is 7.5 °, the experimental data of the drill bit impacting soil are recorded and the drilling trajectory is measured. The results show that the experiment can achieve the expected trajectory trend, as showed in Figure 6(b). Impact body with drilled bit can be drilled in the soil and be able to drill the macro-curve trajectory.

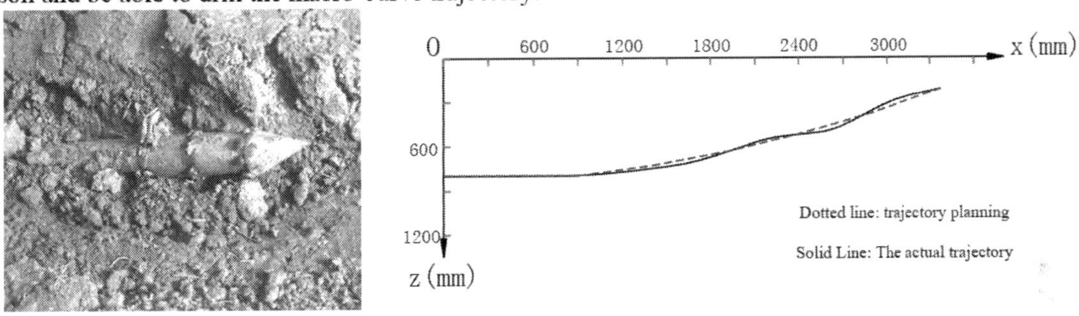

(a) (b)

Figure 6. State of curve prototype drilling out of soil at some curvature

Experiments show that the Steering mechanism realizes the principle of rotation, and proves that the overall design scheme of the robot is feasible.

5. Conclusion

This paper introduces a method of testing the pneumatic impact performance, the impact force is taken as the direct measurement object, and the design of the impact and the steering mechanism is studied. The influence of the impact mechanism parameters on the performance is analyzed by energy conservation method. Experiments demonstrate that the impact force of the robot has reached the performance index of similar products at home and abroad. The designed steering mechanism can drive the rotation of the impacting head in the soil and control the perforation in the soil curve to realize the autonomy movement of the robot.

References

[1] R.L. Sterling 2010 *Journal of GeoEngineering* **5** pp 81-85
[2] J. M. 2014 *Science and Technology Review* **8** pp32-26
[3] W. L. Wang, S.M. Peng, and K.Yin 2004 *Construction Machinery and Equipment* **35** **pp**29-33
[4] W. L. Wang 2005 *Construction machinery* **5** **pp** 32-34
[5] Clarke 2001 *Equipment for Geotechnical Engineering* **1** pp55-56
[6] R. Yach 2002 *Trenchless Technology* **2** pp19-21
[7] S.R. Kramer and J.C. Thomson 2014 *Trenchless Pipeline Projects-Practical Applications, ASCE* pp1235-1242
[8] P.C. Yang, X.G. Jian, Y.F. Wang et al 2012 *Journal of Machine Design* 11 pp1-4.
[9] B.F. Xv, Y. Yao and L.D. Shi 2001 *Construction Machinery*. **3** pp47-50
[10] R.S. Wang, S.M. Peng and K. Yin 2009 *Exploration Engineering* **36** pp338-341
[11] B.F. Xv, M.X. Lu and R.A. Zhang 2000 *Hydraulics Pneumatics and Seals* **82** pp29-31
[12] Z. Liu, J.S Wu, and W. Li 2007 *Chinese Journal of Mechanical Engineering* **18** pp 1769-1772

IOP Conf. Series: Materials Science and Engineering **417** (2018) 012010 doi:10.1088/1757-899X/417/1/012010

Modeling and Validation of New Continuum Robot Backbone Design With Variable Stiffness Inspired from Elephant Trunk

Azamat Yeshmukhametov[1] , Zholdas Buribayev[2,3], Yedilkhan Amirgaliyev[2,3] Rahul R Ramakrishnan[4]

[1] Graduate School of Science and Technology, Tokai University, 4-1-1 Kitakaname, Kanagawa-ken, 259-1292, Japan
[2] Department of Informatics, Kazakh National University, Al-Farabi 71, Almaty, Kazakhstan
[3]Institute of Information and Computational Technologies, Shevchenko28, Almaty, Kazakhstan
[4]Undergraduate School of Science and Technology, SRM Institute, Kattankulathur - 603 203, Tamil Nadu, India

Corresponding author email: yeshmukhametov.coba@gmail.com

Abstract. Attentions to the bio-inspired robots are increasing dramatically in last two decades. Flexible structure, excellent flexibility and possibility of working in constrained environment made continuum robots attractive in robotic community. The purpose of this paper is to introduce, describe and test a novel design of continuum robot backbone design with variable stiffness which using of coil springs. A kinematic model is introduced and explained which could be applied for to a wide range of construction with two pairs of cables per section design. Furthermore, based on the geometry and material property of compliant joint, the robot is verified for justification of the robot construction. Finally, prototype motion analysis tested and repeatability experiments are carried out.

1. Introduction

Continuum robots present a several potential benefits over conventional than rigid link robots in applications that involve reaching through complex trajectories in cluttered environments or where the robot must compliantly contact the environment along its length. Because of inherited flexibility, the shape of continuum robot could be impacted by both actuation and external loading [1]. Moreover, a continuum robot has redundant degrees of freedom, able to move flexibly and has a large work-space [2]. Mostly, continuum robots are increasingly exploited in minimally invasive surgery where is required robot arm with ability to work in confined space [3].

Traditional snake-like manipulators are fully articulated serial robot arms with multiple number of joints. Each joint is actuated by one or two integrated motors [3], depending on the joint type. Unlike rigid articulated robots, continuum robots have flexible structures and hyper-redundant degrees-of-freedom (DOF) that allow for smooth elongation, shortening and bending from force and displacement control [4]. According to the hardness of the joint, it divides into two types: rigid and flexible. Rigid joint is traditionally universal joint or spherical joint, while flexible joints are utilize rubber or helical compression springs [5]. For continuum robots to reach confined and complex trajectory multiple joints are required to achieve necessary bending angle [5,6,7,8,9,10]. However, flexible backbone increases the difficulty in kinematic modelling. The section of the continuum robot generally assumed to bend with a constant curvature[11,12]. Also, the kinematics of multi-section continuum robot was presented with the Denavit-Hartenberg parameters [7] . The kinematics of concentric tube robot has also been presented based on a constant curvature . Moreover, variable curvature kinematics has been investigated. However, controlling the robot with variable kinematics curvature is still challenging issue.

In this paper, continuum robot construction design with variable helical compression spring is adopted justified and explained. Moreover, kinematics of the continuum robot would be analysed and presented. This project presents a continuum robot with shortening and elongation properties. Section 1 presents design structure and construction features of the robot. Section 3 presents geometric modelling of the robot and conclusion will discuss and presents future development of the proposed robot.

2. Backbone design

This paper presents a bio-inspired tendon actuated multi-section flexible robot. The structure is similar to traditional snake-like robots, serially connected hard part (discs) via compression springs and actuated by 1mm metal cable. Moreover, cables actuates by dc motors.

The main feature of proposed continuum robot is a backbone structure and used materials. Thus, 3 types of compression springs had been applied with variable hardness constant. Prototype has a 3 segments, nine spacer discs and three types of compression springs. These springs have a variable spring constants: 35.3N/mm, 13.2N/mm, 8.8N/mm , outer diameter-20 mm, length- 30mm.

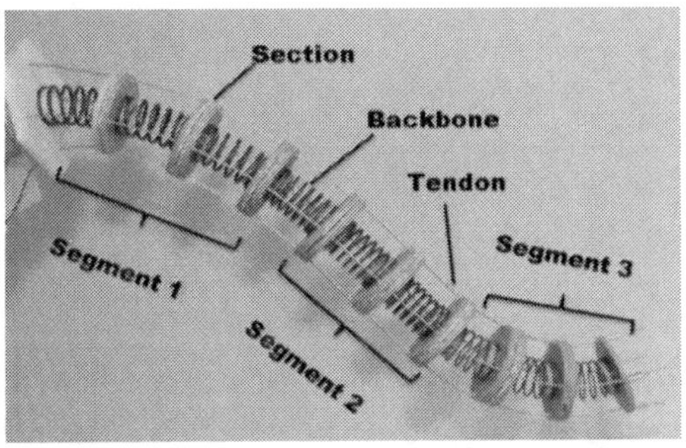

Figure 1. Experimental continuum robot prototype

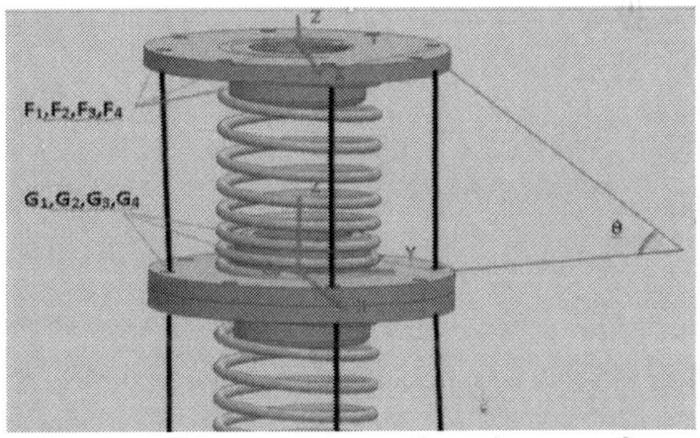

Figure 2. Nomenclature of continuum robot

The reason of using of variable spring constant is comes from a concern about static and natural formation. For this idea an optimal choice drawn came after several experiments in the laboratory. So, the main conclusion is the backbone hardness should be gradually decrease, from, bottom to the top section of the robot. Such kind of variation will increase stability and dexterity of the robot as well [figure4].

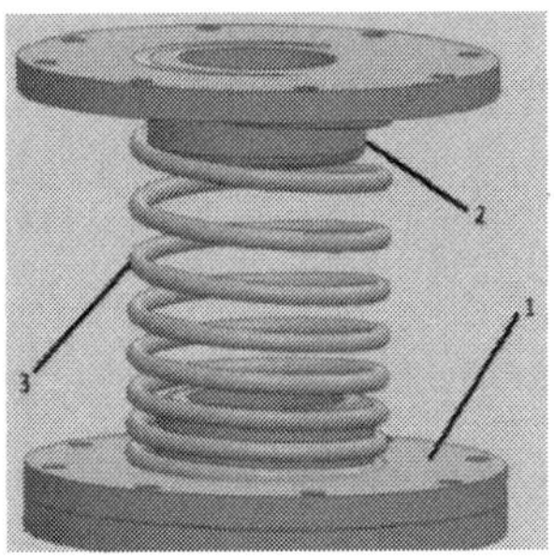

Figure 3. Single section of the robot:. 1-spacer disc, 2-constraint for spring, 3- helical compression spring

Figure 4. Comparison of continuum robot backbone hardness.

Figure 3 illustrates the main difference of the backbone hardness, so in construction of the robot this fact had been taking into an account. Therefore, in prototype there are 3 sections[fig1], so in section-1 mounted the hardest spring WL20-30, in section-2 WF20-30, in section-3 softest spring WR20-30, these customized springs had been ordered from company MISUMI. The reason of using springs for backbone is elongation and shrinking feature. Which gives one more additional degree of freedom. However, using of flexible backbone will accuse buckling of the flexible backbone if the force along the backbone exceeds the bucking load [7]. Therefore, the better solution in such case is to design the backbone which can mechanically minimize the twisting angle.
There are some parameters which requires a justification, for example, number of discs in one segment, length of the section, gap distance, joint length and spacer disc diameter. Gap distance or joint length is mostly depending on output bending of the robot and numbers of the sections. Therefore, the length and the bending of the section should keep the following condition: Above condition could change in terms of final output continuum bending angle. Disc diameter, in this prototype is constant 50 mm, although [4,6] claims the diameter size of the disc should gradually increase from top to the base, same as spring constant, it is because of robot singularity and dynamics. In additional such kind of approach would make simple routing of tendon, because of different size of disc allow actuating separately. Compliant joint hardness is one the crucial part of the continuum robot, so in this prototype springs variable hardness constants applied. According to the conducted experiments, same backbone hardness could increase required motor torque for joint bending

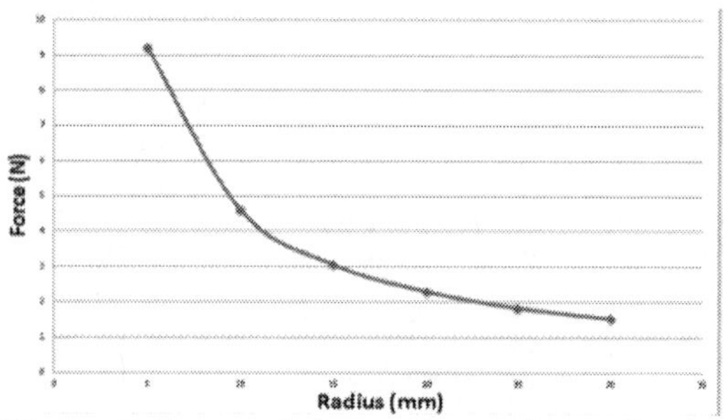

Figure 5. Force relation to the section radius

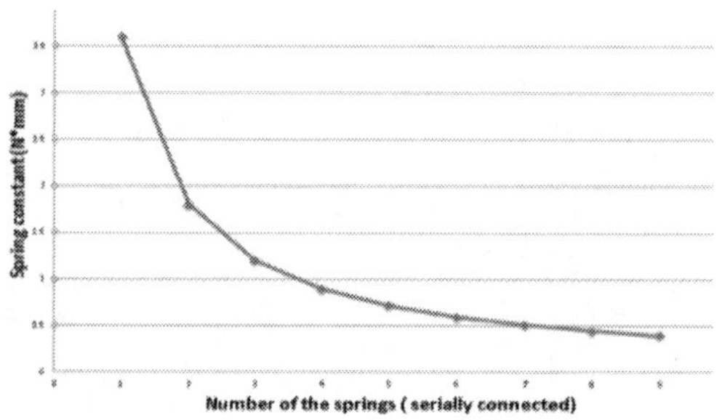

Figure 6. Relation of serially connected number of springs and hardness coefficiency

2.1 Geometric model of tendon driven Continuum Robot Manipulator

The Flexible backbone continuum Robot structure proposed in this paper consists of three modules. Each module contains three sections of individual helical compression springs connected with a 3d printed compliant joints. Each module is divided into three segments and consists of springs with different spring stiffness.

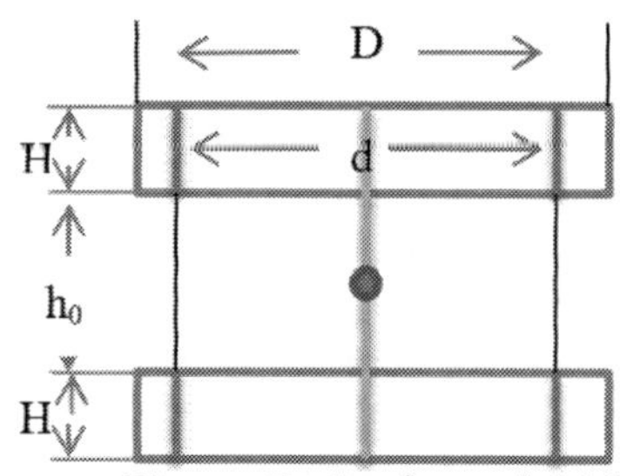

Figure 7. Section at initial state

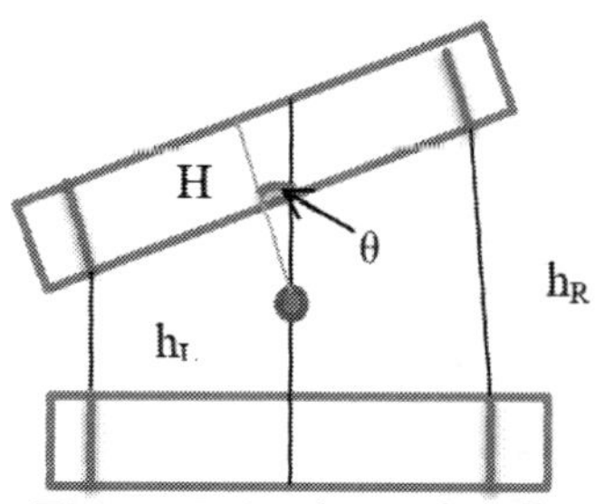

Figure 8. Section at bending state

When each join bends θ leftward, the gap distance and the wire length inside of the section eyelet remains the same. Based on simple geometry, it can be shown that the gap distance after bending is as follows[5].

$$\begin{cases} h_R = h_0 + \left[d \cdot \sin\left(\frac{\theta}{2}\right) - 2h_0 \cdot sin^2\left(\frac{\theta}{4}\right)\right] \\ h_L = h_0 - \left[d \cdot \sin\left(\frac{\theta}{2}\right) + 2h_0 \cdot sin^2\left(\frac{\theta}{4}\right)\right] \end{cases} \quad (1)$$

In summation, the total wire length after bending is shown as below

$$\begin{cases} L_R = L_0 + N\left[d \cdot \sin\left(\frac{\theta}{2}\right) - 2h_0 \cdot sin^2\left(\frac{\theta}{4}\right)\right] \\ L_L = L_0 - N\left[d \cdot \sin\left(\frac{\theta}{2}\right) + 2h_0 \cdot sin^2\left(\frac{\theta}{4}\right)\right] \end{cases} \quad (2)$$

Detailed information about kinematics of the model in [7].

3. Conclusion

In conclusion, design validation backbone with variable stiffness hardness had been tested and analyzed. According to the conducted experiments the following assumption had been confirmed: firstly, structure stiffness, robot became more rigid in control. Secondly, twisting problems were minimized backbone structure mechanically arrange the initial position and desired position of the robot. Variable stiffness backbone became efficient in control as well, for instance, to control cables were used to actuate tip of the robot, so bending was efficient with variable backbone stiffness rather than constant backbone stiffness. Because in case of actuation robot started to bend from the top part of the robot, while constant one started to bent from the middle part.

Reference

[1] D.Caleb Rucker, Robert J Webster III.,Statics and Dynamics of Continuum Robots With General Tendon Routing and External Loading, vol.27.NO.6.December 2011

[2] Guochen Niu, Li Wang and Guanghua Zong, Attitude control based on fuzzy logic for continuum aircraft fuel tank inspection robot, 29 (2015)

[3] Zheng li, Liao Wu, Hongliang Ren, Haoyong Yu, Kinematic comparison of surgical tendon-driven manipulators and concentric tube manipulators, 148-165 pp, Machine and Mechanism (2017)

[4] Kun Cao, et.all. Workspace Analysis of tendon –driven Continuum Robots Based on Mechanical Interference Identification, Journal of Mechanical Design, 2017.

[5] Zheng Li and Ruxu Du, Design and Analysis of a Bio-inspired Wire-Driven Multi-Section Flexible Robot, International Journal of Advanced Robotic Systems, 2013.

[6] Enver T.,Ian D.W. and Darren M.D., Dynamic Modelling for Planar Extensible Continuum Robot Manipulators, Robotics and Automation, Italy, 2007.

[7] Xin Dong, and et.all, A Novel Continuum Robot Using Twin-Pivot Compliant Joints: Design, Modelling, and Validation., Journal of Mechanisms and Robotics, 2016.

[8] K. Suzumori, S. Iikura and H. Tanaka, Development of Flexible Microactuator and Its Applications to Robotic Mechanisms}, pp. 1622-1627, proceedings of the 1991 IEEE International Conference on Robotics and Automation, Sacramento, California, 1991.

[9] M.D. Grissom, I.D Walker, Design and experimental testing of the OctArm soft robot manipulator, Proc. Of SPIE, 2006.

[10] W. McMahan, I.D. Walker field Trials and Testing of the OctArm Continuum Manipulators, pp. 2336-2341, Proceedings of 2006 IEEE International Conference on Robotics and Automation, Orlando, Florida, 2006.

[11] P.E. Dupont, J. Lock, B. Itkowitz and E. Butler, Design and Control of Concentric-Tube robots, Vol.26, No.2, pp. 209-225, IEEE Trans. Robotics., 2010.

[12] P. Sears and P.E. Dupont, Inverse Kinematics of Concentric Tube Steerable needles, pp. 1887-1892, IEEE International Conference on Robotics and Automation, Roma, Italy, April 2007.

Diverting and Sorting Mobile Robotic Table for Motion Control Testing

Héctor Terán[1, a], Oscar Artega[1,b], Eduardo Cárdenas[1, c], Marcelo Ortiz[1, d], Victor H Andaluz[1,e] and Miguel Carvajal[1,f]

[1]Universidad de las Fuerzas Armadas ESPE, Sangolquí, Ecuador

[a]hcteran@espe.edu.ec, [b]obarteaga@espe.edu.ec, [c]mecardenas2@espe.edu.ec, [d]amortiz8@espe.edu.ec, [e]vhandaluz1@espe.edu.ec, [f]macarvajal@espe.edu.ec

Abstract. This article is about the design and construction of a Diverting and Sorting mobile robot table for testing motion control techniques, the robot consists of nine modules with two degrees of freedom each one in three times three matrix configuration; a table with dual rack gear track in bottom move between modules, combined control of each module allows the transport of objects. Computer vision techniques are employed in this case a Pinhole camera model for detection and location of the carry objects. Each of the characteristics of the design and construction of the parts that form the mechanical system are explained in detail. The table design includes high quality and low-cost materials. At finally, the robotic table is verified with different techniques of control, for example follower desired path. The result shows a difference between the real path and the desired path that will be calculated under the quadratic error function.

1. Introduction

The evolution of robotics to move forward in huge strides to achieve a new technologic age, through robotic system that is also reliable and meets the requirements of high capacity [1]. Different types and designs of the new technologic age of robotics have been introduced to industry of manufacture. In the industrial area, the robotic arms are multifunctional, programmed to perform a variety of tasks in environments of high risk to humans. [2][3][4].

Nowadays, there are multiple types of ordering systems with software and standard controllers, which operate in storage and distribution stations feeding operations in supply chains. The high product demands require software and high-speed hardware, this increases productivity, quality and performance. Manufacturing industries handle large warehouses of products in carton packs forced to stop the production line. Systems that divert the flow and classify reduce time and deliver products just in time. As well as artificial vision has allowed the generation of new techniques for data processing. [5] [6] [7] [8]

2. Design and construction

It includes different parts that provide data acquisition, drive and control motors. The camera is the primary data-acquisition device by collecting sequential images that will be previously measured using sophisticated machine vision algorithms. A diagram of the Diverting and Sorting robot table electronic system is illustrated in figure 1.

A Logitech 960-C270 performs image acquisition. Captures successive images of the position of the object on the table, these data are interpreted by a computer, where artificial vision algorithms are performed. This data processing is directly linked to the processor speed of the computer. The computer used for this table complies with the following characteristics: Intel i7, 16 GB RAM and Windows operating system with MATLAB software that is responsible for the artificial vision algorithm and the tracking control for the movement of the motors.

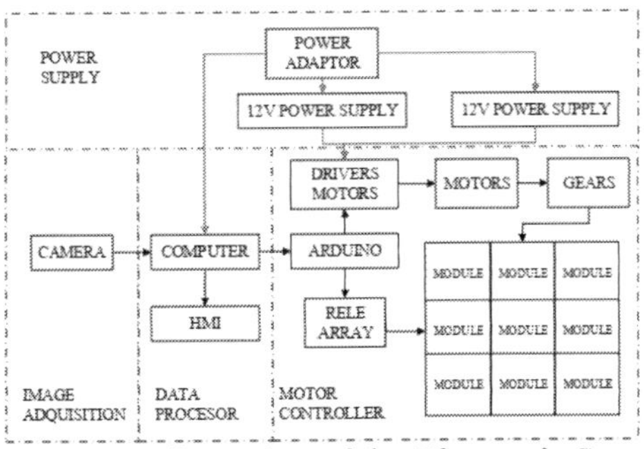

Figure 1. Configuration of the Electronic System.

2.1. Electronic Design

An Arduino Mega board is used to control the motors, sending data through a serial port with 2 information signals for each of the 18 motors in total. The signals represent the direction of rotation and the speed. In addition, a Cytron 10A-25V dual channel DC motor is used to drive and power the Polulo 25Dx54L motors. Finally, each channel to Cytron 10A-25V is connected to a motor controller for each module, two PC switching sources are used to power the entire system.

As actuators for the robotic system, 18 motors are used which are distributed two for each module, to give a total of 9 modules distributed in a 3x3 matrix, one motor for the X axis and the other for the Y axis, which receive their control signal from the Cytron boards. The same ones that manage the energy to drive the motors that will then move the table.

2.2. Image processing

Using the data obtained by means of the artificial vision system, the robot table is controlled, therefore for the process of obtaining images it is carried out by means of a projection of the plane of interest in the image plane where the sensor is located. As a consequence, the perspective is lost and a 2D image is obtained, which provides information on position and speed. In order to compute the coordinates of the object of interest, the pin hole camera model provides a simplified version of the information resulting in a shorter processing time and therefore relief of computational load [9], figure 2.

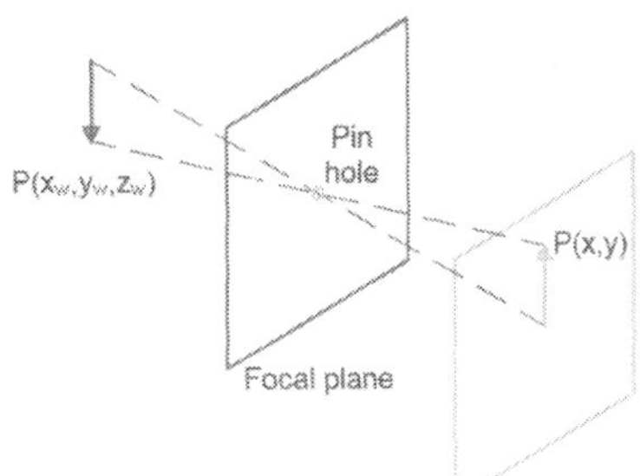

Figure 2. Pinhole Camera Model Illustration.

The pinhole camera model is used for the determination of mathematical equations describing a three-dimensional point $P(X_w, Y_w, Z_w)$ for the case of the image plane is based on figure 3. [10].
To determine the model, the analysis is carried out for each two-dimensional axis, as shown in figure 3. The 3D information of the world is projected onto the plane of the image located at an equal distance known as focal length.

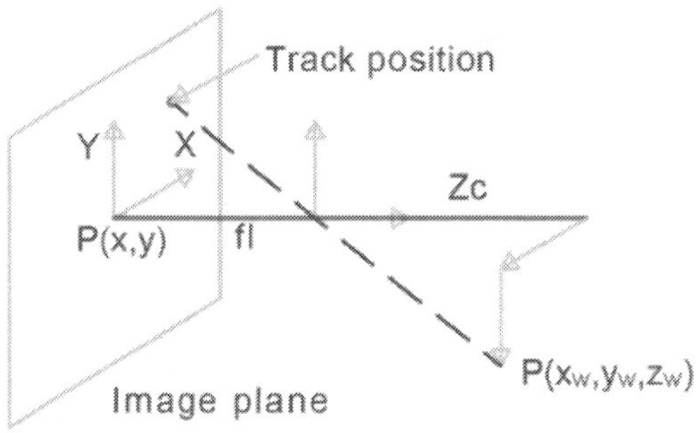

Figure 3. Track position analysis based on the Pinhole Camera Model.

The geometrical relations allow the choice of the position value of each of the pixels as in equation (2) for x axis and equation (3) for y axis.

$$\frac{X}{f} = \frac{x_C}{Z_C} \tag{2}$$

$$\frac{Y}{f} = \frac{Y_C}{Z_C} \tag{3}$$

The following matrix is used to obtain the real value of carry object for this the position of the shown in equation (4).

$$\begin{bmatrix} nx \\ ny \\ n \end{bmatrix} = \begin{bmatrix} 1 & 0 & 0 & 0 \\ 0 & 1 & 0 & 0 \\ 0 & 0 & f^{-1} & 0 \end{bmatrix} \begin{bmatrix} x_C \\ y_C \\ z_C \\ 1 \end{bmatrix} \tag{4}$$

3. Tracking Control

The robotic table is verified with different techniques of control, based in follower desired path Error calculation is being calculated under the quadratic error function see figure 4.

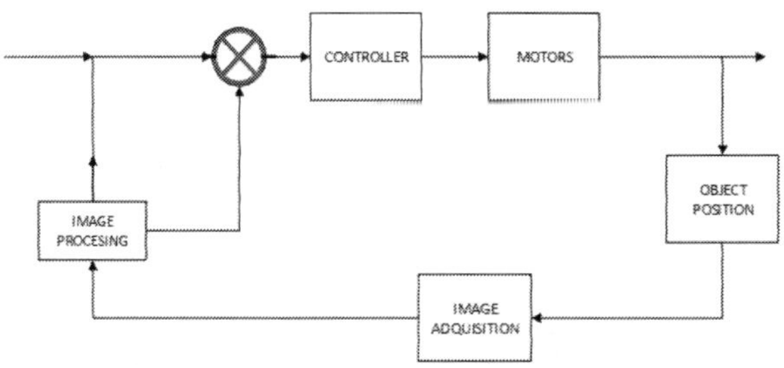

Figure 4. Control System Diagram.

4. Result and Analysis

The tests identify the path of the table and even a red object on it, which is intended to estimate its distance, a situation that will depend on the segmentation and discretion of the colors. The evaluations of the obtained results are specifically the location starting from a real path and another one that references the images acquired in reference points, estimating the position with the algorithm. Both paths are contrasted on the basis of an error criterion, the actual path has 4 points in the plane whose positions are displayed in the figure 5.

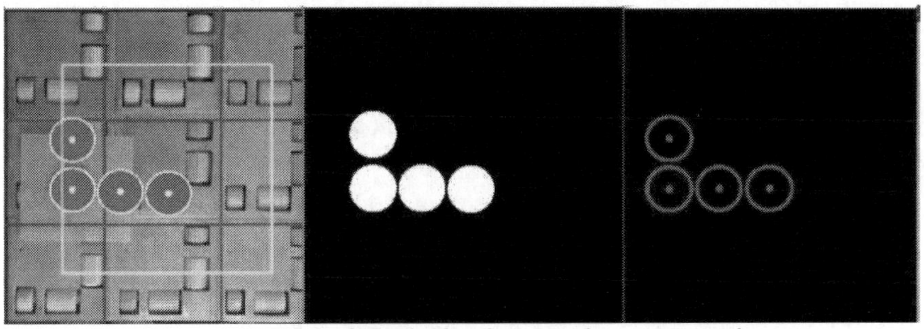

Figure 5. Displays the implemented route control.

Table 1. Real tracking position based in data acquisition.

No. points	xr	yr	Object
P1	220	60	Table – Red Object
P2	138	56	Table – Red Object
P3	130	130	Table – Red Object
P4	200	120	Table – Red Object

A figure. 6 shows the image of the real Path with defined criteria of identification, location, there are not two objects in the same image and its acquisition was made clockwise.

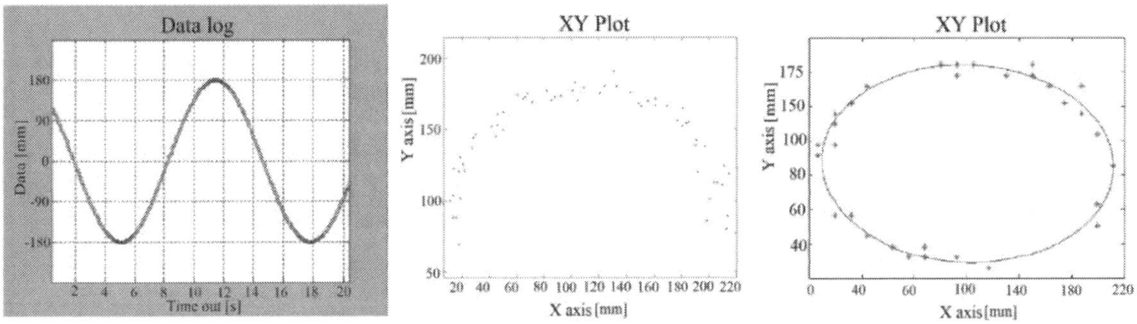

Figure 6. Trajectory data acquisition.

4.1. Error Verification

The result shows a difference between the real path and the desired path that will be calculated under the quadratic error function is show in equation (5).

$$E = \frac{1}{N}\sum_{k=1}^{N}[(x_r(k) - x_d(k))^2 + (y_r(k) - y_d(k))^2]^{1/2} \quad (5)$$

where:

$[x_r(k), y_r(k)]$=real position

$[x_d(k), y_d(k)]$= position is deducted by the mobile robotic table

k = instant where the distance measurement is taken in the field

N= total number of points where a distance measurement was made.

5. Conclusions

After the analysis of the quadratic error, it was found that the table has enough capacity to follow the desired path through the control algorithm with artificial vision, the combined control of each of the motors has been necessary to ensure a smooth movement for the transfer of the double rack on each of the modules.

In addition, the selected materials comply with the requirement which ensures a long service life. The path analysis is observed through the artificial vision system, which efficiently follows the proposed path using the singularity to have an abstraction of the system that facilitates the control algorithm with an error of ±2.4 mm.

References

[1] N. Ackovska and V. Kirandziska, 2017 "The importance of hands-on experiences in robotics courses," *IEEE EUROCON-17th International Conference on Smart Technologies*, Ohrid, Macedonia, p. 56-61.

[2] A Skala et al 2015 IOP Conf. Ser.: Mater. Sci. Eng. 95 012097

[3] H. Cheng, H. Chen and Y. Liu, (2013) "Object Diverting and Sorting using Autonomous Industrial Mobile Manipulator," *IEEE International Conference on Cyber Technology in Automation, Control and Intelligent Systems*, Nanjing, 36-41.

[4] J. A Gwiazda et al 2015 IOP Conf. Ser.: Mater. Sci. Eng. 95 012104

[5] Andaluz, Víctor & Gallardo, Cristian & Santana, Jaime & Villacres, Johana & Toasa, Renato &Vargas, Javier & Reyes, Gabriel & Naranjo, Tannya & Sotelo, Andrea. (2012). Bilatera Virtual Control Human-Machine with Kinect Sensor. 101-104. 10.1109/*Andescon*.2012.32.

[6] W. Jeong, Y. H. Seo and H. S. Yang, (2010) "Effective humanoid motion generation based on programming-by-demonstration method for entertainment robotics," *2010 16th International Conference on Virtual Systems and Multimedia*, Seoul, 289-292.

[7] K. A. Zubair, S. Singha, S. Alam and Kazi Tanvir Ahmmed, (2015) "Humanlike robotic vision system using artificial neural network," *2015 2nd International Conference on Electrical Information and Communication Technologies (EICT),* Khulna, 13-18.

[8] X. Zhang, L. Xue, S. Li and F. J. Kim, (2013) "Digital human modeling for ergonomic evaluation of patient table height," *IEEE Int. Conference on Robotics and Biomimetics (ROBIO)*, Shenzhen, 1480-1485

[9] Víctor Andaluz, Flavio Roberti, Juan Marcos Toibero, Ricardo Carelli, (2012) Adaptive unified motion control of mobile manipulators, *Control Engineering Practice*, **Volume 20**, Issue 12, 1337-1352.

[10] Andaluz V.H., Ricardo Carelli, Lucio Salinas, Juan Marcos Toibero, Flavio Roberti, (2012) Visual control with adaptive dynamical compensation for 3D target tracking by mobile manipulators, *Mechatronics,* **Volume 22**, Issue 4, 491-502.

IOP Conf. Series: Materials Science and Engineering **417** (2018) 012012 doi:10.1088/1757-899X/417/1/012012

Workspace Measurement and Calibration of a Parallel RCM Craniotomy Surgery Robot

Mohammadreza Dehghani, Majid M Moghaddam[1] and Pourya Torabi

School of Mechanical Engineering, Tarbiat Modares University, Tehran, Iran

[1] Corresponding Author: m.moghadam@modares.ac.ir

Abstract. This paper aims to improve the accuracy of a three degree-of-freedom surgery robot by means of calibration. The robot consists of a parallel Remote Center of Motion (RCM) mechanism which is designed for craniotomy surgery. The calibration method is based on the forward kinematic approach, which uses a nonlinear optimization model. The Levenberg-Marquaridt method is used as the optimization technique and the loss function is defined by the least square definition. Sixty experimental data are used for the optimization process, the data are obtained using a Leica total station camera to measure the end-effector position, another sixty measurement data are also obtained to use in the evaluation process. Seven kinematic parameters are considered for the optimization process. After the optimization process the loss function reduced to 0.0002453 square millimeters.

1. Introduction

Robotic bone milling is not science fiction anymore. RIO by Mako Surgical Corp., Robodoc by Curexo Corp. are examples of commercialized systems for knee arthroplasty [1]. Moreover, there are several robotic systems for tooth bone milling and spine surgery [2]. Continuing on this trend, in recent years several studies are conducted on robotic systems for cranial bone milling [2]. Although the concept is the same as the previous systems, the crucial difference is the required high precision.

Pathfinder, RobaCKa and Cranio, are three robots designed and developed for assisting brain operations. The Pathfinder [3] uses CT scan images to find and mark the optimal points on the skull for bone removal and then uses some tools on the robot end effector, for its operations. RobaCka [4] is an industrial CASPAR robot, a 6-DOF serial robot. Cranio [5] has a C shaped arm which holds the tools for the cutting task and the arm is mounted on a Stewart-Gough platform. We [6] proposed a 2-DOF spherical parallel mechanism to keep the milling tool nearly perpendicular to skull surface. To achieve both stability and accuracy needed for bone removal task, this robot uses a parallel mechanism. The proposed design is discussed in the next section. For the skull bone removal task, positional performance is very important. The positional performance of a robot, can be described by three factors; resolution, accuracy and repeatability [7]. Fortunately, most industrial and medical robots have an acceptable repeatability and resolution but they sometimes, lack good accuracy. To improve accuracy, a good way is to use model-based calibration techniques. The calibration processes of parallel robot calibration methods are discussed in [8]. Model-based calibration consists of a step called Measurement [9], which actual end-effector positions must be measured. Different methods and tools are developed and used like cameras, laser trackers, coordinate measuring machines (CMM) and theodolites. In some non-contact methods, they use 2 cameras for position measurement [10]. These methods are based on binocular vision by which the distance can be measured but they lack high accuracy. For enhancing accuracy, some use linear lasers which produce linear vertical and horizontal lines on the surface of the object to help the measurement.

For the present robot, position of the robot's end effector is measured by a Leica TS06plus R1000 total station camera is used. This total station has the ability to measure the location of points from a predefined internal coordinate, by means of laser technology. The points are used during the optimization process of the calibration method. As the kinematic equations are non-linear, Levengerg-Marquaridt is used for optimization process [11].

2. System Description

The Robot System (Figure 1) is a 3-DOF cable actuated mechanism which consists of a 2-DOF spherical parallel mechanism and a tool insertion interface. The active mechanism is installed on a C-shaped base manufactured from 10mm thick aluminum plate. The spherical mechanism has five motion axes, where three of them are dependant of the two actuation joints. The two required motors, control the two shoulder joints via a capstan mechanism with an 18.8:200 transmission ratio. The motors are AC servo drives (ASDA2 DELTA Electronics Inc.). Robot arms are manufactured with 4-axis CNC milling machine with 0.001 mm resolution. Considering the encoder output of 10000 pulse per revolution, the orientation resolution of the end effector is 1.2e-4 rad. The insertion degree is a cable driven rail and cart mechanism which moves a pneumatic drill along the tool joint axis. The theoretical resolution of insertion degree is 4.6e-3 millimeter.

3. Calibration

Accuracy of the robot, can be improved by working with the calibrated model of the robot. One way of finding this model, is to find an ideal robot kinematic equations and then change its parameters by calibration methods. This is called a model-based calibration. A model-based calibrations consists of four steps: Modeling, Measurement, Identification, Compensation[9].

3.1. Kinematic Modeling and Workspace Analysis

The kinematics of the parallel spherical mechanism is described in [12,13], utilizing appropriate coordinate systems according to Denavit-Hartenberg method (Figure 2). To validate the kinematic equations, data obtained from SOLIDWORKS model is compared with analytical equations' output.

The Jacobian of the mechanism was calculated in [6] and used to optimize the mechanism for maximum workspace and isotropy. The optimized link lengths are as follows:

$$\alpha_{13} = 61, \alpha_{35} = 66, \alpha_{24} = 61, \alpha_{46} = 66, \phi = 150$$

To calculate the theoretical workspace of the mechanism, we introduced points covering a unit sphere into the inverse kinematic equations. Points corresponding to valid solutions are considered inside the mechanism workspace.

3.2. Measurement

Methods to measure EE position are different in application, accuracy and price. In this research, a total station camera, Leica TS06plus R1000, is used to measure points with 0.1 mm accuracy.

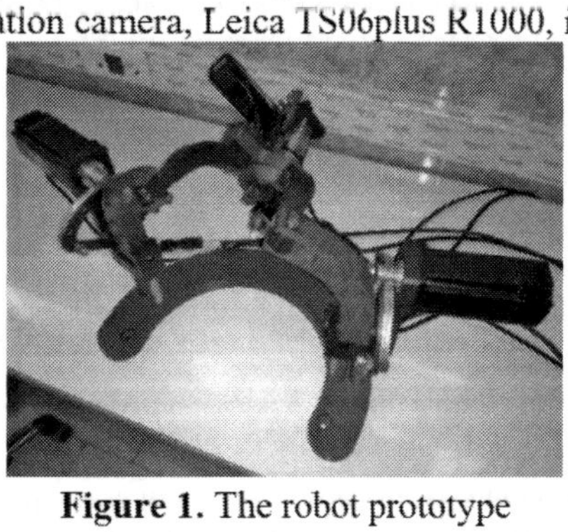

Figure 1. The robot prototype

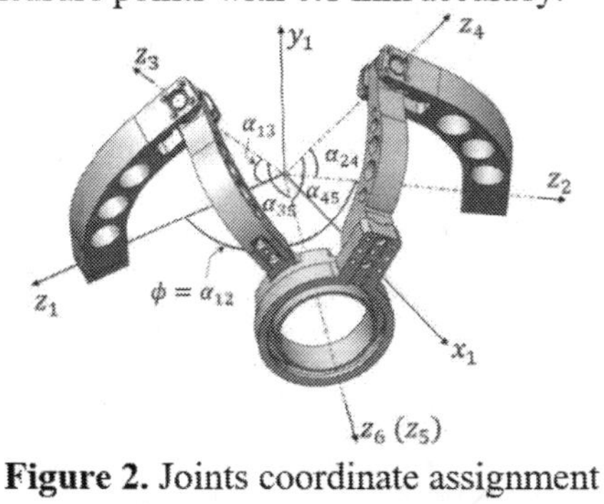

Figure 2. Joints coordinate assignment

3.2.1. Actuation joint calibration
The control input to the mechanism is the motor rotations, while the kinematic equations relate the first link angle to the end-effector position. Therefore, we first need a relation between motor rotation and first link angle. We measured the position of three points (Figure 3) on each arm in four different orientations of the shoulder joints, so that we can pass a plane from each three points and then find the angle between the planes relative to the rotation of each shoulder joints.

We assigned the zero angle of the link to when the shoulder link is completely vertical. We used the Open Controller Z pulse (OCZ) of the controller to synchronize the motor with the zero position. The linear equations (1) and (2) are line fitting of the data of motor rotation vs. arm angle for the right and left motors respectively and equation (3) represents the relation between third motor rotation and tool insertion.

$$\theta_1(r_1) = 0.2785r_1 + 0.1332 \qquad R^2 = 0.9999 \tag{1}$$
$$\theta_2(r_2) = 0.2817r_2 - 0.188 \qquad R^2 = 0.9998 \tag{2}$$
$$I(r_3) = 3.917r_3 + 0.0998 \qquad R^2 = 1 \tag{3}$$

Here θ_1 and θ_2 are right and left arms' angle, I is the amount of the tool insertion in meters and r_1, r_2 and r_3 are motor rotations. R^2 is the yielded Coefficient of determination.

3.2.2. Data acquisition and Workspace measurement
Robot was guided to 120 different actuation angles and the end-effector position is collected. Half of the data is used for calibration and the remaining is used for evaluation. The insertion mechanism was held still and the workspace over a constant radius sphere was covered. Figure 4 shows the recorded EE position alongside the theoretical workspace of the robot.

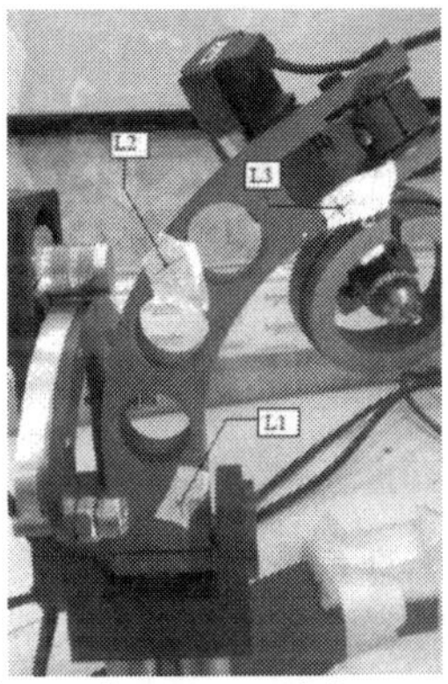

Figure 3. Points of arms used for measurement

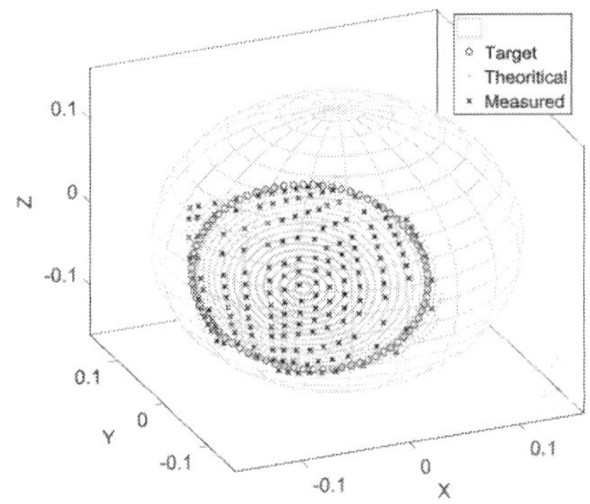

Figure 4. Recorded EE position

3.3. Parameter Identification
7 parameters are chosen to be identified through an optimization technique. Optimization is used to minimize the error between the actual and nominal values. Vector $\overline{V}$, shows the chosen parameters as:

$$\overline{V} = [\alpha_{12}, \alpha_{13}, \alpha_{35}, \alpha_{24}, \alpha_{46}, \delta_1, \delta_2] \tag{4}$$

Where α_{12}, α_{13}, α_{35}, α_{24} and $\alpha_{46}$are the base angle, the right shoulder arm's angle, the right elbow arm's angle, the left shoulder arm's angle and the left elbow arm's angle as shown in Figure 2. Since the kinematic equations are nonlinear, we use the Levenberg-Marquaridt method [11]:

$$\underline{\theta}_k = \underline{\theta}_{k-1} - \eta_{k-1}\left(\underline{J}_{k-1}^T\underline{J}_{k-1} + \alpha_{k-1}\underline{D}\right)^{-1}\underline{J}_{k-1}^T\underline{f}_{k-1} \tag{5}$$

Where $\underline{\theta}_k$ is the vector of robot parameters V in time "k", and $\underline{\theta}_{k-1}$ in time "k-1", α_{k-1} a positive non-zero constant and D is the diagonal value of the matrix $\underline{J}_{k-1}^T\underline{J}_{k-1}$ where J is Jacobian matrix. f is the error between nominal and measured parameters. We used least square method for the loss function as:

$$I(\underline{\theta}) = \sum_{i=1}^{N} f^2(i, \underline{\theta}) \tag{6}$$

IOP Conf. Series: Materials Science and Engineering **417** (2018) 012012 doi:10.1088/1757-899X/417/1/012012

Where, I is the loss function. To achieve the inverse matrix in the Levenberg-Marquaridt, we have:

$$\left(\underline{J}_{k-1}^{T}\underline{J}_{k-1} + \alpha_{k-1}\underline{D}\right)\underline{P}_{k-1} = \underline{J}_{k-1}^{T}\underline{f}_{k-1} \quad (7)$$

So we can write:

$$\underline{\theta}_k = \underline{\theta}_{k-1} - \underline{P}_{k-1} \quad (8)$$

We assume $\eta_{k-1} = 1$. The flowchart of this method is shown in Figure 5:

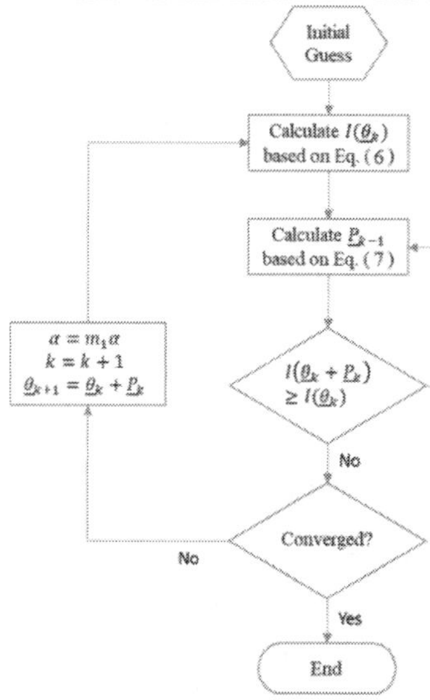

Figure 5. Calibration Flowchart

From [14], the convergence criteria is:

$$I_v = \frac{I}{n - v + 1} < \epsilon \quad (9)$$

Where n is the number of data, v the number of parameters and ϵ a positive value. We inserted 10, 30 and 60 data in the calibration algorithm respectively and the results are shown in **Table 1**.

Table 1. Calibration Results

Number of points	Convergence value	α_{12}	α_{13}	α_{35}	α_{24}	α_{46}	δ_1	δ_2
10	0.0000447	144.25	63.05	68.08	55.82	67.02	0.041	-3.360
30	0.0000920	146.03	62.09	67.47	57.21	65.02	0.149	-2.156
60	0.0000757	146.68	61.48	67.31	58.36	63.81	0.190	-2.036

3.4. Evaluation

For evaluation, first the motor rotations are used as input to the calibrated kinematic and then the output is compared to the relative measured position (Figure 6). The convergence value of the outputs generated using the calibrated kinematics is calculated to be 0.0002453.

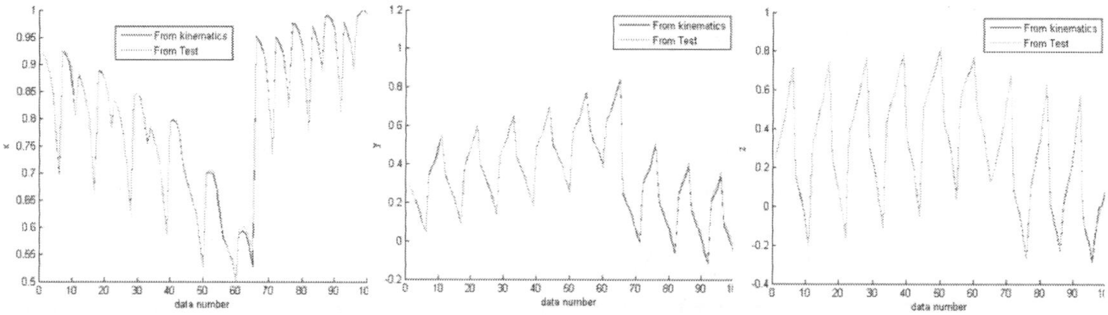

Figure 6. Comparison between the x, y and z of calibrated kinematic output and the measurement

4. Conclusion

To help surgeons in the craniotomy surgery, we developed a robot to ease the task of skull bone removal process. As the bone removing required a tough drilling process, we used a parallel mechanism for the robot and to have back-drivability, we used a capstan driving system. Since in the brain surgery operations, the accuracy is important, after the robot is manufactured and assembled, we used some calibration techniques to improve accuracy. For calibration process, first the kinematic equations were obtained. We considered 7 parameters to be calibrated. The kinematic equations were nonlinear so for the optimization process the Levenberg-Marquaridt was used. With the help of a Leica total station camera, we measured 120 points of the robot end effector in different positions. 60 points were used in calibration and the others were used in evaluation process. Based on the defined convergence criteria, the convergence value of the calibrated kinematics outputs (for the same input points) is 0.0002453.

References

[1] Hoeckelmann, M. et al. 2015. Current capabilities and development potential in surgical robotics. *International Journal of Advanced Robotic Systems*. 12, 5 (2015), 61.

[2] Mattei, T.A. et al. 2014. Current state-of-the-art and future perspectives of robotic technology in neurosurgery. *Neurosurgical Review*.

[3] Deacon, G. et al. 2010. The Pathfinder image-guided surgical robot. *Proceedings of the Institution of Mechanical Engineers, Part H: Journal of Engineering in Medicine*. 224, 5 (2010), 691–713.

[4] Korb, W. et al. 2003. Development and first patient trial of a surgical robot for complex trajectory milling. *Computer Aided Surgery*. 8, 5 (2003), 247–256.

[5] Cunha-Cruz, V. et al. 2010. Robot-and computer-assisted craniotomy (CRANIO): From active systems to synergistic man—machine interaction. *Proceedings of the Institution of Mechanical Engineers, Part H: Journal of Engineering in Medicine*. 224, 3 (2010), 441–452.

[6] Dehghani, M. et al. 2017. Analysis, optimization and prototyping of a parallel RCM mechanism of a surgical robot for craniotomy surgery. *Industrial Robot: An International Journal*. (2017).

[7] Shiakolas, P.S. et al. 2002. On the accuracy, repeatability, and degree of influence of kinematics parameters for industrial robots. *International journal of modelling and simulation*. 22, 4 (2002), 245–254.

[8] Majarena, A.C. et al. 2010. An overview of kinematic and calibration models using internal/external sensors or constraints to improve the behavior of spatial parallel mechanisms. *Sensors*. 10, 11 (2010), 10256–10297.

[9] Roth, Z. et al. 1987. An overview of robot calibration. *IEEE Journal on Robotics and Automation*. 3, 5 (1987), 377–385.

[10] Grosso, E. et al. 1996. Robust visual servoing in 3-D reaching tasks. *IEEE Transactions on Robotics and Automation*. 12, 5 (1996), 732–742.

[11] Shi, J. et al. 2011. Kinematic model identification of planar delta manipulator using Random Levenberg-Marquardt algorithm. *Intelligent Control and Automation (WCICA), 2011 9th World Congress on* (2011), 1097–1102.

[12] Ouerfelli, M. and Kumar, V. 1994. Optimization of a spherical five-bar parallel drive linkage. *Journal of mechanical design*. 116, 1 (1994), 166–173.

[13] Wu, C. et al. 2010. Optimal design of spherical 5R parallel manipulators considering the motion/force transmissibility. *Journal of Mechanical Design*. 132, 3 (2010), 31002.

[14] Gavin, H. 2011. The Levenberg-Marquardt method for nonlinear least squares curve-fitting problems.

IOP Conf. Series: Materials Science and Engineering **417** (2018) 012013 doi:10.1088/1757-899X/417/1/012013

CVCER: Robot to Learn Basics of Computer Vision and Cryptography

A Yeleussinov[1,2], T Islamgozhayev[2], M Satymbekov[2] and A Kozhagul[3]

[1]al-Farabi Kazakh National University, 71 al-Farabi Avenue, Almaty, Kazakhstan
[2]Institute of Information and Computational Technologies, 125 Pushkin str. Almaty, Kazakhstan
[3]Institute of Mathematics, Physics and Informatics, Kazakh National University named after Abai, Dostyk ave. 13, Almaty, Kazakhstan

arman.04081991@gmail.com

Abstract. This paper describes the design of CVCER (Computer Vision and Cryptography Educational Robot) - the working model of the mobile robot used as a training stand for students. We describe the hardware and software of the robot. Hardware part of the robot can be used to teach the basics of robot architecture and software part can be used to teach the basics of computer vision and cryptography. The main goal of the project was to create a low cost, re-configurable and re-programmable robotic platform that could be used to teach the basics of the subjects described above and to test the algorithms written by students having internships in our institute and partner university.

1. Introduction

Research and development in the area of robotics during past decades brought new subjects into an educational program of schools and universities. Highly motivated scientists and hobbyists are creating and developing new platforms, educational programs, and methods of teaching STEM subjects like physics, mathematics, engineering, etc [1-3]. However, most of robots are vendor-dependent, without source code and expensive. Aim of this paper is to describe the process of continual development of a robotic platform, which can be used in teaching STEM subjects and basics of Robotics, Control Theory, Machine Vision, Information Security, and building blocks of which can be bought or 3D printed and used to test robotics projects or in competitions. To reach this aims we developed CVCER - cheap, flexible and multi-functional robotic platform. CVCER's hardware part is built using Arduino platform, which is most popular hardware-platform that gives open architecture and lot of compatible sensor sets.

CVCER is an educational robotic platform (Figure 1) that was firstly developed as a platform for practical realization of Machine Vision algorithms [4], now the system evolved and we wanted to be able to teach Information Security basics with the help of the system. From this field we wanted to teach the basics of information retrieval between the operator and robotics system, that is why to ensure the privacy of sent commands we developed a system with an information security using and NPNs algorithm described in [5]. In our work the main aim is not to teach the basics of constructing physical robot by joining different 3D printed objects but to foster the teaching of development of control, pattern recognition, and information security algorithms and immediate testing of results on the platform because the real-life tests drive more attention in training process than simulation or theoretical results [6-8]. To make the process easier, OpenCV library was used for the purpose of implementation of computer vision teaching software.

2. Platform

CVCER body is made from 3D printed chassis with modification - all wheels were replaced by omni-wheels that adds flexibility [4]. Body frame is made from perforated aluminum that makes platform flexible to any modification and attachment of additional equipment.

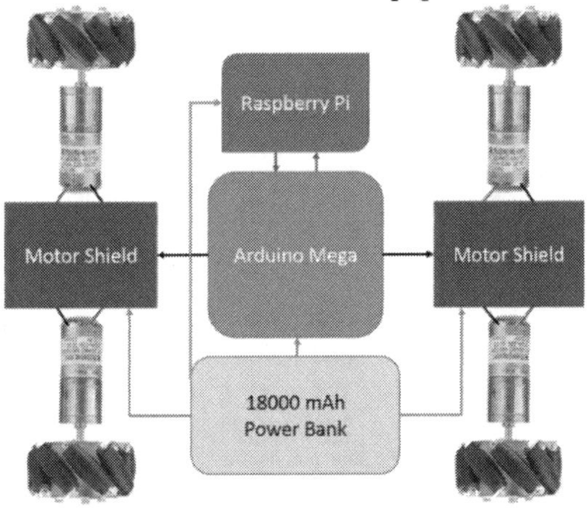

Figure 1. Platform Architecture

For now, 18000 mAh power bank with 5-20V outputs gives the system 80-100 minutes of life to power the platform. Parts can be replaced and new parts can be easily adjusted to the base that has drilled holes.

Movement of the platform is done using four motors: two motors on each side of the platform. Each motor can be programmed to move on its own, which gives more flexibility in motion planning tasks, because you can control almost any behavior of the platform by using the torque from each of them.

Figure 2. CVCER robotic platform

Raspberry Pi 3 Model B and Arduino Mega 2560 are selected as the brain and motion control unit of the robot. Raspberry Pi includes 802.11n Wi-Fi, Bluetooth 4, and Quad Core 1.2GHz Broadcom BCM2837 64bit CPU with 1Gb RAM [9]. Such characteristics and capability of using Linux-family operating system makes it one of the most commonly used single-board computers for robotics projects. Moreover, it makes possible the usage of OpenCV and programmable GPIO's - functions that perfectly suit the aim of our project. Because we have to run object recognition and data decryption programs, as well as sending control commands to Arduino to perform motion.

Arduino is very popular platform for robotics. Its flexibility and availability makes it ideal for fast prototyping and experiments. Arduino Mega 2560 is used as the main controller. This board is based on the ATmega2560 chip and has 54 digital input/output pins, 16 analog pins [10]. Description of mechanical and electronic parts you can read from our previous work [4].

Platform makes use of mecanum wheels to add the flexibility and fast side-to-side movements [11,12]. It can surge (forward/backward), sway (left/right) without turning to sides, and yaw around the Z-axis like crawler which gives it two more controllable degrees-of-freedom, thus, making the system holonomic.

In the given system, four wheels are placed like standard wheels of four-wheeled car, it gives the system stability and simplifies control approach [13].

Motion control is written using IICTv3.h library used in previous version of the system. Additional functionality was added including the motion control of the 2DOF camera mount. Camera mount is a pan and tilt system driven by two Hitec HS-785HB servos capable of working with maximum payload of 2.5 kgs.

Tests were done by using the combination of two Arecont Vision cameras with optical zoom lenses (this approach requires router and an additional source of power to be placed, because AV cameras are network cameras which consume a lot of energy) and paired webcams, availability of USB ports on Raspberry PI makes replacement by any affordable USB webcams possible.

We estimate the cost of the entire system to be around 500 US dollars (with simple 5MP webcam pairs).

3. Software

As mentioned earlier, the robot is composed of several parts. Therefore, the latest complete software consists of a combination of machine vision, data decryption, and motion control blocks (Fig. 3 software architecture). We believe that last year students are familiar with the basics of programming languages.

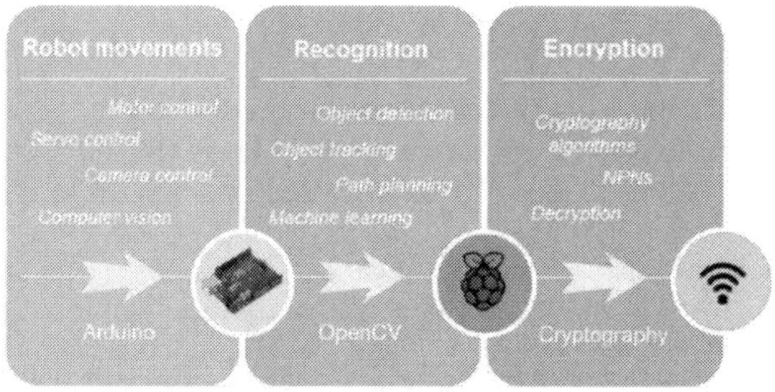

Figure 3. Architecture of the software

Students can use the programming language based on their choice, programming environment and methods, different algorithms, and, if necessary, additional devices (motors, sensors, etc.). But the introductory software is written as the console application in C++ language using OpenCV library, and several System classes, our main aim was not to build "put and go" system, but to build a system that can help to teach full stack of the development process - from combining the hardware part to programming the basic behaviors, to learning the object detection, tracking, and at the end, to learn to encrypt and decrypt incoming and outgoing messages. Figure 4 illustrates packages included into CVCER library.

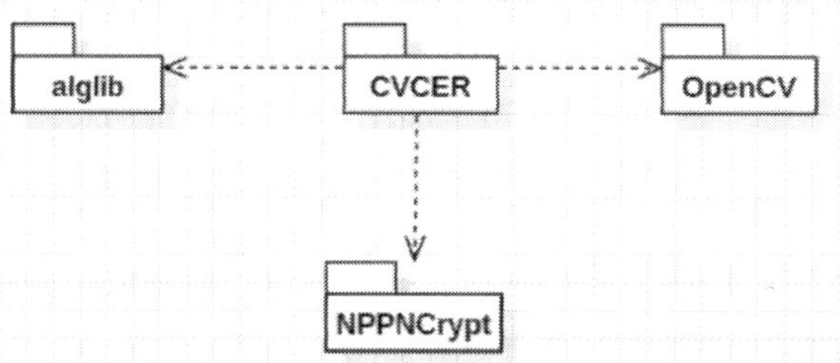

Figure 4. Packages diagram

3.1. Programming basic behaviours of the robot
At this stage, user needs to move actuators connected to the Arduino platform through motor drivers (originally 4 motors and 2 servomotors). Code should be written in C/C++ language. For this purpose, user can program on his own or call our IICTv3.h library key functions for motion programming (like forward(seconds), backwards(seconds), left(seconds/angle), right(seconds/angle), etc). The main purpose of this stage is to carry out all possible movements of the robot and complete the motion control task. Main library (CVCER) includes two modules - IICTv3 (responsible for motion control), NPPNCrypt (encryption/decryption). OpenCV and alglib libraries are ready to use simply by "including" them.

3.2. Object recognition algorithms and the robot movement based on machine vision
Since the platform is designed to be used at different levels of learning, the connection of robot motion and computer vision parts should be made to meet this differentiation (it should be made to make everyone familiar with the connection of these two parts). A combination of two cameras makes it possible to use stereo vision approaches and camera mount with 2 DOF allows user to learn basics of trajectory generation, motion planning and perform object tracking and mecanum wheels makes the object following simpler by making it possible without turning the whole platform. OpenCV's basic template matching methods (square difference matching, correlation matching, correlation coefficient matching, normalized square difference matching, normalized correlation matching, and normalized correlation coefficient matching) were used to perform the task of recognition of the object, the corresponding functions were written to be able to test each method by using just single line of code. Object tracking and following tasks were made to give simple two variable outputs that denotes the distance and angle to object from the optical center of the cameras.

3.3. Application of the encryption algorithm for robot control
There is a need to encrypt the security of data exchange between the robot and other robots (in multiagent systems) or the operator. To that end, the user can use the encryption algorithm based on non-positional polynomial notations developed by our colleagues [5]. Keys of different lengths are required for encryption and decryption between the operator and the robot. It is generated in a special program on the personal computer used by the operator and is sent to the computer on a mobile robot through WiFi. Of course, sending simple control commands (forward/backward, left/right, etc) is not the true usage of the algorithm, because messages sent could be easily caught in the middle and copied by intruders, the main idea of this part was to encrypt combination of commands or complex tasks.

4. Conclusion
The article gives brief description of the motion control, computer vision strategies and encryption algorithm used in the development of mobile educational robot. It describes the achieved results, ideas used and in future report will be described the ready-to-use software package with friendly GUI, rebrained and finished backend and polished body. As the platform has the ability to hold different sensors and additional hardware, by adding them and including additional functions in combination with our library users can modify and enhance the capabilities of the system. Presented library includes commands for setting up pins, driving motors, turning the platform, pre-written commands for combination of joystick buttons (if user is going to use them), basic computer vision algorithms and encryption algorithm. The mobile platform was developed to help beginners – enthusiasts, school and university students – to become familiar with the basics of development of robot systems, which involves different fields of study. We picked up trending areas of robotics and computer science fields and tried to make the robotics software development and motion programming easier, more understandable by providing live examples of how things made and what processes robot developers are undergoing during the development cycle. We also used most popular computer vision library, encryption algorithm and motion control library developed in our institution to make it easy to get the documentation (in case of OpenCV) and help from community as well as to describe the technologies from first person view (as we have the direct connections with each of the developers of the given

IOP Conf. Series: Materials Science and Engineering **417** (2018) 012013 doi:10.1088/1757-899X/417/1/012013

system). Future report will also be about the development of the described system, but more detailed insights will be given to each part.

References

[1] Bellas F. et al. *The Robobo Project: Bringing Educational Robotics Closer to Real-World Applications.* J. Adv. Intell. Syst. 630 (2018), 226-237.

[2] W. Lepuschitz et al. *OTO – A DIY Platform for Mobile Social Robots in Education.* J. Adv. Intell. Syst. 630 (2018), 257-262.

[3] Filippov, S., Ten, N., Fradkov, A., Shirokolobov, I. *Robotics Education in Saint Petersburg Secondary School.* J. Adv. Intell. Syst. 630 (2018), 38-49.

[4] Islamgozhayev, T.U., Mazhitov, Sh.S., Zholmyrzayev, A.K., Toishybek, E.T. *IICT-bot: Educational robotic platform using omni-directional wheels with open source code and architecture.* Proceedings of IEEE International Siberian Conference on Control and Communications (SIBCON) (2015).

[5] Biyashev, R.G., Nyssanbayeva, S.E., Begimbayeva, Y., Magzom, M.M. Modification of the cryptographic algorithms, developed on the basis of nonpositional polynomial notations. *New Developments in Circuits, Systems, Signal Processing, Communications and Computers* (2015), 170-176.

[6] Gaba, D. 1999. *Human work environment and simulators*, In Anaesthesia 5, R.D. Miller, Ed. Churchill Livingstone, London, United Kingdom, 18–26.

[7] Staranowicz, A., Mariottini G.L. 2011. A survey and comparison of commercial and open-source robotic simulator software. *In Proceedings of the 4th International Conference on PErvasive Technologies Related to Assistive Environments* (PETRA '11). ACM, New York, NY, USA, Article 56, 8 pages.

[8] Kumar, A., Mittal, A., Arya, R., Shah, A., Garg, S., Kumar, R. 2017. Hardware in the loop based simulation of a robotic system with real time control and animation of working model. International Conference on Inventive Systems and Control (ICISC), Coimbatore (2017), 1-5.

[9] Raspberry Pi. (2018). Raspberry Pi 3 Model B - Raspberry Pi. [online] Available at: https://www.raspberrypi.org/products/raspberry-pi-3-model-b/ [Accessed 11 Mar. 2018].

[10]Store.arduino.cc. (2018). Arduino Mega 2560 Rev3. [online] Available at: https://store.arduino.cc/usa/arduino-mega-2560-rev3 [Accessed 11 Mar. 2018].

[11] Asama, H., Sato, M., Bogoni, L., Kaetsu, H., Mitsumoto, A., Endo, I. Development of an omni-directional mobile robot with 3 DOF decoupling drive mechanism. *Proceedings of 1995 IEEE International Conference on Robotics and Automation* (1995).

[12] Diegel, O., Badve, A., Bright, G., Potgieter, J., Tlale, S. Improved mecanum wheel design for omni-directional robots. *Proceedings of Australian Conference on Robotics and Automation* (2002).

[13] Tătar, M.O., Popovici, C., Mândru, D., Ardelean, I., Pleşa, A. Design and development of an autonomous omni-directional mobile robot with Mecanum wheels. *Proceeding of 2014 IEEE International Conference on Automation, Quality and Testing, Robotics* (2014).

IOP Conf. Series: Materials Science and Engineering **417** (2018) 012014 doi:10.1088/1757-899X/417/1/012014

Design and Implementation of Hardware's Firefighting Legged Robot

Lindawani Siregar

School of Electrical Engineering and Informatics, Bandung Institute of Technology Jalan Ganesa No 10 Bandung, Indonesia

Email: lindawanisiregar@gmail.com

Abstract. This paper discusses about hardware design for firefighting legged robot. This research aims to design and make a robot control system hardware of firefighting legged robot. This research realize a firefighting legged robot contest, which using microcontroller of ATMEGA128. Robot search for fire and extinguish the fire in the changing space condition. It uses many kinds of sensor. Sound activation (sound sensor) is used to start all the process to be performed by the robot. Robot start navigation after sound activation has been activated. Sensor ultrasonic used to detect the distance to the wall. Line sensor is used to detect the white line when robot will enter the room and detect candle location. To detect light of candle, it used UVTron flame detector. When robot crashes the furniture so limit switch will be used for moving away from the crashed object. Robot must blow on a fire candle at candle location (white color). The tools used to turn off the flame is the fan (DC motor). The result show that firefighting legged robot has been function and working of the various sensors and controller. Overall, the system can work well and stable according desired requirement.

1. Introduction

Firefighting legged robot contest refers to Trinity College International Robot Contest (TCIRC). The objective of the contest is to build a robot, which will extinguish fire in the quickest time possible [1]. In some cases, the rules of TCRIC are adjusted to condition in Indonesia. The Firefighting robot is designed to search for a fire in a small floor plan of a house of the specific dimension, extinguish the fire by using the fire extinguisher, and then return to the front of the house [2].

The robot will be placed in the field of fire modeled in a labyrinth with 4 rooms. One room contained a candle representing the point of fire. Robot must be able to find the fire and then extinguish it. The better locomotion performance of hexapod robot in uneven floors may be obtained by use of a great number of different kinds of sensors. The signals generated by these sensors may be used to represent the real state of the robot and its interactions with the environment [3].

There are four different types of system unit is use 1) Locomotion system 2) Fire detection system 3) Extinguishing system 4) Communication system. The Locomotion system is used for obstacle detection and ultrasonic range finder to find the distance between obstacle and system. Fire detection system is used for the detection of fire in this the gas sensor is used. Extinguishing system is for successfully extinguish the fire [4]. Extinguisher detects and extinguishes the fire sensed by a sensor.

Microcontroller is the heart of the project. Microcontroller controls all the parts of the robot by the use of programming. In this robot as the fire sensor senses the fire, it sends the signal to microcontroller [5]. This paper discusses about electronic system or hardware controller which has an important role in a robotic design. The electronic system in the robot consist of the main controller circuit, the sensor circuit, the motor controller circuit for moving six legs.

IOP Conf. Series: Materials Science and Engineering **417** (2018) 012014 doi:10.1088/1757-899X/417/1/012014

2. System Architecture

The goal of the contest is to build a legged robot which will extinguish a fire in the quickest time possible. A candle will represent the fire, the robot must find and then extinguish fire. The last, it returns to the start position. Design of hardware controller must be compatible with the rules. Figure 1 represent all hardware controller needed.

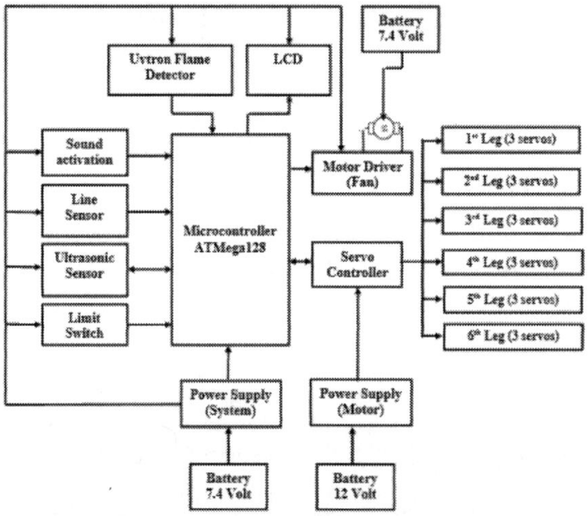

Figure 1. Block diagram of hardware controller

ATMega 128 is the brain of firefighting legged robot. All program control is executed there. The input microcontroller, consist of sound activation, line sensor, fire detector sensor (UVTron), ultrasonic sensor, limit switch, servo controller. The output microcontroller is connected to the fan motor driver, ultrasonic sensor, servo controller and a character LCD to display the condition of robot. Servo controller able to control 18 servos. Power supply is used separately between motor and supply system.

3. Hardware Design

3.1 Minimum system circuit

Minimum system circuit is designed by using 128. Schematic of minimum system firefighting legged robot can be seen in figure 2. Input voltage of minimum system is 5V in VCC. Schematic of minimum system shows pin usage for ultrasonic sensor, UVTron sensor, sound activation, servo controller, line sensor, motor driver, limit switch, stop button and LCD.

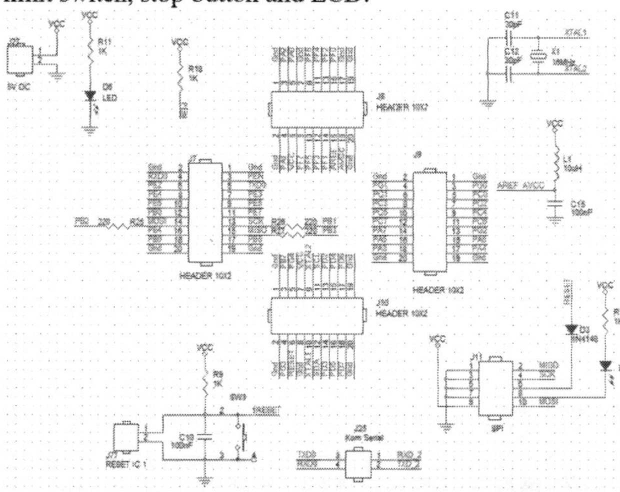

Figure 2. Schematic of minimum system ATMega 128

3.2 Sound Activation Circuit

3.2.1 Transmitter Circuit. Transmitter can generate sound from 3 kHz — 4 kHz. Transmitter circuit used to decrease voltage is shown in figure 3. Transmitter requires 5V DC power supply. The output of this power supply will be connected to transmitter as VCC. Transmitter circuit can be seen in figure 4.

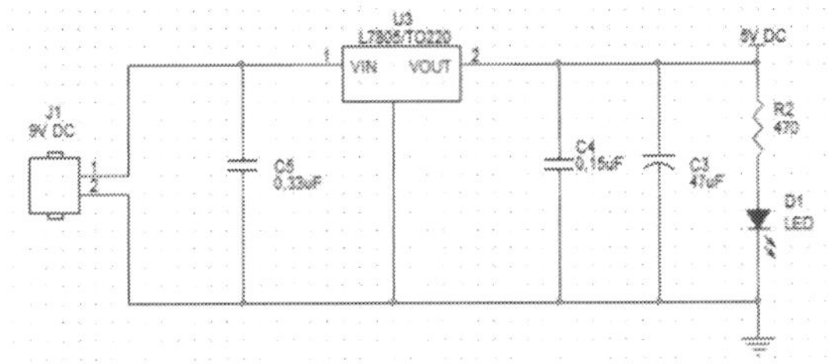

Figure 3. Schematic of minimum system ATMega 128

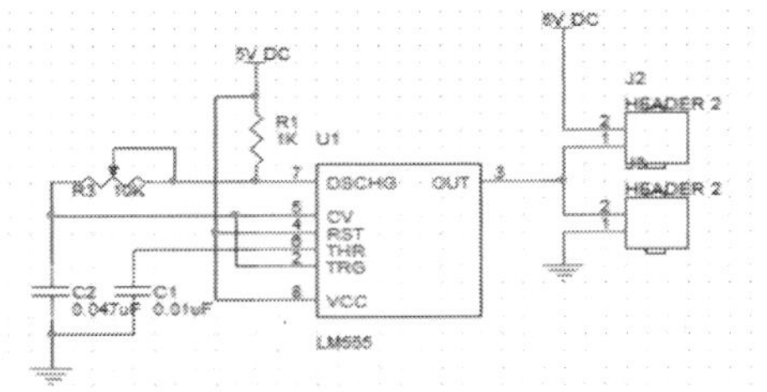

Figure 4. Schematic of transmitter circuit

To design sound generator (transmitter) is required IC LM 555 and buzzer. According to rules of robot contest, frequency that should be emitted is about 3.8 kHz.

3.2.2 Receiver Circuit. Receiver circuit is used as a sound receiver from specific frequency that can be seen in figure 5. Sound transmitted by the sound activation is received and converted to electrical energy by the condensor mic, then pass signal to the signal amplifier component (using LM358 as amplifier), from signal amplifier data will be filtered by IC tone decoder LM567. IC will filter the frequency between 3-4 kHz.

3.3 Line Sensor

Line sensor is used as a robot marker when robot entering the home space and candle location for existence of fire. To identify the entrance of the room and candle location, robot will use a line sensor placed at the bottom of the robot. Schematic of line sensor can be seen in figure 6.

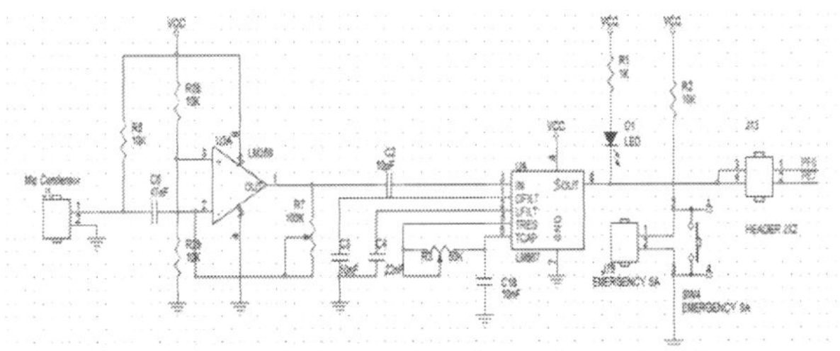

Figure 5. Schematic of receiver circuit

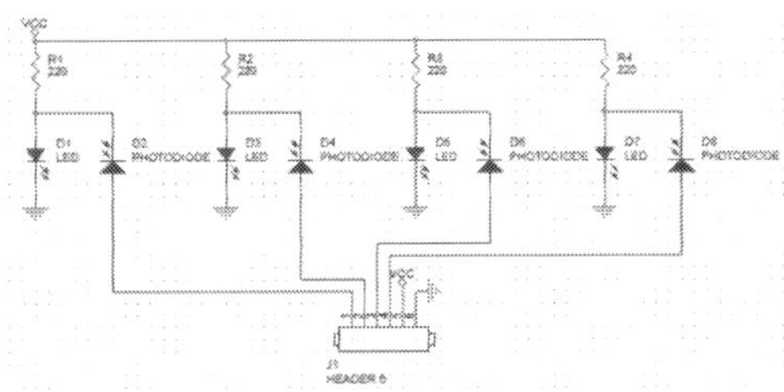

Figure 6. Schematic of line sensor circuit

3.4 Layout of Minimum System

In designing PCB, the first is make the schematic design of electronic. The schematic circuit is created using OrCAD software [6]. Overall layout of hardware controller in layout can be seen in figure 7. Figure 8 shows firefighting robot.

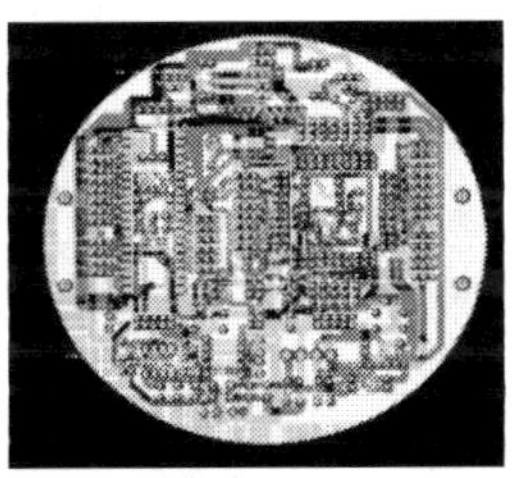

Figure 7. Minimum system layout

Figure 8. Firefighting leged robot

4. Testing Result

Hardware controller designed will be tested and analyzed to know the output in each circuit. The testing result represent that the circuit can be used for firefighting legged robot.

4.1 Ultrasonic Sensor Testing

In the navigation system testing robot, data taken based on the condition of room. Every existing condition in the room, data is taken and used as a reference robot to determine the direction of the movement. The robot uses 8 ultrasonic sensors: front, diagonal right front, diagonal left front, right, left, diagonal back right, diagonal left back, and back. Ultrasonic selection is adjusted to the program needs when reading the room.

Ultrasonic is commonly used in the distance measurement, because of high precision, low energy consumption, long dissemination distance [7]. Ultrasonic sensor testing is related to the level of accuracy of the ultrasonic sensor used by the robot to detect the distance of the robot with the wall. It is done by comparing the measured value on theultrasonic sensor with the actual distance. Table 1 shows the result of distance measurement ultrasonic sensor.

Table 1. Result testing of ultrasonic sensor

Actual Distance	Detected Distance
3 cm	3 cm
10 cm	11 cm
50 cm	54 cm
100 cm	107 cm
200 cm	211 cm
250 cm	265 cm
300 cm	338 cm

There are error or difference between the actual distance and the measured distance but the measured distance is still close to the actual distance. It is from the delay in the program in the microcontroller and from the sensor response itself.

4.2 Uvtron Flame Detector Testing

This sensor works with ultraviolet transmissions. Fire detected by UVtron while the signal processing unit using the module interface Hamamatsu UVtron. The advantages of this sensor is able to detect ultraviolet light until a distance of 5 meters. In order to test UvTron sensor, it uses external interrupt function by microcontroller. If there is a transition up on this external interrupt pin, then the microcontroller will display the result of fire detection as counter on the LCD. Table 2 shows result testing of UVtron sensor.

Table 2. Result testing of UvTron sensor

The distance of sensor to fire	Fire condition	Sensor Respon
50 cm	On	detected
100 cm	On	detected
300 cm	On	detected
500 cm	On	detected
550 cm	On	detected
575 cm	On	not detected

The distance testing is about 10 cm to 5.5 m. Through the above data, the UVTron can detect candle light well up to a distance of 5.5 m.

4.3 Servo Motor Testing

Firefighting legged robot uses 18 servos on the legs. Servo controller is used to control servos. Robot can move and work according to the working principle designed. Table shows result testing of servo.

Table 3. Result testing of servo

Pulse (uS)	Degree of Servo (°)
500	0
600	10
800	30
1100	60
1400	90
1900	140
2300	180

This testing is done by giving pulse to the servo and then measuring the angle on the servo. This result is used for moving legged.

4.4 Minimum System Testing

Minimum system testing is done by testing the minimum system circuit with downloader to find out whether the minimum system can receive program from compiler. The output voltage and input voltage are obtained when the microcontroller is given program. The result shows the high voltage level of 4.9V and the low voltage level of 0.1V. In conclusion that microcontroller can be function.

Robot try to navigate in getting the extinguishing's duration. Candle is placed in a position of the room, the robot scan it. The robot will move closer to the candle and extinguish it. The result is the robot successfully scanning the fire and extinguish the fire. Robot take time around 90 to 120 seconds for moving and extinguishing.

5. Conclusion and Future Work

The goal of this research is to develop design and implementation of hardware's firefighting legged robot. Firefighting legged robot has been designed by using ATMega128. In developing hardware's firefighting legged robot, this paper has proposed several stages of designing, creating, testing and analyzing. The result testing of hardware controller shows that is capable for firefighting robot. Designing hardware of control system must define the robot strategy. Using the data described at this work, an optimization hardware controller based on functional analysis must be developed.

Acknowledgment

The author would like to thank LPDP (Indonesian Endowment Fund of Education) for their support and sponsor the project presented in this paper. This research was supported by Dadang Lukman Hakim as a supervisor.

References

[1] Lee W, Perez R. *2017 Trinity College Firefighting Robot Competition*. EE495 Senior Project Seminar University of Evansville

[2] Mathew S, Sushanth, KR. Vishnu, Kumar V. *Fabrication of Fire Fighting Robot*. International Journal of Innovation and Scientific Research. 2016; Vol. 22 pp. 375-383.

[3] Bachega R P, Pires R, Campo A B. *Hardware Configuration of Hexapod Robot to Force Feedback Control Development*. 44th IEEE Southeastern Symposium on System Theory University of North Florida, Jacksonville, FL. March 11-13 2012.

[4] Kadam K, Bidkar A, Pimpale V, Doke D, Patil R. *Fire Fightimg Robot*. International Journal Of Engineering And Computer Science. 2018; page no. 23383-23485.

[5] Pooja, Singh R, and Rai S, *A Review on RF based Fire Extinguishing Robot*. Imperial Journal of Interdisciplinary Research (IJIR). 2016; volume-2 issue-7.

[6] Wrdhana, Lingga S A. *Pembuatan Skematik Rangkaian Elektronik dan Layout PCB Menggunakan OrCAD Release 9.1*. Yogyakarta, 2006.

[7] Luo Q, Hui Z, Han B. Circuit Design for the SMA Hexapod Walking Biorobot. ICIEA 2009; 978-1-4244-2800-7.

IOP Conf. Series: Materials Science and Engineering **417** (2018) 012015 doi:10.1088/1757-899X/417/1/012015

Time-domain Response Simulation Research on a Certain Type of Ammunition Transport Vehicle

Liu Yuelong, Li Tianpeng, An Zhentao and Liu Qiwei

Army Engineering University, Shijiazhuang, China 050003

3163304178@qq.com

Abstract. A 1/2 seven degree-of-freedom vehicle dynamic model was built for exploring the excitation that was obtained from a certain type of ammunition transport vehicle, which running on standard highway at the speed of 20m/s. According to the power spectral density function advised by national standard, the harmony superposition method (HSM) and pseudo white noise method (PWNM) were both used for simulating the road roughness. And the roughness of generated by these two methods was analysed and compared by MATLAB. Moreover, the vertical acceleration responses spectrum of grade A; B; C road were calculated. The numerical results reflect that the HSM is better than PWNM in road roughness simulation with the same time step size, and the root mean square value (RMS) of car centre of mass will grow largely with the decrease of road grade, the calculation results show the RMS of grade C road is above 300% than that of A.

1. Introduction

A better ride comfort of military automobile, particularly of a certain type of ammunition transport vehicle, is vital to guarantee the routine training and ammo delivery safe and timely. An ammunition transport vehicle would ride on complicated road conditions under some certain circumstances, which included standard level road, rural road and field road et al. Because of road roughness affection, the automotive vibration, with stochastic, stationary, and ergodic properties, was generated[1]. The energy of vibration would reduce the service lifespan of vehicle, and lower the handling stability of drivers. On the other hand, transmitted vibration that through suspension would undermining the structural integrity of ammo, that is, the combat effect might fall short of a requirement, and more severely, it could cause dangerous accidents[2-3]. The dynamic response of ammo carried by vehicle could be the criteria to predict its service life. The research of dynamic response of ammo, of course, became the hotspots.

The road transportation of ammo was a multiple degree-of-freedom (DOF), nonlinear system that contained vehicle, package and ammo. A precise dynamic model of vehicle was the foundation of analysing and assessing the response, and had direct effect on the credibility of numerical results. In a large number of investigations of analysing dynamic response, a typical 1/4 nonlinear two-DOF model was developed to discuss the influence of different parameters[4]. And a more complex 13-DOF dynamic model, with the consideration of suspension systems and tires, was used to evaluate the corresponding parameters affections on the dynamic interaction[5]. But a linear two-DOF suspension has been proved could not be described the vibration performance of vehicle on different road conditions, commendably[6]. The practicability of five freedom half vehicle system was confirmed by absorbed power method in a recent study[7]. In addition, the simulation results of a specific heavy vehicle based on 15 kinds of roads provided basic references on improving the ride comfort performance[3]. And a chaotic response has been found in a full-vehicle model[8]. Overall, the more detailed of the dynamic model, the more accurate of the acquired simulation results, yet the computational complexity would grow excessively.

The reasonably simplified 1/2 seven DOF vehicle dynamic model was built in this paper, and the road roughness of grade C was modelled by the harmony superposition method (HSM) and pseudo white noise method (PWNM), respectively. Subsequently, the dynamic response of carriage was computed by an improved four-order precise Runge-Kutta integration, which achieved by MATLAB[9]. Compared with the of the root mean square value of field tests, the HSM proved to be more effective and feasible than the PWNM with the same time-step. Moreover, it also revealed the simulation adopted by nonlinear dynamic model was available, and the excitation obtained from numerical calculation could be used for the structural integrity assessment of the ammo.

2. 1/2 Seven DOF vehicle dynamic model

Measuring the dynamic response by accelerometers in field tests did provide some reliable results for calculating the distribution of stress and strain in ammo. However, this method was tremendously limited by the huge cost, long experimental period, and poor repeatability[10-11]. Reference[12] suggested that excitations of carriage but without ammo under different road conditions collected by several accelerometers could be used in ambient shaking table tests. This method will no doubt be universal and safer. While, drawing support from refined mathematical models of road roughness and improved numerical simulation methods, combining with the relative parameters of transport vehicle, more practical results could be obtained by simulating on MATLAB. By means of this way, the deficiencies of field tests would be made up for.

On the basis of relevant parameters of ammo transport vehicle, and with reasonable simplifications[13], Figure.1 shows a 1/2 seven degree-of-freedom vehicle dynamic model. The vehicle body is represented by a rigid bar with mass m_0, which considered heave motions x_0 and roll motions , were masses, dampers elements and spring equivalent stiffness of sprung, respectively. were dampers elements and spring equivalent stiffness of tires. The road roughness excitations of different tires were defined as $s_1 \ldots s_5$, and $l_1 \ldots l_5$ were the lengths between the sprung to and the centre of gravity of vehicle mass.

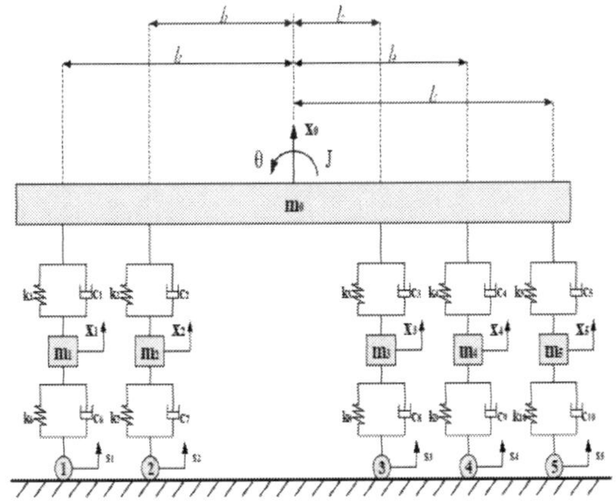

Figure 1. 1/2 seven-degree dynamic model of ammo transport vehicle.

To make it easier to compute the dynamic response, several basic postulates are listed as follows:

(1) The vehicle was symmetrical to the middle plane that cross the centre of mass, its body, car frame and their connections were rigid.

(2) The vehicle kept uniform motion in a straight line, only micro-amplitude vibration happened on equilibrium position. And tires always kept the point contact with road.

(3) The road roughness was random, there were time differences among the excitation of tires.

(4) Only heave and roll motion of vehicle were considered, the vibrations caused by other factors were ignored.

3. Models of road roughness

Road roughness was the shifting over ideal reference plane, and was determined by the displacement of power spectrum density, the general form was revealed as follows[14]:

$$G_s(n) = G_s(n_0)(\frac{n}{n_0})^{-w} \tag{1}$$

Where n the spatial frequency, m^{-1}, n_0 is the reference spatial frequency, usually is $0.1m^{-1}$, and $G_s(n_0)$ is the power spectrum density value at the reference spatial frequency, changed with the road condition, also defined as the road irregularity coefficient. ω is the frequency index, affects the structure of road spectrum, usually is 2.

The frequency-domain model was usually adopted to describe the road surface roughness, however, this analysis method is not applicable to the realistic nonlinear vehicle system[15]. That is, time-domain model was much more suitable for calculating the vertical displacements of vehicle that changed with time. Simultaneously, the accumulative damage and service life prediction became easier. Then, it is necessary to employ the time-domain model to solve the dynamic response of nonlinear vehicle system. Translating spatial frequency into time frequency is the key to carry out the time-domain analysis, transformation relationship as below, where f is the time frequency, and v is the transport speed.

$$f = vn \tag{2}$$

The random road roughness in time-domain was generated by HSM and PWNM, respectively. The time-domain excitation expressions of these two methods were listed as below:

HSM[16]:

$$\dot{s}(t) + 2\pi f_1 s(t) = 2\pi n_0 \sqrt{G_s(n_0)v} w(t) \tag{3}$$

PWNM[17]:

$$s(t) = \sum_{i=1}^{m} \sqrt{2G_s(f_{mid-i})\Delta f_i} \sin(2\pi f_{mid-i} t + \theta_i) \tag{4}$$

Where f_1 is the lower cut-off frequency, $\omega(t)$ is the standard Gauss white noise sequence, m is the interval number of f, m=5000, f_{mid-i} is the centre frequency of the ith interval, and θ_i is the random number of [0, 2π] that satisfies normal distribution.

4. Kinetic equations of transport vehicle

The kinetic equations of the seven DOF ammo model are put forward in equations (5) - (11) using Lagrange method.

Heave kinetic equation of carriage

$$\begin{aligned} &-m_0\ddot{x}_0 + k_1(x_0 - x_1 - l_1\theta) + c_1(\dot{x}_0 - \dot{x}_1 - l_1\dot{\theta}) + k_2(x_0 - x_2 - l_2\theta) + c_2(\dot{x}_0 - \dot{x}_2 - l_2\dot{\theta}) + k_3(x_0 - x_3 + l_3\theta) + \\ &c_3(\dot{x}_0 - \dot{x}_3 + l_3\dot{\theta}) + k_4(x_0 - x_4 + l_4\theta) + c_4(\dot{x}_0 - \dot{x}_4 + l_4\dot{\theta}) + k_5(x_0 - x_5 + l_5\theta) + c_5(\dot{x}_0 - \dot{x}_5 + l_5\dot{\theta}) = 0 \end{aligned} \tag{5}$$

Roll kinetic equation of carriage

$$\begin{aligned} &J\ddot{\theta} - k_1 l_1(x_0 - x_1 - l_1\theta) - c_1 l_1(\dot{x}_0 - \dot{x}_1 - l_1\dot{\theta}) - k_2 l_2(x_0 - x_2 - l_2\theta) - c_2 l_2(\dot{x}_0 - \dot{x}_2 - l_2\dot{\theta}) + k_3 l_3(x_0 - x_3 + l_3\theta) + \\ &c_3 l_3(\dot{x}_0 - \dot{x}_3 + l_3\dot{\theta}) + k_4 l_4(x_0 - x_4 + l_4\theta) + c_4 l_4(\dot{x}_0 - \dot{x}_4 + l_4\dot{\theta}) + k_5 l_5(x_0 - x_5 + l_5\theta) + c_5 l_5(\dot{x}_0 - \dot{x}_5 + l_5\dot{\theta}) = 0 \end{aligned} \tag{6}$$

Heave kinetic equation of tire 1

$$m_1\ddot{x}_1 - k_1(x_0 - x_1 - l_1\theta) - c_1(\dot{x}_0 - \dot{x}_1 - l_1\dot{\theta}) + k_6(x_1 - S_1) + c_6(\dot{x}_1 - \dot{S}_1) = 0 \tag{7}$$

IOP Conf. Series: Materials Science and Engineering **417** (2018) 012015 doi:10.1088/1757-899X/417/1/012015

Heave kinetic equation of tire 2

$$m_2\ddot{x}_2 - k_2(x_0 - x_2 - l_2\theta) - c_2(\dot{x}_0 - \dot{x}_2 - l_2\dot{\theta}) + k_7(x_2 - S_2) + c_7(\dot{x}_2 - \dot{S}_2) = 0 \tag{8}$$

Heave kinetic equation of tire 3

$$m_3\ddot{x}_3 - k_3(x_0 - x_3 + l_3\theta) - c_3(\dot{x}_0 - \dot{x}_3 + l_3\dot{\theta}) + k_8(x_3 - S_3) + c_8(\dot{x}_3 - \dot{S}_3) = 0 \tag{9}$$

Heave kinetic equation of tire 4

$$m_4\ddot{x}_4 - k_4(x_0 - x_4 + l_4\theta) - c_4(\dot{x}_0 - \dot{x}_4 + l_4\dot{\theta}) + k_9(x_4 - S_4) + c_9(\dot{x}_4 - \dot{S}_4) = 0 \tag{10}$$

Heave kinetic equation of tire 5

$$m_5\ddot{x}_5 - k_5(x_0 - x_5 + l_5\theta) - c_5(\dot{x}_0 - \dot{x}_5 + l_5\dot{\theta}) + k_{10}(x_5 - S_5) + c_{10}(\dot{x}_5 - \dot{S}_5) = 0 \tag{11}$$

In summary, the kinetic model of the ammo transport vehicle could be defined as the following equations.

$$[\mathbf{M}]\{\ddot{x}\} + [\mathbf{C}]\{\dot{x}\} + [\mathbf{K}]\{x\} = \{\mathbf{F(t)}\} \tag{12}$$

Where $[\mathbf{M}] = [m_0 \quad J \quad m_1 \quad m_2 \quad m_3 \quad m_4 \quad m_5]$ is the inertial matrix, $[C] = \begin{bmatrix} C_1 & C_2 \\ C_3 & C_4 \end{bmatrix}$ is the damping matrix, and

$$[\mathbf{C}_1] = \begin{bmatrix} c_1 + c_2 + c_3 + c_4 + c_5 & -l_1c_1 - l_2c_2 + l_3c_3 + l_4c_4 + l_5c_5 \\ -l_1c_1 - l_2c_2 + l_3c_3 + l_4c_4 + l_5c_5 & l_1^2c_1 + l_2^2c_2 + l_3^2c_3 + l_4^2c_4 + l_5^2c_5 \end{bmatrix}, \ [\mathbf{C}_2] = \begin{bmatrix} -c_1 & -c_2 & -c_3 & -c_4 & -c_4 \\ l_1c_1 & l_2c_2 & -l_3c_3 & -l_4c_4 & -l_5c_5 \end{bmatrix},$$

$$[\mathbf{C}_3] = \begin{bmatrix} -c_1 & l_1c_1 \\ -c_2 & l_2c_2 \\ -c_3 & -l_3c_3 \\ -c_4 & -l_4c_4 \\ -c_5 & -l_5c_5 \end{bmatrix}, \ [\mathbf{C}_4] = \begin{bmatrix} c_1 + c_6 & 0 & 0 & 0 & 0 \\ 0 & c_2 + c_7 & 0 & 0 & 0 \\ 0 & 0 & c_3 + c_8 & 0 & 0 \\ 0 & 0 & 0 & c_4 + c_9 & 0 \\ 0 & 0 & 0 & 0 & c_5 + c_{10} \end{bmatrix},$$

$[K] = \begin{bmatrix} \mathbf{K}_1 & \mathbf{K}_2 \\ \mathbf{K}_3 & \mathbf{K}_4 \end{bmatrix}$ is the stiffness matrix, and

$$[K_1] = \begin{bmatrix} k_1 + k_2 + k_3 + k_4 + k_5 & -l_1k_1 - l_2k_2 + l_3k_3 + l_4k_4 + l_5k_5 \\ -l_1k_1 - l_2k_2 + l_3k_3 + l_4k_4 + l_5k_5 & l_1^2k_1 + l_2^2k_2 + l_3^2k_3 + l_4^2k_4 + l_5^2k_5 \end{bmatrix}, \ [K_2] = \begin{bmatrix} -k_1 & -k_2 & -k_3 & -k_4 & -k_4 \\ l_1k_1 & l_2k_2 & -l_3k_3 & -l_4k_4 & -l_5k_5 \end{bmatrix},$$

$$[K_3] = \begin{bmatrix} -k_1 & l_1k_1 \\ -k_2 & l_2k_2 \\ -k_3 & -l_3k_3 \\ -k_4 & -l_4k_4 \\ -k_5 & -l_5k_5 \end{bmatrix}, \ [K_4] = \begin{bmatrix} k_1 + k_6 & 0 & 0 & 0 & 0 \\ 0 & k_2 + k_7 & 0 & 0 & 0 \\ 0 & 0 & k_3 + k_8 & 0 & 0 \\ 0 & 0 & 0 & k_4 + k_9 & 0 \\ 0 & 0 & 0 & 0 & k_5 + k_{10} \end{bmatrix},$$

And the matrix of road roughness excitation is $\{\mathbf{F(t)}\} = \{\mathbf{F}_K\mathbf{(t)}\} + \{\mathbf{F}_C\mathbf{(t)}\}$.

Where the damper force that generated by road roughness is

$\{\mathbf{F}_C\mathbf{(t)}\} = [\ 0 \quad 0 \quad c_6\dot{q}_1 \quad c_7\dot{q}_2 \quad c_8\dot{q}_3 \quad c_9\dot{q}_4 \quad c_{10}\dot{q}_5]^T$,

and the sprung force that generated by road roughness is

$\{\mathbf{F}_K\mathbf{(t)}\} = [\ 0 \quad 0 \quad k_6q \quad k_7q_2 \quad k_8q_3 \quad k_9q_4 \quad k_{10}q_5]^T$,

the matrix of displacements response is $\{\mathbf{x}\} = [x_0\ x_1\ x_2\ x_3\ x_4\ x_5)]^T$,

the matrix of speeds response is $\{\dot{\mathbf{x}}\} = [\dot{x}_0\ \dot{x}_1\ \dot{x}_2\ \dot{x}_3\ \dot{x}_4\ \dot{x}_5)]^T$,

the matrix of accelerations response is $\{\ddot{\mathbf{x}}\} = [\ddot{x}_0\ \ddot{x}_1\ \ddot{x}_2\ \ddot{x}_3\ \ddot{x}_4\ \ddot{x}_5)]^T$.

The compute time starts from tire 5, and the excitation of tire 4could be derived from equation (13), and the others could be deducted in the same way.

$$s_4(t) = s_5(t + \frac{l_5 - l_4}{v}) \tag{13}$$

The state space formulation of equation (12) was built for effective use of the improved four-order precise Runge-Kutta integration solving dynamic response.

$$\{\dot{\mathbf{u}}\} = [\mathbf{H}]\{\mathbf{u}\} + \{\mathbf{P(t)}\} \tag{14}$$

And $\{\mathbf{u}\} = \{x\ \dot{x}\}^T$, $\{\mathbf{P(t)}\} = \{0\ \mathbf{M}^{-1}\mathrm{F(t)}\}^T$, $[\mathbf{H}] = \begin{bmatrix} 0 & I \\ -\mathbf{M}^{-1}\mathbf{K} & -\mathbf{M}^{-1}\mathbf{C} \end{bmatrix}$.

5. Numerical simulation and results analysis

The feasibility study of HSM and PWNM adopted in this paper has been verified[18-19].The displacement and acceleration response of the carriage barycentre on grade A; B; C road were calculated utilizing MATLAB. The road irregularity coefficients of grade A; B; C road are 16、64、256 mm^2/m^{-1}, respectively. And the range of spatial frequency is 0.011~2.830 m^{-1}. The corresponding range of time frequency was calculated by equation (2), and that range could totally overlap the natural frequency of the vehicle body and tires[14].

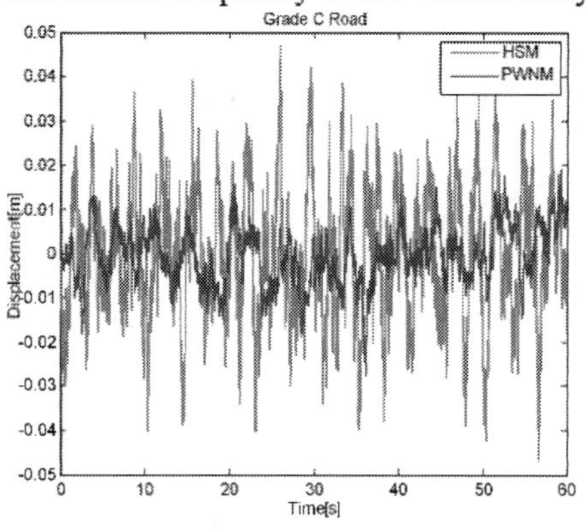

Figure 2. Comparison results of grade C road surface roughness simulated by HSM and PWNM.

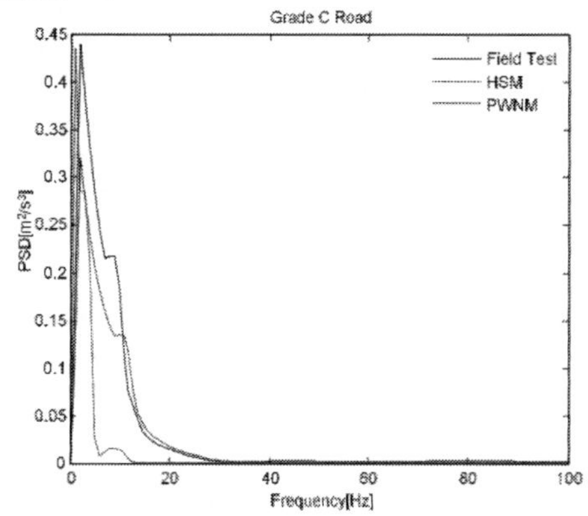

Figure 3. Acceleration response of carriage barycentre on grade C road simulated by HSM and PWNM.

Figure 2 shows the comparison results of grade C road surface roughness simulated by HSM and PWNM, and figure 3 gives the acceleration power spectrum density of carriage barycentre on grade C road, which also provides the comparison result combined with field test. It is not hard to catch sight of that the road roughness simulated by HSM is closer to the field test than PWNM under the same time-step. There are still some deviations in several parts frequency band that attribute to the moment of forces generated by different road roughness on tires of both sides, and the discrepancies between the analogue and complex actual signals. However, the peak of main frequency band simulated by HSM is essentially in agreement with the field test, which could reflect the vibration characteristics, and also illustrates the rationality of the vehicle dynamic model. The following discussion, therefore, is based on the results that simulated by HSM.

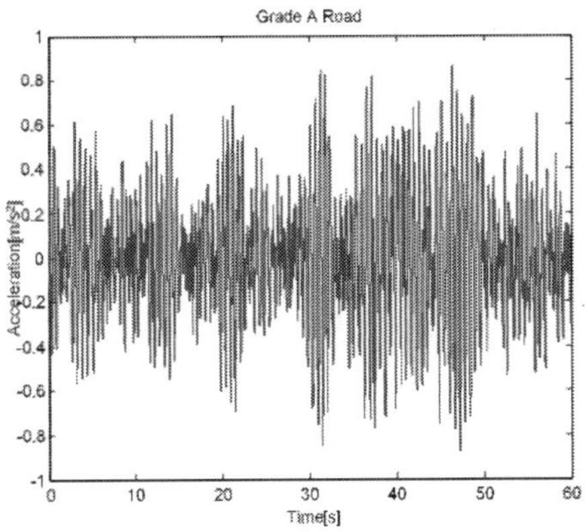

Figure 4. Acceleration response of carriage barycentre on grade A road simulated by HSM.

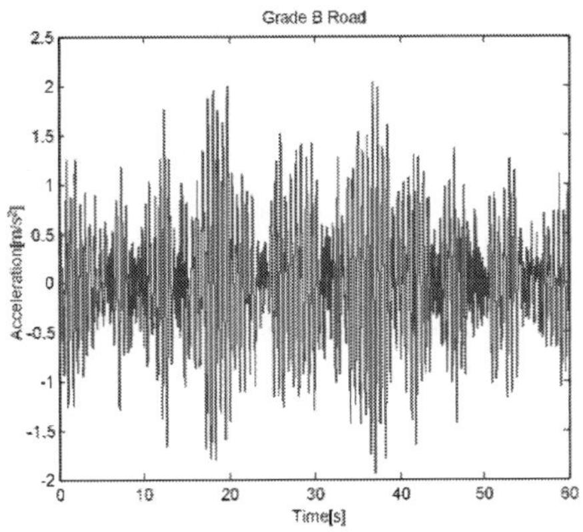

Figure 5. Acceleration response of carriage barycentre on grade B road simulated by HSM.

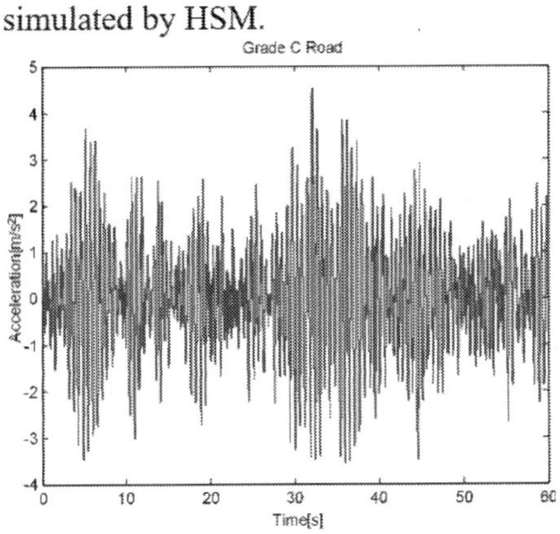

Figure 6. Acceleration response of carriage barycentre on grade C road simulated by HSM.

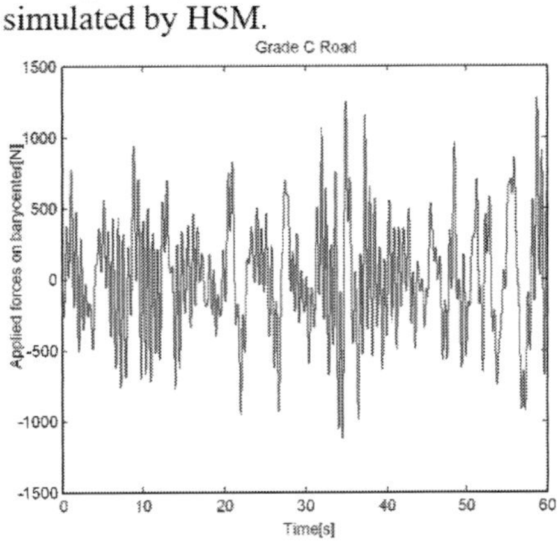

Figure 7. Applied force of carriage barycentre on grade C road simulated by HSM.

The simulation results from figure 4 to 6 indicated that the heave acceleration response of carriage barycentre changes with the road condition, and gets higher when road grades become worse, which accords with the facts. And root mean square (RMS) increased from 0.3177 m/s^2 of grade A to 0.7493 m/s^2 grade B, finally to 1.3005 m/s^2 of grade C. The forces applied on the barycentre showing in figure 7 indicates that the indispensable research on the destructiveness of ammo road transportation, particularly the accumulative damage of ammo caused by long-distance transportation.

6. Conclusion

Each figure should have a brief caption describing it and, if necessary, a key to interpret the various lines and symbols on the figure. This paper presents a reasonably simplified half seven DOF nonlinear dynamic model, according to the propositional expression of road roughness coefficients, and mainly using the HSM to simulate the road roughness with different grades. The displacement and acceleration response of dynamic model under different road excitations were calculated by an improved four-order precise Runge-Kutta integration, and also obtained the applied forces of carriage barycentre on grid C road changed with time.

The deviations of carriage barycentre response under the road roughness simulated by HSM, which compared with that under the field test, strongly indicates that there is a necessity to build a full-vehicle model. The numerical results also reveal that the applied forces on ammo could not cause the strength failure, but the accumulative damage could not be neglected in the next study.

In the research, we can conclude that the method of generate road roughness and the vehicle dynamic model need to be improved, particularly a full-vehicle model that considers pitch motions should be built. Moreover, the applied forces could be taken as the excitations of the vehicle on the ammo model, and the simulation that analyses the affections on the structural integrity of ammo could be carried on.

Acknowledgments
The authors gratefully acknowledge the reviewers who give pertinent proposals about this paper, and also should acknowledge my senior fellow apprentice who teach me how to use MATLAB to analyse the calculations.

References
[1] Song Yifan and Chen Rongfeng 2007 Analysis method of vehicle vibration response caused by pavement roughness *Journal of traffic and transportation engineering* **7(4)** 39–43
[2] Wu Shaoqing, Ai Hongxin and Guo Yingzheng 2013 Coupled vibration analysis of missile-vehicle system under road excitation *Journal of southeast university (natural science edition)* **43(5)** 1055–61
[3] Zhu Yanlei 2014 Simulation and evaluation of vibration performance for multi-axis heavy-duty special automobile under different road models *Jilin*
[4] Lv Shibao 2013 Analysis and calculation of vehicle's random vibration response *Lanzhou*
[5] Le Van Quynh, Zhang Jianrun and Liu Xiaobo 2011 Nolinear dynamic analysis of interaction between vehicle and road surfaces for 5-axle heavy truck *Journal of southeast university (English edition)* **27(4)** 405–09
[6] Marek Borowiec, Asok K. Sen and Grzegorz Litak 2012 Vibrations of a vehicle excited by real road profiles *Forschung im ingenieurwesen* **74(2)** 99–109
[7] Li Junyan, Wan Yiqiang, Zhao Qi and Li Jie 2010 Vibration performance simulation of military automobile *Science paper online* **5(10)** 807–11
[8] Zhu Qin and Mitsuaki Ishitobi 2006 Chaotic vibration of a nonlinear full-vehicle model *International journal of solids and structures* **43(3)** 747–59
[9] Fu Minghui and Liang Huali 2009 An improved precise Runge-Kutta integration *ACTA scientiarum naturalium universitatis sunyasen I* **48(5)** 1–5
[10] Xu Cheng, Hou Zhiqiang and Li Shangao 2002 Calculation on dynamic characteristics of the fuselage of winged missiles during their transportation *ACTA armament AR II* **23(2)** 219–23
[11] Fu Libiao, Hu Mingshu and Yao Jungang 2008 Based on SIMO modal experiment and analysis on frame and bottom-board of military transport vehicle *Science technology and engineering* **8(23)** 6421–23
[12] Han Baohong, Qi Jingli and Yan Qinghua 2012 Random vibration test vehicle in ammunition transportation and theoretical analysis *Machine tool and hydraulics* **40(1)** 78–82
[13] Li Haiguang 2014 Research on ammunition vibration characteristics and safety assessment under highway transportation *Shijiazhuang*
[14] Yu Zhisheng 2006 *Automobile theory* (Beijing: Mechanical Industry Press)
[15] Zhang Yonglin 2010 Study on numerical modelling and computer simulation of stochastic road of vehicles *Wuhan*
[16] Zhao Qi, Li Jie and Cui Naidong 2012 Time field simulation of vibration performance for military automobile under road random input *China science paper* **7(11)** 862–65
[17] Chang Zhiquan, Luo hong and Chu Zhigang 2004 Building model of road roughness *Journal of Chongqing university (natural science edition)* **27(12)** 5–8
[18] Xia Junzhong, Ma Zongpo and Bai Yunchuan 2012 State of the research on model for road roughness excitation *Noise and vibration control* **(5)** 1–5
[19] Xiang Huarong, Zhang Kaibin and Zhang Zhuoyu 2009 Simulation of road roughness based on pseudo white noise and study of vehicle ride comfort *Journal of Chongqing institute of technology (natural science)* **23(9)** 11–15

The Process of Engine Reliability Fuzzy Prediction

Chuan Zhang[1, a], **Shaokun Wang** [1, b], **Qiang Zhang**[*1, c], **Zhongzheng Liu** [1, d]

[1]School of mechanical engineering, Shandong University, Jinan 250061, China
[2]Shandong Labor Vocational and Technical College

[a]627156601@qq.com,[b]iwangshaokun@sina.com,[c]zhangqiangd@163.com,
[d]37381666@qq.com

Abstract. Since reliability is one of the most important indicators of engines, reliability research has been increasing used in engines widely. These studies are based on the failure data which are obtained during working time, and the distribution can be summarized by using computer aided technology, hence improving the product reliability. With the help of computer aided technology, the reliability fuzzy prediction method of the engine system, for example, is used in supercharging system to verify the feasibility.

1. Introduction

Reliability issues came out at the occurrence of the Second World War. During the process when Germany research and develop the V-2 rocket, as well as the aircraft engine improvement in England, due to the complexity of the system structure and the harsh requirement for its usage, reliability requirement was come out in order to guarantee the functional feasibility of the system.

Generalizing reliability technology in automobile manufacturing industry is very significant because the development for both two are complementary. The incidents of recalling cars often happened before the launch of generalizing quality control and reliability index. The report of America automobile index has pointed out that the average value of faulted cars in each 1000 cars have been declined from 99 to 20 from the year of 1980 to 2000, and the reliability of engines have been improved distinctively. Reliability research started late in China compared with some other countries, and the early reliability research was applied to electronic products. Nevertheless, there are still gaps existing between China and some other countries who has already possessed advanced technology about it in the engine reliability field.

In order to study the system reliability, the system structure needs to be defined first. The system structure model should be set up first in the system reliability evaluation: The whole system is expressed in its parts; and then collect data for each part. Finally, the reliability of the system is evaluated with the value of the components and the reliability model of the system[1].The fault tree is a common system model, which is used for display critical event of system failures, the logical graph model between the causes of the parts malfunction. Building a fault tree is also a connection between the state of the system and the state of the parts[2].

After verifying the fuzzy prediction feasibility of universal confidence level, the related applications of reliability fuzzy prediction in failure associated problem can be found, following by selecting subsystem unite prediction, figuring out probability of subsystem unite faults by using triangular fuzzy numbers range, obtain and eventually obtaining the reliability prediction consequence.

2. The method of engine system reliability fuzzy prediction

2.1. Selecting the influence factors of subsystem reliability

The general system can be divided into multiple subsystems which contains multiple parts. The engine for example, the valve system and fuel feed system of which has numerous failure modes, and factors that influences its reliability vary as well. Without conducting the selection process, it will not only reflect no research value on main influence factors but also impact the rationality of the fuzzy prediction. Therefore, it is necessary to take the categories of influence factors as the reference when screening the factors, which can be done by dividing some factor into groups and constituting the factor compilations, following by refining into the sub-factors and constituting the factor subsets.

The inherent attributes of the subsystem can be neglected when selecting the factors since they have already been considered in the process of design and production.

2.2. The wight of influence factors

The ability of exerting influence for the factors need to be measured by the weight, which contains the weight of all kinds of factors, as well as the weight of the sub-factors within each of them. The degree of hazard needs to be identified by the hazard analysis when defining the weights of the factors in terms of vital, very important, important, unimportant, relatively unimportant, very unimportant and extreme unimportant in sequence. Considering that the fuzziness of the reliability information when it is designed, fuzzy numbers are used to express.

$$\tilde{A} = \{\tilde{1},\tilde{2},\tilde{3},\tilde{4},\tilde{5},\tilde{6},\tilde{7},\tilde{8}\} \tag{1}$$

The stated factors weight and sub-factors weight respectively are:

$$\tilde{Q} = (q_1, q_2, q_3) \tag{2}$$

$$\begin{cases} \tilde{Q}_1 = (q_{11}, q_{12}, q_{13}) \\ \tilde{Q}_2 = (q_{21}, q_{22}, q_{23}) \\ \tilde{Q}_3 = (q_{31}, q_{32}, q_{33}, q_{34}) \end{cases} \tag{3}$$

2.3. Comprehensive assessment

The comprehensive assessment is a multi-objective decision actually. It is the comprehensive fuzzy assessment that evaluated the incidents and faults that influenced by multi-factor which include fuzzy factors or replaced by fuzzy factors. The fuzzy comprehensive evaluation has one and two grades, and generalized further could be multistage. All the sub-factors of each factor obtain the comprehensive assessment that is fuzzy comprehensive assessment. The degree of reliability of sub-system might achieved is impressed by alternative domain $Y = \{y_1, y_2, ..., y_3\}$. Take the number j sub-factor x_{ij} in the sub-factor $X_i = (i = 1,2,3)$ as an example, the fuzzy probability of fetching the alternative factor Y_k is $\tilde{P}_{ijk}$, and the one grade fuzzy comprehensive assessment matrix is:

$$\tilde{P}_i = \begin{bmatrix} \tilde{p}_{i11} & \tilde{p}_{i12} & \cdots & \tilde{p}_{i1m} \\ \tilde{p}_{i21} & \tilde{p}_{i22} & \cdots & \tilde{p}_{i2m} \\ \cdots & \cdots & \cdots & \cdots \\ \tilde{p}_{in1} & \tilde{p}_{in2} & \cdots & \tilde{p}_{inm} \end{bmatrix}, \quad i = 1,2,3 \tag{4}$$

Then the fuzzy comprehensive assessment each factor $X_i(i = 1,2,3)$ could obtain the judging aggregate

$$\tilde{O}_i = \tilde{Q}_i \circ \tilde{P}_i = (\tilde{o}_{i1}, \tilde{o}_{i2}, ..., \tilde{o}_{im}), i = 1,2,3 \tag{5}$$

The "$\circ$"means fuzzy operational symbol. If regard the $\tilde{O}_i$ of the one grade fuzzy comprehensive assessment as comprehensive estimated matrix $\tilde{P}$, reconsider the weight $\tilde{Q}$ of each factor could obtain the judging aggregate of second grade fuzzy assessment:

$$\widetilde{R} = \widetilde{Q} \circ \widetilde{P} = (\widetilde{r}_1, \widetilde{r}_2, \ldots, \widetilde{r}_m) \tag{6}$$

Doing the weighted average when predict by using fuzzy assessment. And using the weight $\widetilde{R}_k$ to weighted average figure up the alternative factors, the predicted consequence is shown by fuzzy numbers:

$$\widetilde{P}_{ss} = \sum_{k=1}^{m} \widetilde{r}_k y_k / \sum_{k=1}^{m} \widetilde{r}_k \tag{7}$$

2.4. Fuzzy prediction

The consequence of fuzzy comprehensive assessment prediction is made up with the fuzzy numbers. The degree of reliability that figured out represents the interval named confidence interval of confidence level. According to the possible theory of the probability statistics, this interval in the fuzzy aggregate $\widetilde{P}_{SS}$ has more possibility, and the higher confidence level λ, the littler ambiguity. Generally speaking, confidence level λ rather approaching to 1 than equate, because the fuzzy aggregate is determined value when confidence level is 1.

3. The Application of Engine Fuzzy Prediction

The malfunction assessment and reliability prediction of automobile engine system have been more complex due to the updating of engine materials, structure and technology. Applying the fuzzy reliability prediction is an efficient path. It is a gradual change process from the engine first been used to first break down, far from "entire" order changed into "entire" failure state. It would appear the failure symptom in the pilot process. Therefore, the malfunction and omen are fuzzy.

Take the pressurization system as application object, According to the setting initial each factor and sub-factor in the last section, the degree of reliability of pressurization system average fault-free period is 0.97 around by consulting the statistical data of references[3-4]. The design of pressurization system went wrong if the predicted value was divergent by using the aforesaid way. Increase the predicted range somewhat could get the alternative set:

$$Y = \{0.95, 0.96, 0.97, 0.98, 0.99\} \tag{8}$$

The weight of the each factor that was represented by triangular fuzzy numbers is shown in table 1. The first number of triangular fuzzy numbers represents the grade $\{\widetilde{1},\widetilde{2},\widetilde{3},\widetilde{4},\widetilde{5},\widetilde{6},\widetilde{7},\widetilde{8}\}$ of weight that divided in above. And the other two numbers are lift and right distributed parameters of $\widetilde{A}$, that means greater than or less than the fuzziness of this grade, and bigger subtraction of the two numbers represent the bigger possibility.

Table 1. Weight of the each factor that was represented by triangular fuzzy

Each factor	Group one	Group two	Group three	Group four
Design	(7,0.5,0.5)	(8,0.4,0.4)	(8,0.6,0.2)	(8,0.7,0.1)
Usage	(6,0.4,0.7)	(7,0.6,0.6)	(7,0.5,0.3)	(6,0.2,0.8)
Production	(8,0.4,0.3)	(8,0.5,0.2)	(7,0.2,0.3)	(7,0.7,0.3)

During the counting process in last section could get the weight aggregate

$$\widetilde{Q} = (0.35, 0.31, 0.34) \tag{9}$$

And the weight aggregate of sub-factors

$$\begin{cases} \widetilde{Q}_1 = (0.334, 0.41, 0.256) \\ \widetilde{Q}_2 = (0.375, 0.301, 0.324) \\ \widetilde{Q}_3 = (0.296, 0.241, 0.272, 0.191) \end{cases} \tag{10}$$

The triangular fuzzy numbers are particularity in the one-level comprehensive fuzzy assessment prediction. That means the lift and right distributed parameters of $\widetilde{A}$ are equal. In order to simplified representation, using the letters replace the lift and right distributed parameters directly. And divide into three grades "very possible 0.04", "rather possible 0.08", "common possible 0.15", table 2 aims at the fuzzy weight of design objective sub-factors. According to the sequencing computation consequence of alternative elements aggregation in the above, obtain the one-level fuzzy comprehensive assessment matrix.

Table 2. Sub-factors weight were represented by triangular fuzzy numbers

alternative elements	Group one	Group two	Group three	Group four
0.95	(0.30,0.08)	(0.25,0.08)	(0.20,0.04)	(0.20,0.15)
0.96	(0.85,0.04)	(0.80,0.08)	(0.75,0.08)	(0.70,0.08)
0.97	(1.00,0.08)	(0.95,0.04)	(0.90,0.08)	(0.95,0.08)
0.98	(0.90,0.08)	(1.00,0.08)	(1.00,0.15)	(1.00,0.15)
0.99	(0.70,0.15)	(0.75,0.08)	(0.60,0.08)	(0.90,0.15)

$$\widetilde{p}_{11}=[(0.24,0.09)\ (0.78,0.07)\ (0.95,0.07)\ (0.98,0.12)\ (0.74,0.12)] \tag{11}$$

Figure out the two-level comprehensive fuzzy assessment matrix according to the above next:

$$\widetilde{P}=\begin{bmatrix}(0.25,0.07) & (0.81,0.09) & (0.95,0.08) & (0.97,0.10) & (0.77,0.10)\\(0.41,0.11) & (0.76,0.10) & (0.97,0.08) & (0.93,0.10) & (0.75,0.11)\\(0.64,0.08) & (0.92,0.09) & (0.98,0.07) & (0.93,0.09) & (0.57,0.10)\end{bmatrix} \tag{12}$$

During the reliability prediction in the end the confidence level need to be ensured. The confidence level λ generally value 0.95 in the mathematical statistics, and approach to the initial predicted value 0.97 and all belong to the range of the alternative aggregation. So it is observed that the consequences are reasonable though figure out by reliability fuzzy prediction. Today, with the continuous improvement of human's thinking consciousness, we have gradually utilized this non quantitative model to break the shackles of traditional knowledge, and become a scientific theory with distinctive characteristics and complete system, and have been applied successfully in dozens of fields[5].

4. Conclusion

Using the ways of engine system reliability fuzzy prediction, briefly verify the feasibility of fuzzy prediction under the common confidence level by taking the pressurization system in the FAT as example. And come up with the malfunction relevance application of reliability fuzzy prediction. Proved the feasibility of reliability fuzzy prediction. It provides theoretical and data support for the test of automotive production design stage, and draws practical guidance. Starting from the product design stage, it improves reliability and prolongs the service life of products.But still need the optimized analysis.

References

[1] Mosleh, Ali, Bier, Vicki M. Uncertainty about probability: a reconciliation with the subjectivist viewpoint[J].. IEEE Transactions on Systems, Man, and Cybernetics Part A: Systems and Humans. 1996(25): 979-987.

[2] Hamada. Bayesian Reliability[M]. National Defence Industry Press, 2014:1-2.

[3] Zhao Dezi. Fuzzy Prediction and Allocation Techniques for Mechanical System Reliability in Early Design Stage[M]. National Defence Industry Press, 2010:103-234.

[4] Li Hailong, Ma Dengwu, Liu Xiao. Application of Bayesian network Theory in Equipment Fault Diagnosis[M]. National Defence Industry Press, 2009,07: 14-18.

[5] Guozhang Zhan. Reliability Prediction and Fault Analysis Technology of CNC Vertical Lathe Based on Fuzzy Theory[D]. Changchun: Jilin University, 2014.

IOP Conf. Series: Materials Science and Engineering **417** (2018) 012017 doi:10.1088/1757-899X/417/1/012017

Finite Element Model of Multi-Body Brake Quarter Car for Braking Analysis

M J Han[1], C H Lee[1], T W Park[1], S P Jung[2] and Y I Lee[3]

[1]Department of Mechanical Engineering, Ajou University, Suwon 16499, Korea
[2]Korea Automotive Technology Institute, Jeollanam-do 58463, Korea
[3]SB&F Eng., Jeollanam-do 58463, Korea

park@ajou.ac.kr

Abstract. A finite element multi-body model was developed for evaluating the dynamic characteristics of the brake quarter car. All parts of the brake quarter car were modeled as a finite element considering the elastic deformation. Each part was connected by a joint and has a degree of freedom by kinematic constraint. After stabilization by preload, friction braking analysis was performed. The vertical acceleration of the lower control arm was confirmed. Vibration effects due to deformation of the flexible body motion not occurring in the rigid body motion were confirmed at the acceleration level. The coupled thermo-mechanical analysis considering the frictional heat generation was performed using the result of dynamic displacement assuming small deformation in the elastic region. The temperature results of the brake disc were confirmed.

1. Introduction

The quarter car model is the basis for analyzing the dynamic behavior of the full car system. Conventional quarter car model, modelled as a rigid body, was used primarily to analyze the static and ride vibration of mass, spring, and damper systems [1, 2]. However, the rigid body cannot consider the local deformation, so only the low frequency mode analysis is possible for the whole system. This is a simple assumption except for the effect of high frequency vibration mode due to the local deformation of each node point. As computer performance is improved and efficient numerical methods are developed, rigid body models can be replaced with flexible body models [3]. However, static stress analysis and natural frequency analysis are mostly performed for single parts, and there is almost no multi-physics analysis composed of flexible multi-body system.

Braking is basically a phenomenon of energy conversion. A kinetic energy of vehicle speed is converted into thermal energy by friction contact between the brake disc and the pad during braking. The frictional heat generation causes sudden temperature increase on the brake disc and the pad surface. Thermal expansion due to temperature increases thermal stress. Therefore, a thermo-mechanical coupling which can consider structural deformation and thermal deformation is essential in frictional braking analysis.

In this study, the finite element multi-body brake quarter car system was developed to evaluate the local vibration effects and the thermoelastic behavior. The thermo-mechanical coupling conditions are modeled for the heat generation by frictional contact of different parts. Stabilization analysis by gravity and preload of chassis suspension was performed before braking analysis. After the stabilization of the dynamics model, friction braking analysis is performed around the equilibrium position. The vertical acceleration of the lower control arm due to the suspension vibration was confirmed. Also, the temperature results of the brake disc were confirmed through the thermal model.

2. Modeling of brake quarter car

2.1. Multi-body joint modeling

The brake quarter car analysis model has a McPherson suspension and consists of a brake disc, a caliper, a lower control arm and a hub. Each part is connected by bushings, revolute joints and fixed joints that enable kinematic motion. A schematic diagram of the brake quarter car model is shown in Figure 1. The suspension was modeled by spring-dashpot connector element. The inertial effect by gravity was given to the whole model. The angular velocity was applied to the revolute joint of the brake disc. Table 1 shows the degree of freedom and mechanical property of joints.

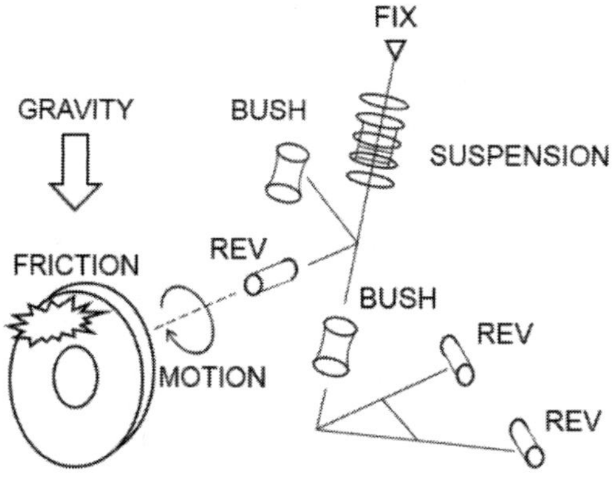

Figure 1. Schematic diagram of brake quarter car model.

Table 1. Degree of freedom and mechanical property of joint.

Joint	Type	D.O.F	Stiffness (N/m)	Damping (Ns/m)
SUSPENSION	Trans.	1	1000	100
BUSH	Trans., Rot.	6	2000	200
REV	Rot.	1	-	-
FIX	-	0	-	-

2.2. Finite element model

In this study, the commercial software ABAQUS Ver. 6.13 was used. All parts are modeled as flexible bodies that allow for deformation. The elements of the caliper, the brake disc and the pad have additional temperature degrees of freedom for calculating the thermal deformation. The suspension is a rigid body model consisting of an equivalent spring and damper system. The finite element model of the brake quarter car is a linear hexahedral mesh with a total of 35,285 nodes and 20,491 elements. Table 2 shows the information of the finite element model.

2.3. Thermo-mechanical coupling condition

Heat generation by frictional contact causes thermoelastic deformation [4]. To calculate the thermal energy and simplify the thermal model, the following assumptions are made: (i) The materials are all homogeneous isotropic and their thermal properties are independent on temperature. (ii) The friction coefficient is constant. (iii) Convective heat transfer is applied by film coefficient. (iv) Radiation heat transfer is not considered.

The transient heat transfer equation described in the Cartesian coordinate system is given as follow:

$$\rho c\left(\frac{\partial T}{\partial t}\right) = k\left\{\frac{\partial}{\partial x}\left(\frac{\partial T}{\partial x}\right) + \frac{\partial}{\partial y}\left(\frac{\partial T}{\partial y}\right) + \frac{\partial}{\partial z}\left(\frac{\partial T}{\partial z}\right)\right\}$$

Where ρ, c and k are the density, specific heat and thermal conductivity, respectively.

Table 2. Information on finite element models

Part	Type	No. of nodes	No. of elements	Elastic modulus (GPa)	Density (kg/m^3)
Disc		9,937	4,902	120	7,200
Pad		1,067	667	3	2,252
Piston		436	206	190	7,400
Hub		5,669	3,261	210	7,215
Caliper		9,208	6,197	180	6,850
Knuckle		5,220	3,215	90	4,570
Lower control arm		3,748	2,043	90	4,570

The heat flux coupling conditions are calculated by the energy conversion ratio, the heat distribution ratio, the friction coefficient, the contact pressure, the effective radius and the angular speed between two different friction-contacting parts. All kinetic energy is assumed to be converted to thermal energy and there is no energy loss during the energy conversion process. The heat flux input to the friction surface of the brake disc and the pad is as follows.

$$q_{disc} = (1-\gamma)\mu\sigma(x,y,t)r\omega(t)$$
$$q_{pad} = \gamma\mu\sigma(x,y,t)r\omega(t)$$

where q is the heat flux, γ is the heat partition ratio, μ is the friction coefficient, σ is the contact pressure, r is the friction radius, and ω is the angular velocity of the brake disc. The heat partition ratio can be obtained as follows [5]:

$$\gamma = \frac{1}{1+\sqrt{\rho_p k_p c_p / \rho_d k_d c_d}}$$

IOP Conf. Series: Materials Science and Engineering **417** (2018) 012017 doi:10.1088/1757-899X/417/1/012017

2.4. Boundary condition

The boundary conditions of the finite element model are shown in Figure 2. In the initial analysis step, the stabilization analysis of the brake quarter car by gravity ($g = 9.81$ m/s^2) is performed. Stabilization analysis is the process of converging the static position due to suspension stiffness and damping. The angular velocity (ω = 30 rad/s) of the brake disc, which has reached the maximum speed after stabilization, is reduced by the braking pressure (p = 5 MPa). The braking pressure during braking is constantly applied to the brake pads.

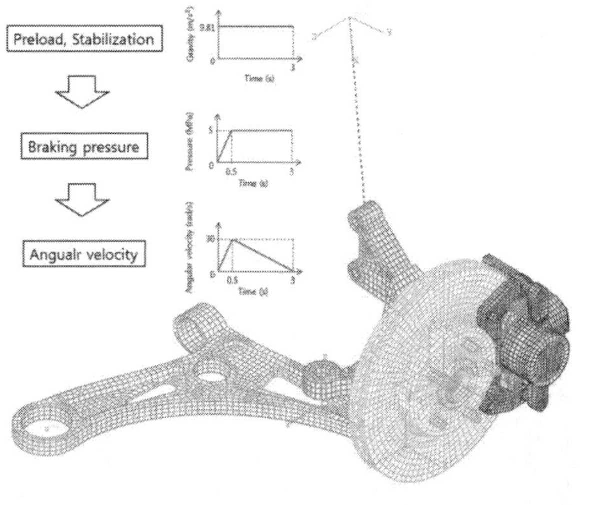

Figure 2. Boundary conditions of the finite element model.

3. Results of braking analysis

3.1. Acceleration.

The vertical vibration acceleration of the lower control arm is shown in Figure 3. The vertical acceleration is gradually reduced by spring stiffness and damping of the suspension. In addition, the local vibration can be confirmed in the acceleration result. This is because the local deformation mode of the flexible body is considered. This local vibration can be easily confirmed by Fast Fourier Transform (FFT) results. Figure 4 shows the FFT for vertical acceleration. The peak of the natural mode of the suspension can be seen at about 6.2 Hz. It can be seen that the peak frequencies of the elastic modes of the flexible body appear in different frequency bands.

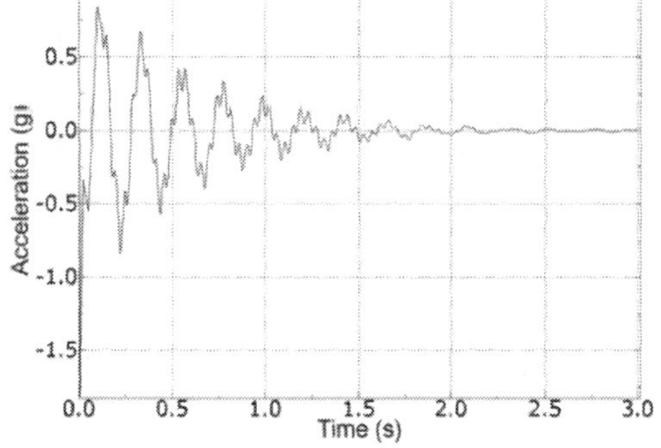

Figure 3. Vertical acceleration of the lower control arm.

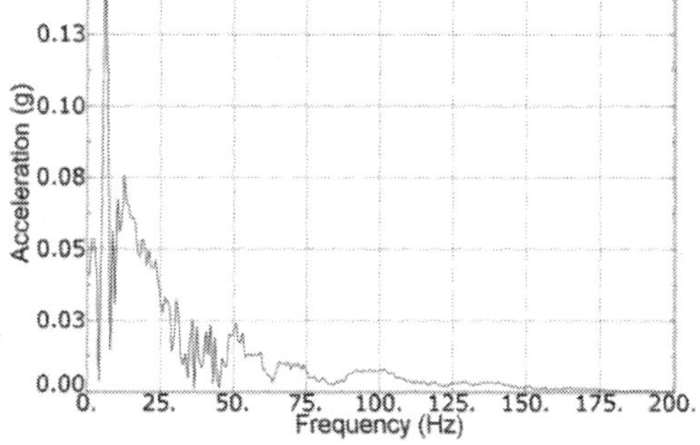

Figure 4. FFT of the vertical acceleration of the lower control arm.

IOP Conf. Series: Materials Science and Engineering **417** (2018) 012017 doi:10.1088/1757-899X/417/1/012017

3.2. Temperature.

The temperature distribution of the brake quarter car is shown in Figure 5. A temperature gradient occurs around the brake disc. Figure 6 shows seven temperature peaks on the disc surface. It can be confirmed that the temperature is reduced by convective heat transfer. The maximum temperature and the end temperature are about 53 degrees and 40 degrees, respectively. Therefore, heat generation by the heat flux coupling condition and convective heat transfer to the ambient temperature are well modeled reflecting the physical heat transfer effect.

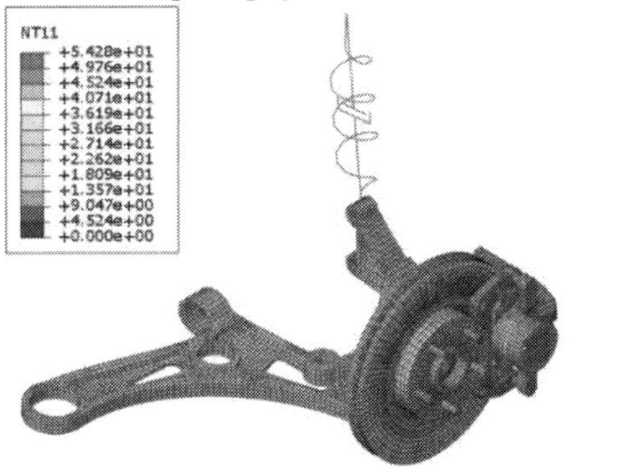

Figure 5. Temperature distribution of the brake quarter car.

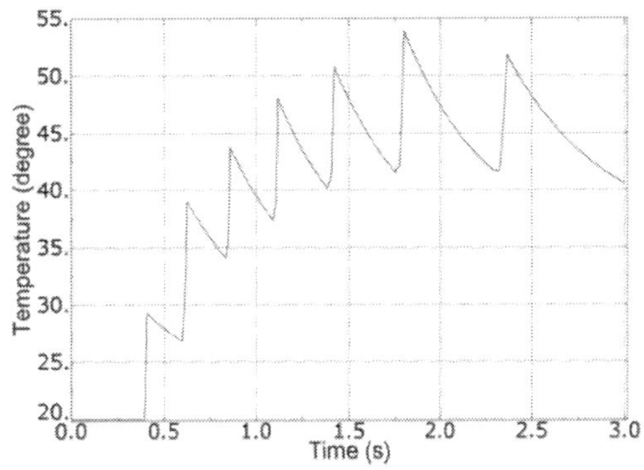

Figure 6. Temperature result of the brake disc surface.

4. Conclusion

The finite element model of the multi-body brake quarter car was developed. The local vibration by the flexible body mode was confirmed from the vertical acceleration acting on the lower control arm. Also, the temperature gradient by frictional heat and convective heat transfer was confirmed. Based on the brake quarter car model, full quarter car model assembled with wheel-tire will be developed. The full quarter car model will be able to analyze not only braking but also ride and handling. In the next study, the validation of the analytical model will be carried out by the brake quarter car dynamo experiment.

Acknowledgments

This research was financially supported by Jeollanamdo Province and Jeonnam Institute for Regional Program Evaluation(JNIRPE) through the Research and Development for Auto Tuning Part Industry, "Developed high-performance tuning brake friction material with improved fade resistance and durability life by using composite materials".

References

[1] Chikhale S J and Deshmukh S P 2013 *Int. Journal of Innovative Research in Science, Engineering and Technology* vol 2 no 8 (Comparative Analysis Of Vehicle Suspension System in Matlab-SIMULINK and MSC-ADAMS with the help of Quarter Car Model) p 4074-4081

[2] Sathishkumar P, Jancirani J, John D and Manikandan S 2014 *Int. Journal of Innovative Research in Science, Engineering and Technology* vol 3 no 1 (Mathematical modelling and simulation quarter car vehicle suspension) p 1280-1283

[3] Gao C H, Huang J M, Lin X Z and Tang X S 2007 *Journal of Tribology* vol 129 no 3 (Stress Analysis of Thermal Fatigue Fracture of Brake Disks Based on Thermomechanical Coupling) p 536-543

[4] Gao C H, Huang J M, Lin X Z and Tang X S 2007 *Journal of Tribology* vol 129 no 3 (Stress Analysis of Thermal Fatigue Fracture of Brake Disks Based on Thermomechanical Coupling) p 536-543

[5] Yevtushenko A and Grzes 2011 *Numerical Heat Transfer* vol 59 no 7 (Finite Element Analysis of Heat Partition in a Pad/Disc Brake System) p 521-542

Sound Reconstruction of an Overdetermined Mixture in a Mini Plant: An Experimental Study

Anindita Adikaputri Vinaya[1], Dhany Arifianto[2], Yupit Sudianto[3], Aulia Siti Aisjah[2]

[1]Department of Engineering Management, Universitas Internasional Semen Indonesia, Gresik, Indonesia 61122
[2]Department of Engineering Physics, Institut Teknologi Sepuluh Nopember, Surabaya, Indonesia 60117
[3]Department of Information System, Universitas Internasional Semen Indonesia, Gresik, Indonesia 61122

Email: anindita.vinaya@uisi.ac.id

Abstract. In this study, we propose a sound-based signal analysis technique emitted by rotating machines. Three water pump machines are set in a row to reflect a mini plant. We raise an overdetermined case with four sensors that are used to record three machines' signal sounds with unbalanced, normal and bearing fault conditions. We use non-negative matrix factorization (NMF) approach to reconstruct the signal sounds of each machine which had previously been mixed. Based on the results, the type of engine damages can be determined based on the fundamental frequency of each machine. The first reconstructed signal $\hat{S}_1$ is similar to the baseline signal of normal machine with LSD of 1 and MSE of 5.84×10^{-9}. The second reconstructed signal $\hat{S}_2$ has a similarity to the baseline signal of the machine with bearing fault with LSD of 1 and MSE of 1.10×10^{-7}, while the third reconstruction signal $\hat{S}_3$ is similar to the machine's baseline signal with unbalanced condition,that is, LSD of 1.02 and MSE of 1.38×10^{-8}

1. Introduction

Blind Source Separation (BSS) is a method that can be used to separate mixed sound and to reconstruct signal without having to know much of the source signal information or the mixing process. This method was originally used as a technique for separating human speech signals (speech) or often called as cocktail party problem [1,2]. From these studies, research related to BSS has been developing with different objects of research; one of them is a machine with the aim to support the fault identification of sound-based machine. Machinery sound separation has already been conducted in previous studies using parallel factor analysis [3], full rank approach [4], and multiway array decomposition on the motor-pump system [5].

In 2011, Yu et al. used BSS with low-order statistical approach in order to separate engine vibration signals [6]. It was used to identify machine damages. Zhong et al.in 2006 separated machine's signal sound in semi-anechoic chambers by reconstructing wavelet transform coefficients in time-frequency domains. In their study, the mixing process was done instantaneously in which the number of sensors exceeded the sound sources [7]. The ICA (Independent Component Analysis) approach has been successfully applied by Atmaja et al. [8] to separate pump signals. However, the

ICA algorithm was bounded by independent and non-gaussian characteristics from its sources' signals [1,2].

In determined mixing (in which the number of sensors is equal to the number of sources) and underdetermined mixing (where the number of sensors is less than the number of sources) – if the mixing matrix is not invertible due to information limitations – failure in signal separation may occur. Overdetermined mixing can help to minimize this. In overdetermined mixing – in which sensor is used more than the source – the mixed signal will generate much information. It can be used to improve the separation performance. One of the advantages of an overdetermined mixture is that the data processing or computation can be done easier than underdetermined mixture [1].

Several methods can and have been used to separate mixed signals to reconstruct the source signal. In separating an overdetermined mixture with ICA, the separation signal is resulted from scaling by constants at each frequency. Thus, additional algorithms are required to improve signal quality [1]. In addition, the ambiguity in the permutation process can cause similar signal to be activated at specific frequencies [2]. One of the assumptions and limitations of BSS is an unknown matrix mixture represented through a square matrix. In overdetermined mixing, the matrix is the result of the redundant mixture. In this case, the redundant mixture can be removed, so that the matrix can be a square matrix. This redundancy removal technique is presented mathematically in [9,10].

In overdetermined mixing, redundancy will persist as the number of sensors exceeds the number of sources. Therefore, reducing the redundancy is required [2,9]. It is expected that the output of the estimation signal is equal to the number of sources. The separation results can be identified according to the original signal of the source. In this study, we use NMF on overdetermined mixing to reduce our matrix dimensions. NMF is also used to increase the sparsity of the data. NMF was introduced in 1994 by Paatero and Tapper [11] and continued to be popularized by Lee and Seung [12]. In this research, the above method will be implemented on a mini plant. This study will act as the initial study before the real implementation in the actual plant. A mini-plant is a lab-scale plant with several pumping machines with varying conditions. In recording signal sounds, we use microphones array as sensors in an overdetermined mixture.

2. Theory

Blind source separation (BSS) is a method used to separate mixed signal x from multiple sources without knowing the information from the source and the mixing process. BSS consists of the sum of several sources S, mixing matrix A, and mixed signal x [9,11]. In BSS, these commonly used linear instantaneous mixing equates as:

$$x(t) = A.S(t) \tag{1}$$

where S=[S1(t),··· Sm(t)]T, x=[x1(t),…xn(t)]T, and A is mixing matrix n × m × L, in which L is length of FIR Filter.

2.1. Factorization method

Non-negative matrix factorization (NMF) is a linear regression technique that can be applied to non-negative data. NMF is part of the linear reduction technique and is able to estimate the representation of existing data. NMF can also be referred to as an iteration optimization algorithm that modifies each of the iterations at the non-negative base function [2,13]. In NMF, the negative matrix V is composed of the non-negative matrix factor W and H with a Gaussian assumption as formulated in Equation (2).

$$V_{Fx\,N} \approx W_{Fx\,K} \cdot H_{Kx\,N} \tag{2}$$

In an overdetermined mixture, the dimensional reduction is required; thus, the concept of principal component analysis will be used. Singular value decomposition will calculate the mixed signal, in which the number of observations I will be equal to the number of sources J [9,10]. In this study, estimates of source spectrograms consisting of non-negative matrices W and H will be used. The number of sources consists of several p components which will be grouped by the number of used sources. . The spectral power resulting from the separation of the audio signal by the NMF method can be reconstituted through estimated likelihood, i.e., W and H values. By minimizing the difference through divergence, the likelihood estimation of the parameters can be obtained. Factorization can be done by minimizing the cost [12,13]. Cost functions, in this study, can be approximated by various methods, e.g., Euclidean distance, Kullback-Leibler divergence, Beta divergence[12].

2.2. Sound Reconstruction

In the early stages of sound reconstruction, Short Fourier Transform (STFT) function will be used to change the domain x (t) with $\omega(t\text{-}\tau)$ as a window function.

$$STFT\,\{x(t)\} = \int_{-\infty}^{\infty} x(t).\,\omega(t-\tau).\,e^{-i2\pi ft}\,dt \tag{5}$$

W and H components will be randomly initialized and estimated according to the factorization method. Through the expectation maximization algorithm, maximum likelihood W and H will be obtained. The presence of phase addition will convert W and H into complex numbers which were initially non-negative. To be able to reconstruct a signal that is the sum of Gaussian components, Wiener filter will be used. The Wiener filter is used to calculate the STFT component which represents the sum of Gaussian components based on the previously obtained W and H factors [2][14]. Inverse Short Time Fourier Transform (ISTFT) will be used to convert the signal back into a time domain.

3. Experimental study

The object of this research is a mini plant in which there are three similar types of water pump machines. In this research, each pump has a different condition, i.e., normal, unbalance, and bearing fault. The distance of each pump between one another is 30 cm. The recording processes are carried out in two stages. The first stage is conducted to generate the baseline signal while the second stage is to obtain the mixed recording signal. For mixed and baseline signal recording, we use four microphones as sound sensors which reflect cardioid characteristics. The experiment setup is shown in Figure 1.

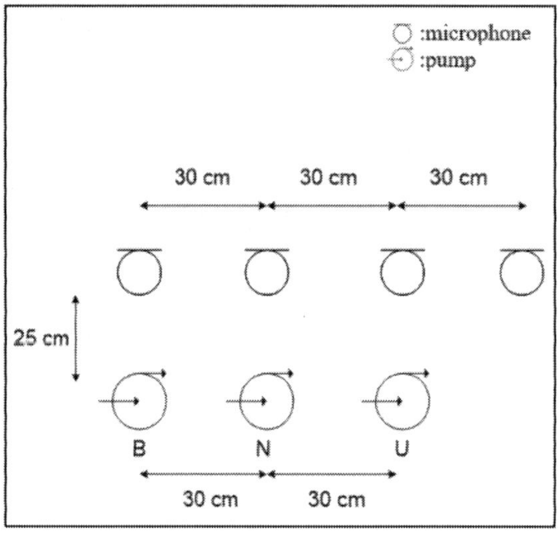

Figure 1. Experimental setup of overdetermined mixing with varying conditions of the pump (B:bearing fault, N: normal, U: unbalance)

In this study, each sound will be modelled with four-p components. The first step of the sound reconstruction (through NMF) is to transform signals from time-domain to time-frequency domain. We use Short Time Fourier Transform, and we define sine window function with an overlapping level of 50%. Then, the likelihood of the dividing matrix and the source matrix will be estimated using Itakura Saito divergence; combined with expectation maximization algorithm with 500 iterations. Subsequently, the estimation of the separation matrix and source will be compared with the mixed signal through the Wiener filtering process.

4. Result

Three source signals are reconstructed using NMF. The results of the reconstruction of the signals shown in the spectrograms reflect distinctive characteristics. Based on Figure 2, the highest signal power is found in signal two which is -49.32 dB at normalized frequency 0.331xπ rad/sample. Signal 1 marks the highest power -73.19dB at the normalized frequency of 0.107xπ rad/sample, while signal 3 generates the highest power of -68.47dB at the normalized frequency 0.106xπ rad/sample. The bright yellow color on the spectrograph indicates the highest power at a certain frequency, while the blue color signifies the low signal power at certain frequencies.

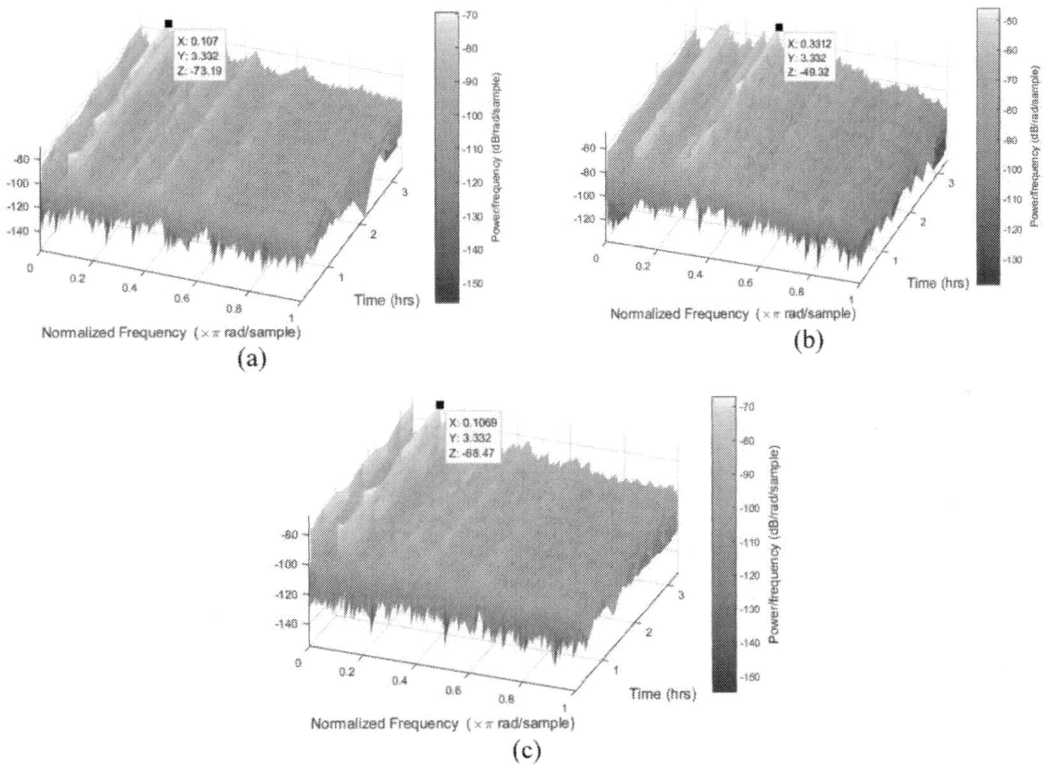

Figure 2. The spectrogram of the reconstructed signals (a) signal 1 (b) signal 2 (c) signal 3

Through the BSS method, the mixed signal can be separated. However, matching the reconstruction signal to the source signal or baseline is a complex task. In order to help finding out the level of similarity at each reconstruction signal with the baseline signal, we use LSD and MSE parameters. These parameters are also used to ascertain the performance of reconstruction signal. In this study, we use log spectral distance (LSD) in the frequency domain and MSE in the time domain. LSD is used to notice the ratio of rand noise signals; in this study, the relative similarity between spectral signal and the baseline will be measured [4]. The smaller the LSD and MSE, the closer the reconstructed signal to the baseline signal. Table 1 and 2 exhibit the MSE and LSD results.

Table 1. LSD between the reconstructed signal and the baseline signal

Reconstructed signals	Baseline signals of the machine		
	Normal	Unbalance	Bearing fault
$\hat{S}_1$	1	1.05	1.36
$\hat{S}_2$	1.191	1.163	1
$\hat{S}_3$	1.07	1.02	1.34

Table 2. MSE between the reconstructed signal and the baseline signal

Reconstructed signals	Baseline signals of the machine		
	Normal	Unbalance	Bearing fault
$\hat{S}_1$	$5.84 x 10^{-9}$	$8.87 x 10^{-9}$	$7.79 x 10^{-8}$
$\hat{S}_2$	$3.86 x 10^{-8}$	$4.15 x 10^{-8}$	$1.10 x 10^{-7}$
$\hat{S}_3$	$1.09 x 10^{-8}$	$1.38 x 10^{-8}$	$8.34 x 10^{-7}$

Based on the above tables, the first reconstructed signal $\hat{S}_1$ is similar to the baseline signal of the normal machine with LSD 1 and MSE $5.84 x 10^{-9}$. The second reconstructed signal $\hat{S}_2$ is also close to the baseline signal of the machine with bearing fault of LSD 1 and MSE of $1.10 x 10^{-7}$. The third reconstruction signal $\hat{S}_3$ is similar to the machine baseline signal with unbalance condition of LSD 1.02 and MSE $1.38 x 10^{-8}$. In addition to the LSD and MSE parameters, we compare the spectrogram of the reconstruction signal and the baseline signal. Below are the comparison results (Figure 3-5):

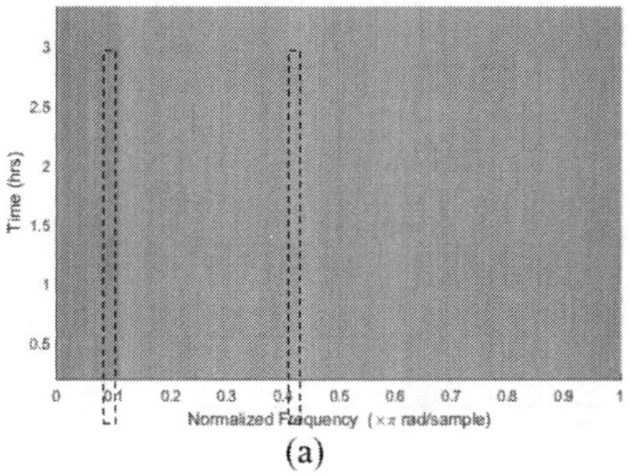

(a)

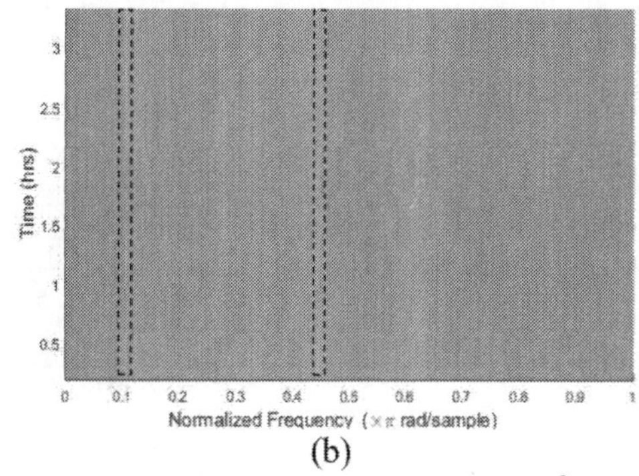

(b)

Figure 3. The spectrogram of (a) baseline signal-normal and (b) reconstructed signal $\hat{S}_1$

As have been mentioned earlier, the bright yellow color signifies the highest signal power on a frequency. Frequency, in its peak level, can be expressed as a normalized fundamental frequency [8]. The dashed black lines in the pictures above reflect the location of the frequency. In Figure 3, the normalized fundamental frequencies of baseline signal of the normal machine and the reconstructed signal are located at $0.1 x \pi$ rad/sample and $0.45 x \pi$ rad/sample.

IOP Conf. Series: Materials Science and Engineering **417** (2018) 012018 doi:10.1088/1757-899X/417/1/012018

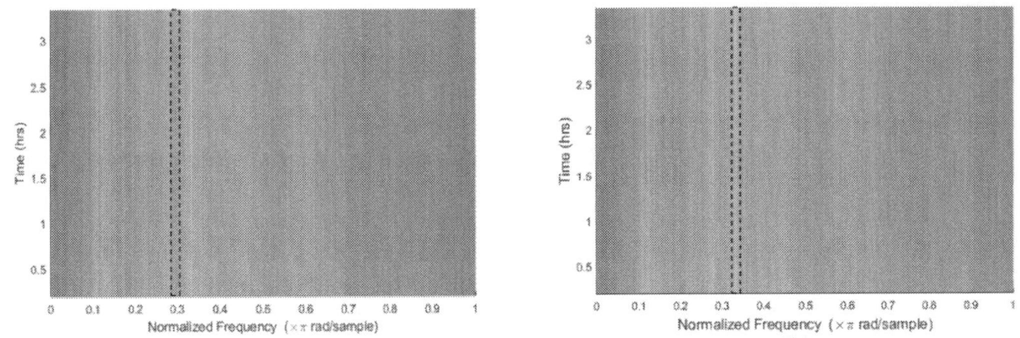

(a) (b)

Figure 4. The spectrogram of (a) baseline signal-bearing fault and (b) reconstructed signal $\hat{S}_2$

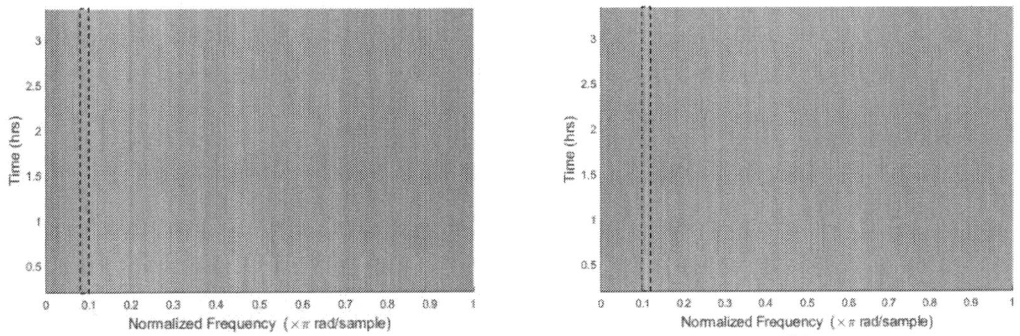

(a) (b)

Figure 5. The spectrogram of (a) baseline signal-unbalance and (b) reconstructed signal $\hat{S}_3$

The location of normalized fundamental frequencies in Figure 5 also tends to lower frequencies similar to the spectrograms in Figure 3, which bases at 0.1xπ rad/sample. In Figure 4, the baseline signal on the machine with bearing fault condition reflects the normalized fundamental frequency similar to the reconstructed signal $\hat{S}_3$ (0.33xπ rad/sample). Based on the spectrogram results, each machine is known to have distinctive characteristic. Therefore, it can be inferred that each machine conditions is distinguishable.

5. Conclusion

In this study, estimation of source spectrograms consisting of non-negative matrices is conducted. Each machine's signal sound in a mini plant has been successfully reconstructed on an overdetermined mixture using the NMF method with varied MSE and LSD. The normalized fundamental frequency of the baseline signal becomes the basis for determining the condition of each machine, i.e., the normalized fundamental frequency in the normal machine is equal to the normalized fundamental frequency of one of the reconstruction signals located at 0.1xπ rad/sample and 0.45xπ rad/sample. Each machine has different characteristics which depend on the machine's condition.

Acknowledgments

We would like to express our gratitude to the Direktorat Riset dan Pengabdian kepada Masyarakat (DRPM), Ditjen Penguatan Riset dan Pengembangan, Kemenristekdikti Indonesia which has provided funding for this research grant program.

References

[1] Osterwise, Christopher, and Steven L. Grant, "On over-determined frequency domain BSS," *IEEE/ACM Transactions on Audio, Speech and Language Processing,* vol. 22, no. 5, pp. 956-966, 2014.

[2] A. Ozerov and C. Févotte, "Multichannel nonnegative matrix factorization in convolutive mixtures for audio source separation," *IEEE Transactions on Audio, Speech, and Language Processing,* vol. 18, no. 3, pp. 550-563, 2010.

[3] A. A. Vinaya and D. Arifianto, "Fault Diagnosis of Rotating Machinery based on Acoustic Emission using PARAFAC-Source Separation," in *Journal of Proceedings Series,* 2015.

[4] A. A. Vinaya and D. Arifianto, "Underdetermined blind source separation based condition monitoring," in *2015 International Conference on Science in Information Technology (ICSITech IEEE*, 2015.

[5] Vinaya, A.A. and Arifianto, D., "Multi-way Array Decomposition on Acoustic Source Separation for Fault Diagnosis of a Motor-Pump System," *International Journal of Electrical and Computer Engineering (IJECE)*, vol. 7, no. 6, pp. 3052-3059, 2017.

[6] Yu, Gang, Zhidong Xiao, and Changning Li, "Blind source separation of rotating machinery fault signals based on fractional lower order statistics," in *Prognostics and System Health Management Conference*, Shenzhen, 2011.

[7] Zhong, Z. M., J. Chen, P. Zhong, and J. B. Wu, "Application of the blind source separation method to feature extraction of machine sound signals," *The International Journal of Advanced Manufacturing Technology*, vol. 28, no. 9-10, pp. 855-862, 2006.

[8] B. T. Atmaja and D. Arifianto, "Machinery fault diagnosis using independent component analysis (ICA) and Instantaneous Frequency (IF), 200," in *International Conference on Instrumentation, Communications, Information Technology, and Biomedical Engineering (ICICI-BME)*, 2009.

[9] Nadal, Jean-Pierre, and Nestor Parga, "Redundancy reduction and independent component analysis: Conditions on cumulants and adaptive approaches," *Neural Computation*, vol. 9, no. 7, pp. 1421-1456, 1997.

[10] Wang, L., Reiss, J.D. and Cavallaro, A, "Over-determined source separation and localization using distributed microphones," *IEEE/ACM Transactions on Audio, Speech, and Language Processing*, vol. 24, no. 9, pp. 1573-1588, 2016.

[11] Paatero, Pentti and Tapper, Unto, "Positive matrix factorization: A non-negative factor model with optimal utilization of error," *Environmetrics*, vol. 5, pp. 111-126, 1994.

[12] Lee, Daniel D. and Seung, H. Sebastian , "Learning the parts of objects by non-negative matrix factorization," *Nature*, vol. 401, pp. 788-791, 1999.

[13] A. Cichocki, R. Zdunek, A. H. Phan and S. I. Amari, Nonnegative matrix and tensor factorizations: applications to exploratory multi-way data analysis and blind source separation, 1 ed., John Wiley & Sons., 2009.

[14] Févotte, Cédric, Nancy Bertin, and Jean-Louis Durrieu, "Nonnegative matrix factorization with the Itakura-Saito divergence: With application to music analysis," *Neural computation*, vol. 21, no. 3, pp. 793-830, 2009.

IOP Conf. Series: Materials Science and Engineering **417** (2018) 012019 doi:10.1088/1757-899X/417/1/012019

Design Considerations for Automated Crane Assignment

Seung-Jun Lee, Yang Koo Lee, and Daesub Yoon

IoT Research Division, Electronics & Telecommunication Research Institute (ETRI), Daejeon, Korea.

lsj0209@etri.re.kr

Abstract. This paper describes a method of multiple cranes assignment to operators who call cranes in a factory. Due to a crane moving on a rail, there is a restriction on the movement. Therefore, the crane cannot be delivered without proper assignment. To solve the cranes assignment problem, we propose two methods. The first is the optimal solution search method which finds the minimum travel distance combination from the number of cases that are derived by calculating the distance between all the cranes and all the operators. The second is the dynamic mapping method which finds the number of operators between each crane. Simulation results show that the dynamic mapping method sometimes performs the same assignment results as the optimal solution search method, but on average, the result of dynamic mapping method is farther than the optimal solution as the number of cranes increases. When partial regions are created by using the dynamic mapping method and an optimal solution search method is applied to each region, optimal solutions can be acquired while reducing the number of calculations.

1. Introduction

Factory machines have been developed with connectivity using industrial IoT technology [1]. The introduction of IoT will eventually lead to a complete automation of the factory. For example, forklifts and automated guided vehicles now are able to be autonomously operated [2]. Also, automation of cranes has been studied in consideration of moving stability [3,4]. At the current stage, the operator can use the controller to move the crane to an arbitrary position when using a crane in a factory. It is inconvenient because the operator must work with the controller and it may be dangerous because it moves together with using the controller. Proper assignment, a method to deliver a large number of cranes under the requisite operators may give operators benefits. Because the crane moves over rails, we must consider that a crane cannot cross a crane and rails beside it.

In this paper, we introduce considerations for automated crane assignment and propose two methods. We also performed simulations using these methods and verified the assignment results.

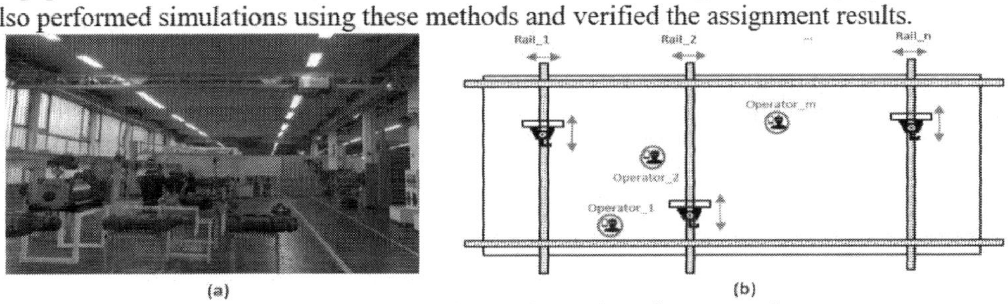

Figure 1. (a)Physical picture and (b) schematics of automated crane system

2. Requirements of Automated Crane System

To properly assign a large number of cranes to operators, the requirements of HW specifications and some functions are described below.

2.1. Location awareness

To send a large number of cranes to the location of any operator, location information of cranes and operators is required. The position of the crane can be measured by the movement of the driving motor installed on the rail. The positions of operators and cranes are generally determined by wireless communication [5, 6].

2.2. Sensors

As for the method of grasping the position of the crane, an error occurs when there is physical interference in the factory or its location cannot be determined in real time. Large or great numbers of objects cannot be recognized by such location awareness, so additional sensors are required to check the presence of obstacles around the crane and their relative location for safe automated operation.

2.3. Communication module

The crane should be equipped with a communication module to exchange information with not only each crane and operator, but also the central control system. The communication module of the crane recommends a protocol that satisfies high reliability and real time communication. Also, pairing confirmation between controller and crane have to be satisfied some mandatory such as European standards (EN 60202-32).

2.4. Computing power

A micro controller unit (MCU) is installed in each crane to drive sensors and the communication module. Depending on the system's specifications, location awareness and crane assignment are calculated by each crane. If the specifications of the MCU are low, it operates based on the information calculated by the central control system and sends the task report there. Operating a crane with low performance, if some crane problems occur during an automated operation, the whole operation is suspended until a decision is delivered by the central control system.

3. Assignment method

This chapter introduces two types of assignment problem solutions. In general, multi agent problem is solved by Munkres Assignment Algorithm to find an optimal assignment [7]. However, Munkres Assignment Algorithm can be applies to a given N x N matrix, which means the number of operators and cranes must be the same. To simplify the calculation, the distance between the operator and crane is calculated using only the x-axis length.

3.1. Method of Optimal solution search

First of all, calculate the distance between all the cranes and operators and fill in the table shown in Figure 2. The order of operators and cranes written on the table are sorted by distance to origin, respectively. Then, the number of cases where a crane is assigned to each operator is calculated.
Considering that the cranes cannot cross the lateral cranes, for example, if crane 1 is assigned operator 2, other cranes cannot be assigned operator 1. Therefore, the number of cases is smaller than common combination problem. Finally, the combination that satisfies the minimum distance from the number of cases is the optimal solution.
Figure 3 shows cases that contain an optimal solution depending on the number of cranes and operators. As the number of cranes increases, the number of cases will increase. Moreover, when the number of operators is half the number of cranes, the number of cases becomes the largest.

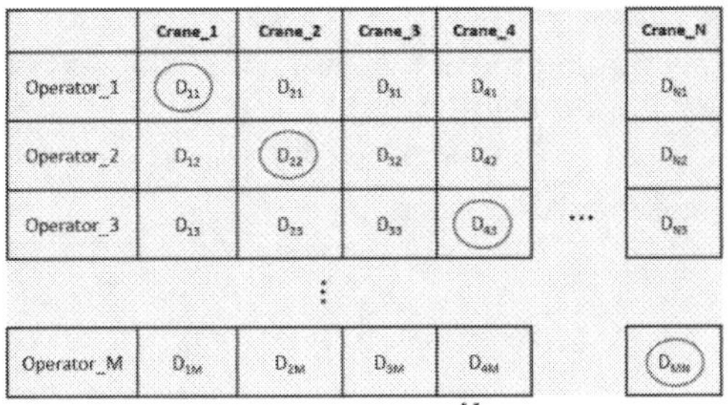

$$Optimal\ solution = \min(\sum_{i=1}^{M} D_{ij})\ j \in N$$

Figure 2. Method of Optimal solution search

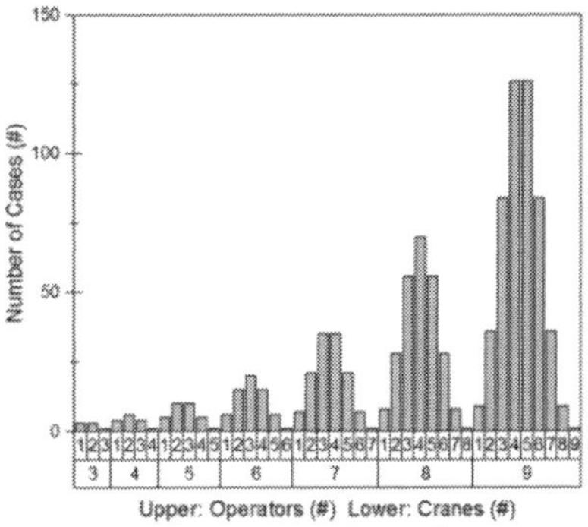

Figure 3. Number of cases

3.2. Method of Dynamic mapping

The dynamic mapping method assigns cranes by using how many operators are placed between each crane. Figure 4 shows how to derive a solution using the dynamic mapping method. First, all cranes and operators are rearranged according to the x-axis of the coordinates. Then, some areas are created between cranes, and the operators in each area are determined. If some operators are placed on the right side of the rightmost crane, those operators are assigned to the rightmost crane. After that process, there are no operators placed on the right side of the rightmost crane. Next step, method finds some areas where two or more operators are placed. If so, a nearby crane moves in a virtual manner so that no one operator exceeds one area. Finally, the right-hand crane is assigned to each operator. In the case of problem shown in Figure 4, the result of allocations are crane2→operator1, crane3→operator2, crane4→operator3 and crane5→operator4.

One of the advantage of this method, it can solve assignment problem with relatively few calculations. This method is not a method for deriving an optimal solution.

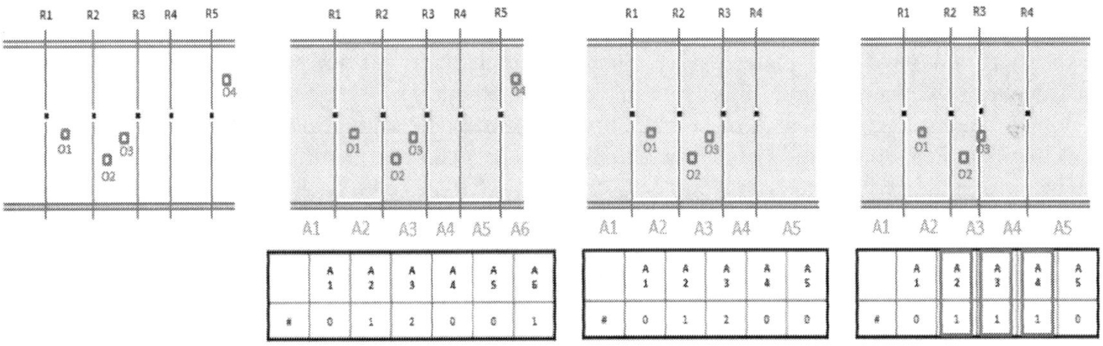

Figure 4. Method of Dynamic mapping

(a)sorting, (b)counting by each area, (c) assigning cranes and operators in the corner, and (d) crane traveling in a virtual manner

4. Simulation results

4.1. Simulation Scenarios

To verify the validity of both methods, we use MATLAB to perform the simulation scenario described below:

- The area of the factory is 100 x 100 m.
- Two operators request the cranes simultaneously.
- When two cranes reach the operators' locations, two other operators are randomly generated and request the cranes simultaneously.

- The operator positions used in both methods are the same.
- The above scenario is repeated 10 times and the sum of the travel distance about all cranes is calculated.
- The number of operators was varied from three to eight, and each simulation was executed 10 times to obtain its average sum of travel distance and standard deviation.

4.2. Results

Figure 6 shows the sum of travel distance results from the crane calls. The larger the number of cranes, the smaller the total distance that all cranes move. The results of optimal solution search method present the shortest traveling distances under given conditions. The result of dynamic mapping method is farther than the optimal solution as the number of cranes increases. On the other hand, assignment solution of dynamic mapping method also performs that all cranes travel without collision of cranes in the iterative simulation.

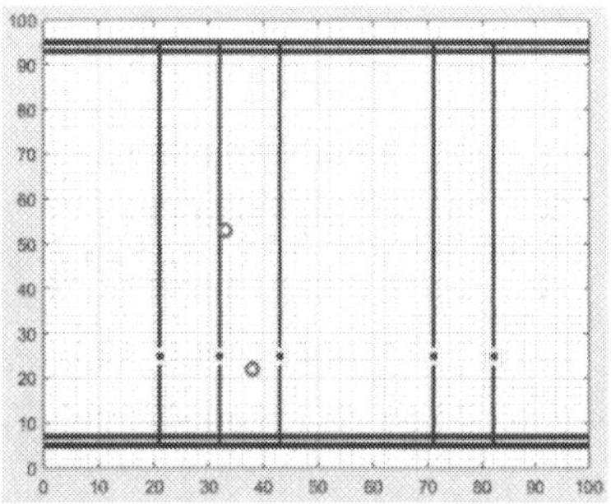

Figure 5. Simulation image

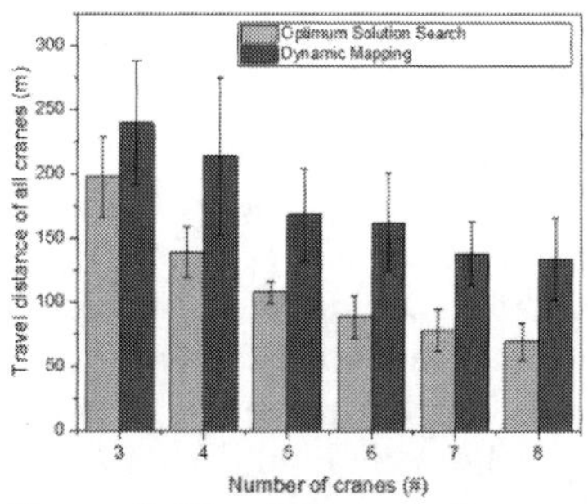

Figure 6. Simulation results

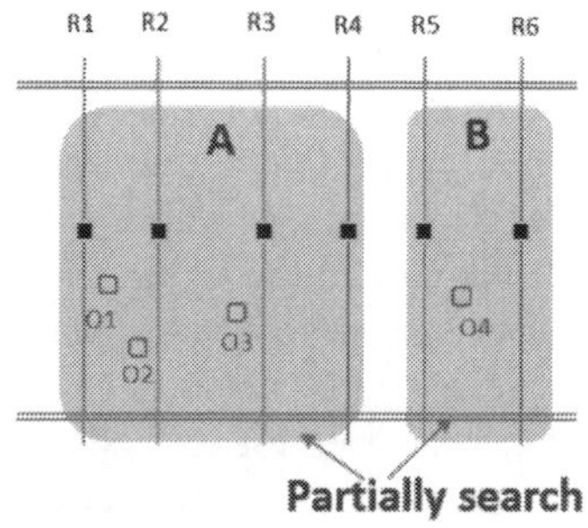

Figure 7. Concept of Partial search method

Figure 7 shows the method of applying the dynamic mapping method to find the position of the operator and then applying the optimal solution search method in each part. For example, the optimal solution is found from the number of 16 cases when 6 cranes and 4 operators are placed in factory. If dynamic mapping method divides operators and cranes into two parts like group A(4 cranes and 3 operators) and group B(2 cranes and 1 operator), each parts has the numbers of 4 and 1 cases, respectively. In that example, this combination of methods lead to reduce the calculation to find optimal solution. However, this combination use method cannot always be solved with few calculations. For example, if there are three operators between one crane and another crane, the crane to be used in the calculation is added to both cranes and two more on each side, so that a maximum of six cranes are covered. In this case, the combination of methods has no effect of reducing the calculation amount. When the larger the number of cranes than the number of operators, this combination of methods presents faster result than only applied to optimal solution search method.

5. Conclusions

We introduce two methods of assigning cranes to operators in a factory, taking into account the limit of movement of the crane. When designing an automation system, it can be calculated by using optimal solution search method if it is operated by a central control system, and it can be operated by transmitting an allocation position to each crane. Also, if it is possible to simulate and allocate by itself in each crane, dynamic mapping can be applied to reduce the number of calculations.

Acknowledgments

This research was financially supported by the Ministry of Trade, Industry and Energy(MOTIE) and Korea Institute for Advancement of Technology(KIAT) through the International Cooperative R&D program(N0002620) and has been achieved in the European ITEA project "Optimised Industrial IoT and Distributed Control Platform for Manufacturing and Material Handling (OPTIMUM)

References

[1] Malte Brettel, Niklas Friederichsen, Michael Keller and Marius Rosenberg 2014 *International Journal of Information and Communication Engineering* **8(1)** 37-44

[2] L. Li, Y. H. Liu, M. Fang, Z. Zheng and H. Tang 2015 *IEEE International Conference on Automation Science and Engineering (CASE)* 264-265

[3] A. Broad, J. Schultz, M. Derry, T. Murphey and B. Argall 2017 *IEEE Robotics and Automation Letters* **2(1)** 239-246

[4] Oliver Sawodny, Harald Aschemann and Stephan Lahresc 2002 *Control Engineering Practice* **10** 1323-1338

[5] Kyungsu Yun, Ji Kyu Park, Jae Young Ahn and Jae Kyun Kwon 2018 *ETRI Journal* 40(1) 101-110

[6] K. J. Flattery, B. Henry, M. Muadi, K. Walters, A. Eroglu and J. Rickman 2016 *IEEE/ACES International Conference on Wireless Information Technology and Systems (ICWITS) and Applied Computational Electromagnetics (ACES)* 1-2

[7] Margaretha Gansterer and Richard F. Hartl 2018 *European Journal of Operational Research* 1-12

IOP Conf. Series: Materials Science and Engineering **417** (2018) 012020 doi:10.1088/1757-899X/417/1/012020

Effect of Tread Scratch on Fatigue Life of Wheels in High - speed EMU under Low Temperature Servicing Condition

Song Hao[1, a], Zhao Deming[1,b], Hu Ming[1,c], Yang Liuqing[1,d], Zhou Xun[1,e], Qin Yudong[1,f], Chen Wenhua[1,g], Zhao Yongxiang[2,h]

[1] National and Local Joint Engineering Research Center of Reliability Analysis and Testing for Mechanical and Electrical Products ,Zhejiang Sci-Tech University, Hangzhou 310018, China;
[2] State Key Laboratory of Traction Power, Southwest Jiaotong University, Chengdu 610031, China;

[a]1921474165@qq.com, [b]miancer@163.com, [c]huming@zstu.edu.cn, [d]409709426@qq.com, [e]5048338@qq.com, [f]602089581@qq.com, [g]chenwh@zstu.edu.cn, [h]yxzhao@home.swjtu.edu.cn

Abstract. In order to study the influence of scratches on the fatigue life of CRH5 EMU wheels under -40°C servicing condition, a dynamic model of the vehicle-rail system is established along with the contact model of the scratched wheels. Considering the irregular and typical working condition of the railway line, a random load spectrum and a cyclical wheel-rail impact load spectrum are obtained. The *S-N* character of the wheel material is obtained according to a series of fatigue test in serving environment -40°C. Using the finite element quasi-static superposition method, the fatigue life of the wheels with and without scratch are compared, and then the fatigue life of the scratched wheel is evaluated. The results show when the scratch length $L_R \leq 20$ mm, the fatigue life of the scratched wheel is less affected by the wheel-rail impact caused by scratch; when the scratch length $L_R > 20$ mm, the fatigue life decrease quickly, less than 1e6km, affecting the safety of the EMU.

1. Introduction
Impact fatigue is one of the important causes of fatigue failure of mechanical components under impact load, and it is commonly found in strong impact cases [1, 2]. A strong wheel-rail impact that caused by the wheel tread scratch can cause EMU wheels impact fatigue damage, reducing EMU train safety [3, 4]. Wheel tread scratch usually generates during the EMU emergency braking or acceleration. Most of the existing researches are focused on the fatigue and fracture of the EMU's Components under ordinary temperature. The lowest servicing temperature for the Harbin-Dalian Line opened in China is around -40 °C, and the train wheels in the low-temperature environment are tend to produce tread scratch.

2. Load spectrum and material properties

2.1. *Wheel-rail random load spectrum*
According to the basic parameters of the CRH5 EMUs, the high interference spectrum in the German orbit spectrum is selected as the simulated orbit spectrum. Due to the extremely complicated road

conditions of the Harbin-Dalian line, it is impossible to analyze all line conditions. Therefore, typical operating conditions are selected for calculation. The conditions are shown in table 1.

Table 1. The typical operating conditions parameters of Harbin-Dalian line

line	Condition	Curve radius (m)	Transition (m)	Straight (m)	Running speed (km/h)	Superelevation (mm)	Deficient superelevation (mm)	Time(s)
Straight	1	—	—	—	200	—	—	20
	2	8000	680	250	190	180	80	20
Curve	3	5500	570	210	180	180	82	20
	4	2000	320	120	160	180	56	20

Applying the above model, the random loading time history of the non-scratch wheels under various typical operating conditions is obtained. The lateral force is much smaller than the vertical force, and the influence can be neglected.

2.2. *Wheel-rail impact load spectrum*

Due to the random disturbance of the rail and other factors, the wheel-rail impact produced may be slightly different for each revolution. In order to simplify the calculation, it is assumed that the wheel-rail impact produced by the scratched wheel is exactly the same in every revolution. The dynamic model of the vehicle-rail system is established, the wheel tread scratch is simulated by the variable wheel diameter method. The wheel scratch length should be controlled within 32 mm [5]. Therefore, the wheels with different scraping length (10mm, 20mm, 30mm and 32mm) are respectively simulated and analyzed. Then the impact loading time histories of each wheel under four typical working conditions are obtained. As shown in figure 1.

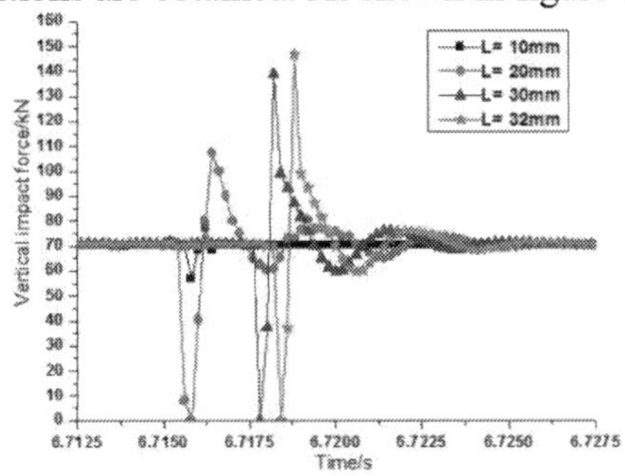

(a) Vertical impact force time history curve of condition 1

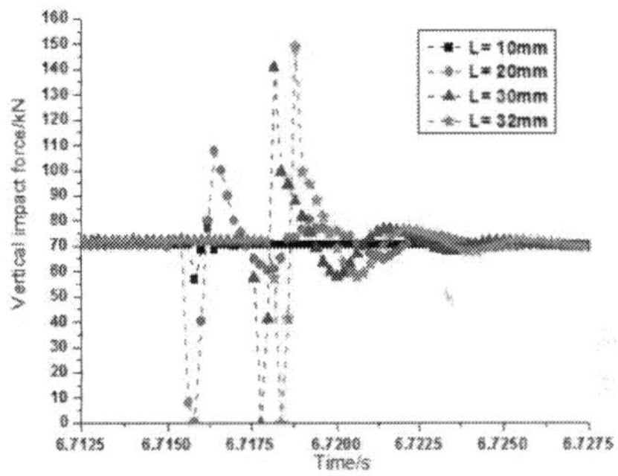

(b) Vertical impact force time history curve of condition 2

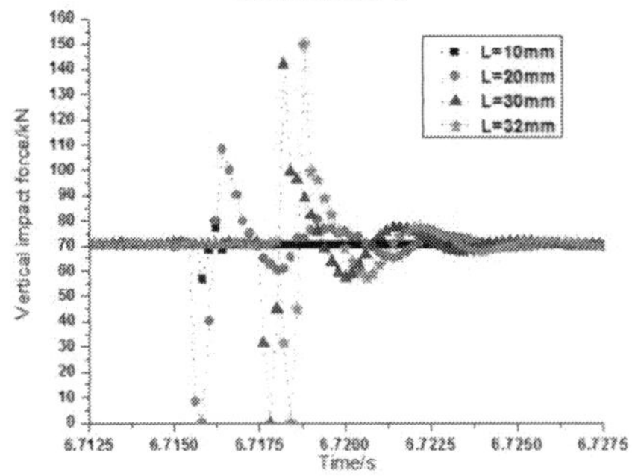

(c) Vertical impact force time history curve of condition 3

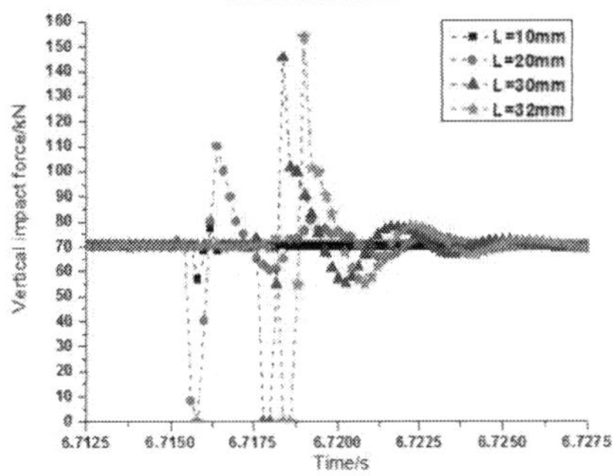

(d) Vertical impact force time history curve of condition 4

Figure 1. Vertical impact force time history curve of four conditions

As can be seen from figure 1, the longer the wheel scratch length under the same working condition, the stronger the wheel-rail impact. Under the same working conditions, the length of the wheel suffered wheel-rail impact is basically the same.

2.3. *The material ER8C properties*

To obtain CRH5 EMU wheel material ER8C properties, the static tensile test and *S-N* character test of material ER8C in temperature -40°C are carried out. The results show that the average rim strength σ_b = 993.25 MPa and the average spoke strength σ_b = 919 MPa. The fatigue limit of the material ER8C is tested by the ascending and descending method. The fatigue limit σ_{-1} = 392 MPa and the fatigue limit of the spoke σ_{-1} = 343.5 MPa. The *S-N* characters of ER8C are shown in figure 2 and figure 3.

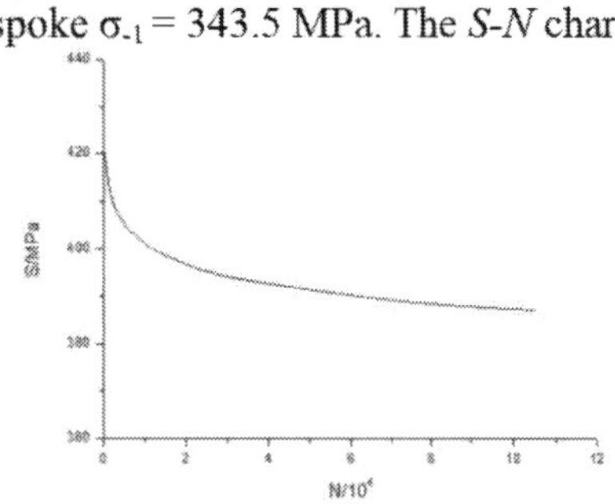

Figure 2. Rim S-N curve

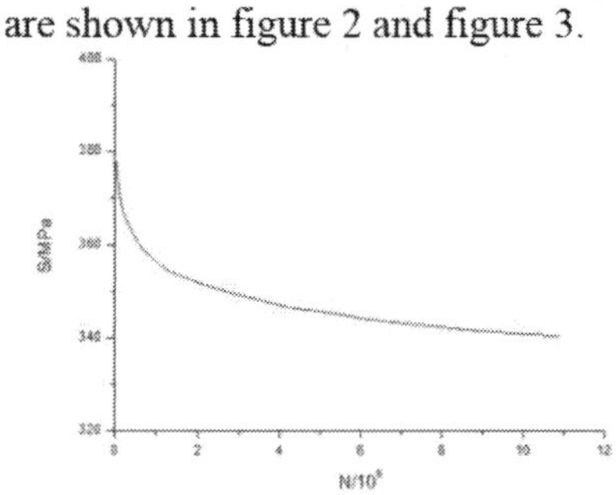

Figure 3. Spoke S-N curve

3. Fatigue life of wheel

3.1. *Fatigue life of the non-scratch wheel*

To simplify the problem, the wheel fatigue damage equivalent calculation method is used to complete the analysis, that is: Assuming that the wheel is stationary, the wheel and rail forces always act on the same contact point, and the dangerous point is equivalent to rotating around the wheel center. So the damage of dangerous point is equal to the mean value of the the damages of nodes on the same radius and the same plane with the dangerous point. This method ignores the force of inertia, will result in the calculated damage slightly less than the actual situation.

According to the UIC510-5 standard, the wheel model is constrained and the boundary conditions are given. CRH5 EMU wheel using XP55 tread, rolling contact patch area of wheel and rail about 100mm^2. Accordingly, a unit vertical load (1N) was applied to a tread having an area of 100 mm^2 and a center of 70 mm from the inside of the wheel. The equivalent stress cloud and equivalent strain cloud of the wheel under the unit vertical load are solved, as shown in figure 4.

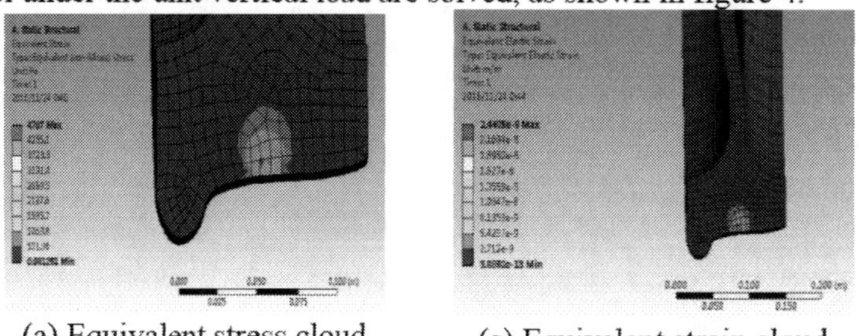

(a) Equivalent stress cloud (a) Equivalent strain cloud

Figure 4. Equivalent cloud image under vertical load of unit

As can be seen from figure 4, the maximum equivalent stress point and the maximum equivalent strain point of the wheel under the unit vertical load are the same nodes, and the point is the dangerous point at a depth of 8mm from the tread surface. The fatigue life of the dangerous point is the smallest, and the fatigue life of all the nodes on the radius circle and the same plane with the dangerous point is extracted. The effect of each node under condition 1 to 4 of the primary load block (load block or load time history curve) fatigue damage are shown in table 2.

Table 2. Fatigue life and fatigue damage of each node on the same radius and same section with dangerous point under condition 1 to 4

Node	Life (load block)				Damage (load block once)			
Condition	1	2	3	4	1	2	3	4
1	14166	18784	21808	1.1027e5	7.0591e-5	5.3236e-5	4.5854e-5	9.0686e-6
2	27273	35335	44331	1.9096e5	3.6666e-5	2.8300e-5	2.2557e-5	5.2367e-6
3	26832	35600	45002	1.7434e5	3.7268e-5	2.8089e-5	2.2221e-5	5.7359e-6
⋮	⋮	⋮	⋮	⋮	⋮	⋮	⋮	⋮
1620	3.5765e7	3.6838e7	3.7911e7	3.8139e7	2.7960e-8	2.7145e-8	2.6377e-8	2.6219e-8
Mean	/	/	/	/	1.3688e-7	1.0936e-7	9.4040e-8	3.9013e-8

Can be seen from the table 2, the damages of dangerous node under condition 1 to 4 are 1.3688e-7, 1.0936e-7, 9.4040e-8 and 3.9013e-8. Based on the above analysis, it can be concluded that the fatigue damage of the wheel under condition 1 to 4 (1km) is 1.23201755e-7, 1.03610611e-7, 9.40401319e-8 and 4.38899514e-8.

Assuming that the conditions 1 to 4 are respectively 50%, 20%, 20% and 10% of the Harbin-Dalian Line, the fatigue damage of the wheel is 1.0552002e-7. According to the Miner damage law, the running distance of the wheel under random loading is 9,476,900 km. If the EMU runs 300,000 km per year, the service life of the wheel is 31.59 years.

3.2. *Fatigue life of the scratched wheel.*

The scratched wheels are subject to cyclical wheel-rail impact loads and are subject to constant impact. In order to simplify the calculation, it is assumed that the point of impact is the point of vertical load under straight-line conditions. It is assumed that the tread of the EMU wheels is scratched when leaving factory and the length of the scratch does not change with the operation of the high-speed EMU. And the fatigue life of the scratched wheel with different scratch length is analyzed and calculated.

First of all, the scratched wheel with a length of 32mm was analyzed. It can be concluded that the fatigue damage (running 1 km) under each working condition (the scratched wheel is affected by random loading of the track and cyclical wheel-rail impact). As shown in table 3.

Table 3. Fatigue damage of 32mm scratched wheel under each working condition

Condition	Damage (single impact load)	Damage(1 km)
1	2.1898e-5	7.8179e-3
2	2.7932e-5	9.9718e-3
3	3.1129e-5	1.1113e-2
4	4.7492e-5	1.6954e-2

Table 4. Fatigue life of the scratched wheel with various length scratch

Length of scratch	Life[km]
LR=32mm	101.8
LR=30mm	156.4
LR=22mm	56.36e4
LR=20mm	947.69e4
LR=10mm	947.69e4
LR=0mm	947.69e4

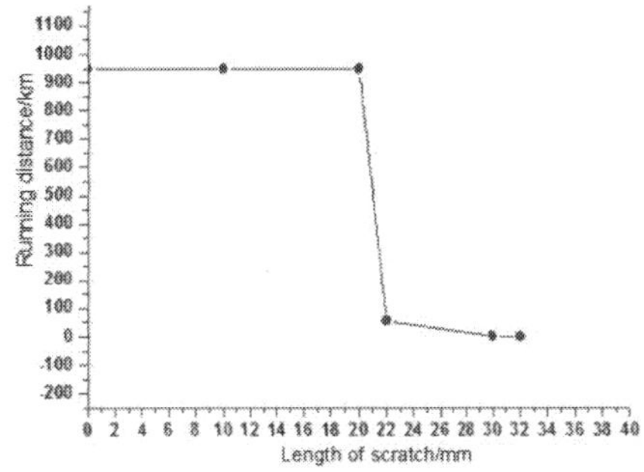

Figure 5. The relationship lines of running distance with length of scratches

Assuming that the operating conditions 1 to 4 respectively account for 50%, 20%, 20% and 10% of the Harbin-Dalian line. The damage of wheel with 32mm scratch is 0.009821482, running distance of 101.8km. Based on this, the fatigue life of the wheel with the scratch lengths of 30 mm, 20 mm, 22 mm, and 10 mm, respectively, is calculated as shown in table 4. The relationship between the scratch length and wheel running distance is shown in figure 5. As can be seen from figure 5, when the scratch length $L_R \leqslant$ 20 mm, the fatigue life of the scratched wheel is less affected by the cyclical wheel-rail impact; when the scratch length $L_R >$ 20 mm, the fatigue life of the wheel is significantly affected by the cyclical wheel-rail impact, the life of the wheel is rapidly decreasing.

4. Conclusion

The S-N character of the wheel material ER8C was obtained based on the fatigue test in servicing environment -40 °C. Combined with the finite element quasi-static superposition method, the wheel fatigue life with and without scratches were compared. The conclusions are as follows:

(1) When the wheel tread is free from scratch, the running distance of the wheel is 9,476,900 km. If the EMU runs 300,000 km per year, the wheel can be used for 31.59 years. This estimation meets the design life of 30 years.

(2) When the scratch length $L_R \leqslant$20mm, the fatigue life of the scratched wheel is barely affected; when the scratch length $L_R >$ 20mm, the fatigue life of the scratched wheel is obviously affected by cyclical wheel-rail impact, the service life of the scratched wheel tend to decrease sharply, which affects the safety of the EMU. The scratched wheel should be technically treated as soon as possible.

The above results provide a basis for further analysis of the fatigue life of EMU scratched wheel.

Acknowledgments

This work is supported by the Program for the National Natural Science Foundation of China (Grant No.U1334204), Zhejiang Provincial Natural Science Foundation of China (Grant No. LZ12E05004), Doctoral Fund of Ministry of Education of China (Grant No.20123318130001), Zhejiang Sci-Tech University "521 Talent Training Plan".

References

[1] TANAKA T. Studies on impact fatigue[J]. Bull.J.S.M, 1973, (16): 1814-1828.

[2] DARLINGTON J F, BOOKER J D. Development of a design technique for the identification of fatigue initiating features[J]. Engineering Failure Analysis, 2006, 13(7): 1134-1152.

[3] ZHAO X, WEN Z, ZHU M, et al. A study on high-speed rolling contact between a wheel and a contaminated rail[J]. Vehicle System Dynamics, 2014, 52(10): 1270-1287.

[4] ZHAO X, LI Z. The solution of frictional wheel–rail rolling contact with a 3D transient finite element model: Validation and error analysis[J]. Wear, 2011, 271(1): 444-452.

[5] YUDONG Q, MING H, LIUQING Y, et al. Mechanics characteristics analysis of wheel-rail impact induced by wheel flats based on SIMPACK[J]. China Mechanical Engineering, 2017, (17): 2029-2035.

Effect of Ceramic Coated Pistons on the Performance of a Compressed Natural Gas Engine

Hafiz Liaqat Ali, Fubai Li, Zhi Wang, and Shijin Shuai*
State Key Laboratory of Automotive Safety and Energy, Tsinghua University, Beijing 100084, China

* Corresponding author: sjshuai@tsinghua.edu.cn

Abstract. An internal combustion (IC) engine with ceramic wall coating is usually utilized to reduce the heat losses and it is referred to as low heat rejection (LHR) engine. While high fuel price and environment hazard is of major concern for the modern world, LHR engine type is gaining huge importance due to its low emissions and efficient fuel consumption. In this article, the simulation study was conducted using commercial software AVL Boost to analyze the effect of thermal barrier coatings (TBCs) on the performance of a single cylinder, naturally aspirated, compressed natural gas (CNG) engine. The results were carried out on a conventional (uncoated) piston, as well as two different thermal barrier ceramic Titanium dioxide (TiO_2) and Yttria-stabilized zirconia (YSZ) insulated pistons, which were coated with thickness of 0.5 mm. The simulation results were validated against the experimental results. The insulated pistons results showed that better performance at all operating conditions over the uncoated piston. The maximum exhaust gas temperature was increased about 57.35 ^{0}C, improvement in indicated specific fuel consumption (ISFC) up to 9.1%, and maximum 9.78% improvement in indicated thermal efficiency (ITE) were predicted in the insulated pistons, as compared to the conventional piston.

1. Introduction

An IC engine rejects about two third of fuel energy to the environment either through its walls or through the exhaust valves while merely about one third of it is utilized as a power output. The concept of LHR engine eliminates this heat rejection towards the coolant and improves the energy utilization for work output as well as decreased emissions. The use of ceramic TBCs in engines improves the emissions, high piston surface temperature, efficient fuel consumptions and thermal fatigue protection. It also defends piston from thermal stress, corrosion attack, great heat emissions as well as it decreases heat flux into the piston [1].

Furthermore, a vast number of simulation studies on LHR engines have been carried out by different researchers [2-5]. The researchers have reported enhanced exhaust gas temperatures as well as enhanced performance in certain conditions. Chan et al. [6] used thin YSZ ceramic insulation schemes on a piston crown. When the engine was loaded at lower power, the insulated piston provided a better fuel economy of about 6%, as compared to the uncoated engine. Li et al. developed a one dimensional simulation code for a direct injection (DI), four-stroke, 1.9-liter diesel engine by means of commercial software GT-Power [7]. When the engine cooling temperature was varied from 56.5°C to 100°C, the simulated brake fuel conversion efficiency and brake specific fuel consumption was enhanced by 19.14% and 16.07%, respectively. Moreover, Parker [8] developed an air gap piston and found that the heat reduction was 41%. At the piston crown the maximum temperature of the air gap piston and conventional piston were predicted by 610 ^{0}C and 222^{0}C, respectively. Morel et. al established a thermodynamic cycle code for a four-stroke, 14-litre diesel engine [9]. The piston crowns were coated with 0.5 mm of ceramic partially stabilized zirconia (PSZ). The results showed that the thermal efficiency (TE) increased by 7%, heat losses fell by about 30%, and turbine inlet temperature increased by 30 °K.

Herein, the numerical one dimensional simulation was performed on a single cylinder CNG engine by using a commercial engine simulation software AVL Boost. As piston is the hottest part and mainly it contributes toward heat losses. Piston is coated with two different types of thermal barrier coating (TBC) materials, Titanium dioxide (TiO_2) and Yttria-stabilized zirconia (YSZ) with thickness of 0.5mm of each. The simulation was operated at five different excess air ratios (EARs) from 1 to 1.5 with a step of 0.1, different ignition times, and at a constant engine speed of 1000 rpm. The performance results of coated pistons engine were compared with the uncoated piston engine. With the anticipated performance data, the experimental results are carried out to validate the simulation model, as well as study the potential gains produced by the application of the LHR concept. For this simulation the following assumptions were made.

1) A perfect mixture in the combustion chamber
2) Start of combustion (degree) taken at 5% of the mass fraction burned
3) CNG properties similar to Methane (CH_4)
4) In-cylinder heat transfer occurs through forced convection
5) Complete combustion befall inside of combustion chamber

The properties of Titanium dioxide (TiO_2) [10] and YSZ [11] that were used as TBC materials for this investigation are given in Table 1.

Table 1. Properties of Titanium dioxide (TiO_2) and YSZ materials.

Material	Density (g/cm^3)	Melting temperature (°C)	Thermal conductivity (W/mK)	Poisson's ratio	Modulus of elasticity (GPa)	Compressive strength (MPa)	Fracture toughness ($MPa.m^{1/2}$)
TiO_2	4.0	1870	11.7	0.27	230	680	3.2
YSZ	5.7-6.13	2800	2.0	0.3	205	2200	8

2. Modeling and methodology

Commercially accessible engine simulation AVL Boost software was used, which is based on the one dimensional gas dynamics. It implies that the uniform flow is considered along the cross sections. Built-in templates together engine parts and sub models (such as gas exchange model, combustion model, heat transfer model, scavenging model, and emission models) were used to established the engine simulation model, that describe the physical processes taking place in the engine. In this model, SB is the system boundary, E is the engine, C is the cylinder, I is the injector, PL is the plenum, R is the restriction, MP is the measuring point, and 1-9 numbers is the pipes. Figure 1 shows the single cylinder CNG engine simulation model that was adopted for this analysis.

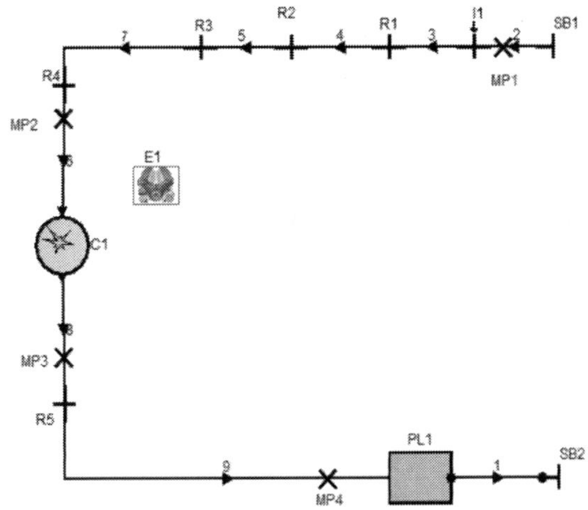

Figure 1. AVL Boost simulation CNG engine model.

2.1 Vibe combustion model

For present study, the Vibe model was selected for the combustion analysis. To calculate the actual mass burning rate and heat release characteristics of an engine the Vibe model is often used. Equation 1 and Equation 2 represent the rate of heat released and mass fraction burned, respectively [12].

$$\frac{dx}{d\alpha} = (m+1)\cdot\frac{a}{\Delta\alpha_c}\cdot y^m \cdot e^{-a\cdot y^{(m+1)}} \tag{1}$$

$$x = \int \frac{dx}{d\alpha}\cdot d\alpha = 1 - e^{-a\cdot y^{(m+1)}} \tag{2}$$

Where $dx = \frac{dQ}{Q}$, $y = \frac{\alpha - \alpha_0}{\Delta\alpha_c}$, Q is the total fuel heat input *(J)*, $\Delta\alpha_c$ is the combustion duration (degree), α_o is the start of combustion (degree), and α is the crank angle (degree). The Vibe parameter a, it was assumed to be 6.9 for the complete combustion and m is the shape parameter.

2.2 In cylinder heat transfer through Woschni model

From the in-cylinder gas to the walls of the combustion chamber (like cylinder head, cylinder liner, and piston), the heat transfer is computed through Equation 3 [13].

$$Q_{\omega i} = A_i \cdot \alpha_\omega \cdot (T_c - T_{\omega i}) \tag{3}$$

Where $Q_{\omega i}$ is the convective heat flux to combustion chamber surroundings, α_ω is the convective coefficient of heat transfer, A_i is the surface area (piston, cylinder head, and liner), $T_{\omega i}$ and T_c is the in-cylinder wall temperature and gas temperature, respectively. The Woschni model [14] was published in 1978. Equation 4 is often used to estimate the convective heat transfer coefficient.

$$\alpha_\omega = 130\cdot T_c^{-0.53}\cdot p_c^{0.8}\cdot D^{-0.2}\cdot\left[(p_c - p_{c,o})\cdot\frac{V_D\cdot T_{c,1}}{p_{c,1} - V_{c,1}}\cdot C_2 + C_1\cdot c_m\right]^{0.8} \tag{4}$$

Where $C_1 = 0.308\ (c_u/c_m) + 2.28$, c_u is the circumferential velocity, and c_m is the mean piston speed. For the indirect injection (IDI) engine $C_2 = 0.00622$. While, V_D is the cylinder displacement, D is the bore diameter of cylinder, $p_{c,o}$ is the cylinder pressure of the motored engine, $p_{c,1}$ and $T_{c,1}$ are the pressure and temperature of in-cylinder at inlet valve closing, respectively.

2.3 Model calibration strategy and validation

This section describes the developed simulation model calibration strategy and validation process. The general model calibration strategy for the present investigation is depicted in Figure 2.

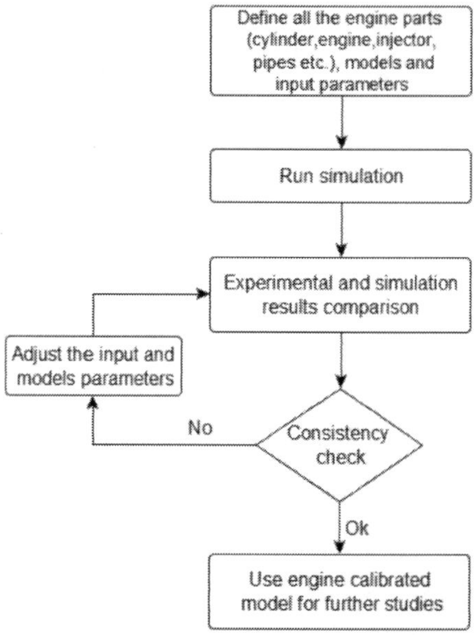

Figure 2. Model calibration strategy.

The simulation results are validated against the experimental data at 4.11 bar and 8.42 bar of indicated mean effective pressure (IMEP). The pressure trace at different engine loads with constant engine speed of 1000 rpm and it compared with the experimental data is illustrated in Figure 3. The results comparison between simulation and experiment show a good agreement because of minor deviation.

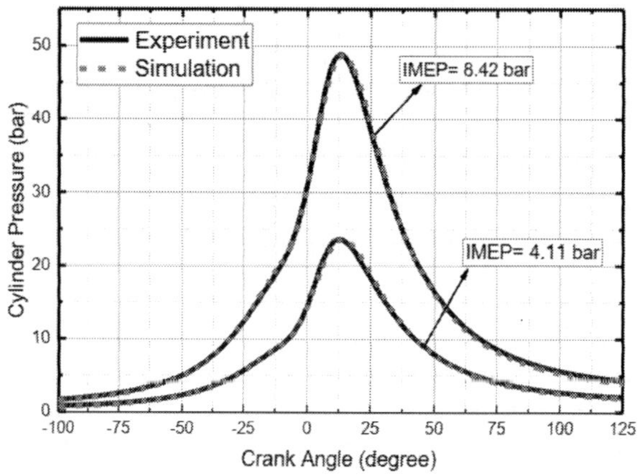

Figure 3. Validation of simulation results using in-cylinder pressure.

2.4 Experimental setup and operating conditions

A four stroke, single cylinder, indirect injection (IDI), naturally aspirated, water-cooled CNG engine is used for the present analysis. The schematic diagram of the single cylinder CNG engine test bench is illustrated in Figure 4. The engine was operated at different engine loads (4 bar, 8 bar and 12 bar), ignition times (21, 24 and 27-degree crank angle before top dead centre), excess air ratios (1 to 1.5) and at a constant engine speed of 1000 rpm. To reduce the intake and exhaust pressure variations two 100L surge tanks are used. The intake air mass flow rate is measured and control by using a six-channel sonic nozzle system. The back pressure valve is used to control the exhaust pressure. Kistler

6125C pressure transducer in combination with a Kistler charge amplifier is used to measure the cylinder pressure of the engine. The general CNG engine specifications are shown in Table 2.

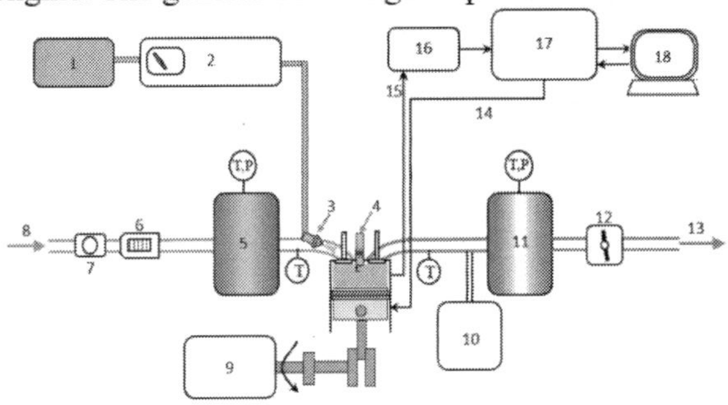

Figure 4. Schematic diagram of CNG engine test bench.
1. CNG 2. CNG flow meter 3. CNG injector 4. Spark plug 5. Intake surge tank 6. Sonic nozzle 7. Pressure regulator 8. Compressed air 9. Horiba LI250 10. Horiba MEXA-7200 11. Exhaust surge tank 12. Exhaust pressure valve 13. Exhaust gas 14. Control signal 15. Pressure signal 16. Charge amplifier 17. NI Control/combustion analysis system 18. PC

Table 2. CNG engine specifications.

Engine type	CNG Engine
Number of cylinders	1
Displacement volume	1.85 (liter)
Stroke	156 (mm)
Bore	123 (mm)
Connecting rod	228 (mm)
Compression ratio	10.5:1
Maximum speed	2300 (rev/min)
Maximum torque	360 (N.m)
Number of strokes and valves	4

3. Results and discussion

The numerical investigation of a conventional piston engine and two ceramic thermal barrier (YSZ and TiO_2) coated piston engines were conducted, as well as the performance results were compared. The conventional and insulated engines performance characteristics are measured and compared in terms of thermal efficiency, volumetric efficiency, indicated specific fuel consumption, and exhaust gas temperature at different excess air ratios. These results are denoted in the Figure 5 to 8. The results and observation of present numerical investigation of the effects of ceramic coated pistons on the performance of a CNG engine are discussed.

3.1 Thermal efficiency

Thermal efficiency (TE) is the type of efficiency through which the chemical energy input in the mode of fuel is transformed to valuable work. The in-cylinder reduction of heat losses is the main intention of LHR engine investigation which has potential be enhanced TE. With increase in EAR the ITE increases for both conventional and insulated (YSZ and TiO_2) pistons. The comparison of the ITE results of uncoated and insulated pistons is shown in Figure 5. In case of improving ITE, the YSZ coated piston have more effect than the TiO_2 coated piston. The increase in ITE of TiO_2 coated piston

(40.79 %) and YSZ coated piston (41.56%) is higher than that of the conventional piston (37.85%). The maximum improvement in ITE was predicated by 9.78% (YSZ coated piston) and 7.75% (TiO_2 coated piston). This may occur in the engine due to larger accessibility of heat in combustion chamber and reduced loss of heat to the coolant.

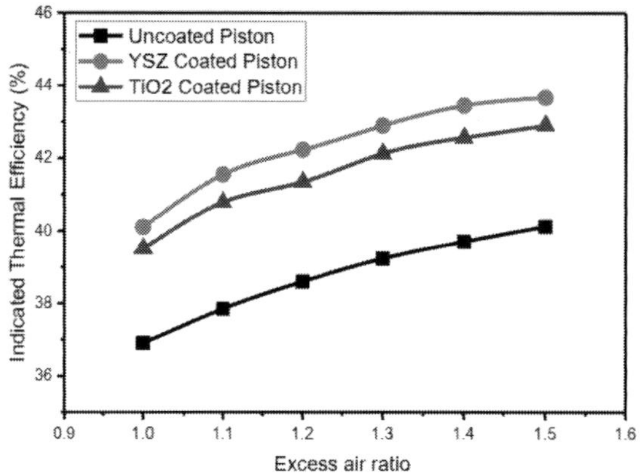

Figure 5. Indicated thermal efficiency against the EAR.

3.2 Volumetric efficiency

Volumetric efficiency is a sign of an engine's breathing capacity. It depends on an engine's operating and ambient conditions. Figure 6 shows the variation in volumetric efficiency with increasing excess air ratio (EAR) for conventional and insulated pistons. For all the cases the volumetric efficiency is lower for thermal barrier insulated piston engines. With increase in EAR, the volumetric efficiency increases for both the conventional and insulated pistons. It is found that the volumetric efficiency of insulated pistons (TiO_2 and YSZ) is 4.32% to 8 % less than the uncoated piston. Because the coated engines operating under high temperature, which cause of rise the piston speed and as a result of smaller amount of air was imbibed in the combustion chamber. This deficiency of air can be accomplished by using the turbocharger.

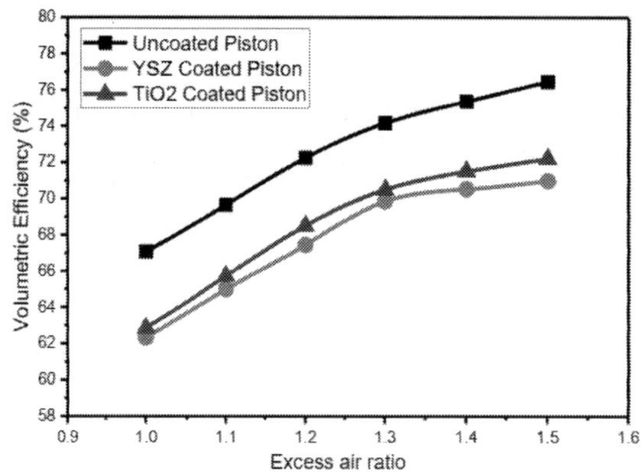

Figure 6. Volumetric efficiency against the EAR.

3.3 Specific fuel consumption

The results comparison of indicated specific fuel consumption (ISFC) of the uncoated and insulated pistons at different excess air ratios is illustrated in Figure 7. For all the cases the specific fuel consumption is lower for thermal barrier insulated piston engines. With increase in EAR, the ISFC decreases for both insulated (TiO_2 and YSZ) and conventional piston engines. The improvement in fuel economy for an LHR engines within the range of 11.2 g/kW·h to 16.8 g/kW·h were predicted over the conventional engine.

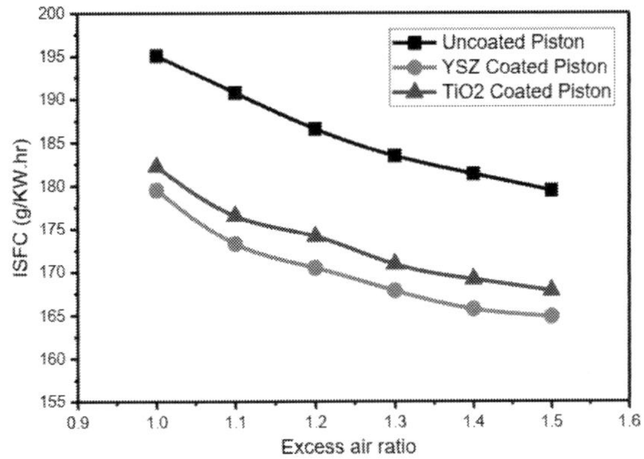

Figure 7. Indicated specific fuel consumption against the EAR.

3.4 Exhaust gas temperature

The comparison of the exhaust gas temperature results of conventional piston and thermal barrier coated (TiO_2 and YSZ) pistons is illustrated in Figure 8. In general, the insulated pistons engine shows higher exhaust gas temperature. The exhaust gas temperature decreases as the EAR increases for both uncoated and coated pistons. The maximum exhaust gas temperature of ceramic YSZ and TiO_2 insulated pistons was enlarged about 57.35 °C and 41.98 °C respectively over the uncoated piston. As the insulated pistons hamper the heat losses inside the combustion chamber in that way rising the exhaust gas temperature is a considerable.

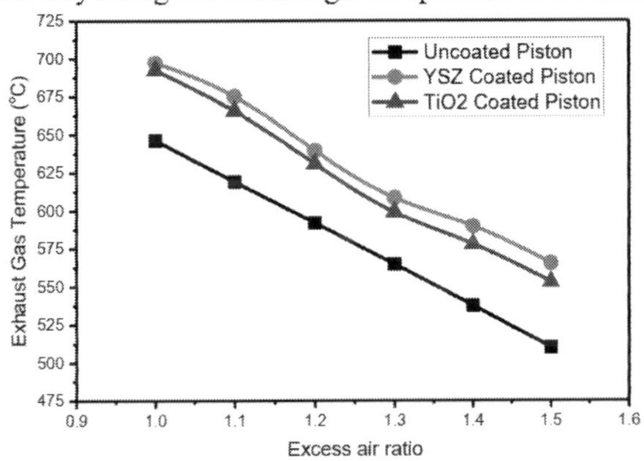

Figure 8. Exhaust gas temperature against the EAR.

4. Conclusion

The numerical investigation was carried out to analyze the effects of ceramic thermal barrier coatings on the performance of a water-cooled, four stroke, single cylinder CNG engine. One conventional (uncoated) piston and two thermal barrier (TiO_2 and YSZ) coated pistons (with thickness of 0.5 mm) were used for the present investigation. Simulations of uncoated piston and insulted pistons were operated under the similar conditions. The conventional engine as well as insulated engines were run at diverse excess air ratios from 1 to 1.5 through a step of 0.1. The following conclusions are drawn.

1) There is maximum about 9.78 % increase in the ITE was predicted with coated pistons (TiO_2 and YSZ) over the conventional (uncoated) piston.

2) The ISFC in case of insulated pistons (TiO_2 and YSZ) is reduced up to 16.8 g/kW·h (9.1%) than the uncoated piston under the same operating conditions.

3) Volumetric efficiency decreased about 4.32 % to 8% in coated pistons over the conventional piston.

4) For ceramic insulated pistons TiO_2 and YSZ, the maximum exhaust gas temperature was increased around 41.98 °C and 57.35 °C respectively, as compared to the conventional piston.

References

[1] S.G. Pandian, S.P. Rengarajan, T.P. Babu, V. Natarajan, H. Kanagasabesan. Thermal and Structural Analysis of Functionally Graded NiCrAlY/YSZ/Al_2O_3 Coated Piston, SAE Int. J. of Mater. Manuf. **8**, p.578-585, (2015)

[2] R. Kamo, W. Bryzik. Adiabatic Turbocompound Engine Performance Prediction, SAE Technical Paper 780068, (1978)

[3] W. Bryzik, R. Kamo. TACOM/Cummins Adiabatic Engine Program, SAE Technical Paper 830314, (1983)

[4] S. Caputo, F. Millo, G. Cifali, F.C. Pesce. Numerical Investigation on the Effects of Different Thermal Insulation Strategies for a Passenger Car Diesel Engine. SAE Technical paper 2017-24-0021, (2017) compared to the conventional piston.

[5] K. Hoag, M. Brands, W. Bryzik, and U. Army. Cummins/TACOM Adiabatic Engine Program, SAE Technical Paper 850356, (1985)

[6] S.H. Chan, K.A. Khor. The effect of thermal barrier coated piston crown on engine characteristics, J. Mater. Eng. Perform. **9**(1), p.103-109, (2000)

[7] T. Li, J. Caton, T. Jacobs. Use of an Engine Simulation to Study Low Heat Rejection (LHR) Concepts in a Multi-Cylinder Light-Duty Diesel Engine, SAE Technical Paper 2016-01-0668, (2016)

[8] D.A. Parker, G.M. Donnison. The development of an air gap insulated piston, SAE Technical Paper 870652, (1987)

[9] T. Morel, E.F. Fort, P.N. Blumberg. Effect of insulation strategy and design parameters on diesel engine heat rejection and performance, SAE Paper 850506, (1985)

[10] S.K. Rupangudi, C.S. Ramesh, K. Veerabhadhrappa, V.R. Rohit. Study of Effect of Coating of Piston on the Performance of a Diesel Engine, SAE Int. J. of Mater. Manuf. **7**(3), pp.633-637, (2014)

[11] G. Sivakumar, S.S. Kumar. Investigation on effect of Yttria Stabilized Zirconia coated piston crown on performance and emission characteristics of a diesel engine, Alex. Eng. J. **53**(4), pp.787-794, (2014)

[12] I.I. Vibe, F. Meißner. Brennverlauf und Kreisprozess von Verbrennungsmotoren, Verlag Technik, (1970)

[13] J.B. Heywood. *Internal combustion engine fundamentals*, p. 672, (1988)

[14] G. Woschni. A Universally Applicable Equation for the Instantaneous Heat Transfer Coefficient in Internal Combustion Engines, SAE Technical Paper 6700931, (1967)

Coupling Faults Diagnosis of a Gear-Shaft-Bearing System with Crack and Tooth Wear Faults

Li Cui, Qingsheng Wang

Faculty of Engineering, Shanghai Polytechnic University, Shanghai, 201209, P.R. China

cuili@sspu.edu.cn, mechcui@163.com

Abstract. Considering tooth wear, crack failure of a gear-rotor-bearing system, multiple faults diagnosis model are established. The gear rotor bearing system with coupled multiple faults is identified by using model-based signal-based identification method, and type of system fault are distinguished in order to raise more efficient and accurate fault identification algorithm. It verifies the theoretical correctness and reliability of the method by making the experiment of gear-rotor-bearing system on a gear-rotor test bench. This study has an important guiding significance and application value to accurately predict the faults of gear-rotor-bearing system, and improves the reliability of the gear rotor system.

1. Introduction

Gear rotor system is an important part of mechanical system, which often work under high speed and heavy load conditions. Fault diagnosis of gear system are very important for a whole machine[1]. Recently, the research on diagnosis of gear system has attracted great attention. Model of rotor dynamics, rotor crack, rule of crack propagation have been proposed and improved[2]. Tooth wear and tooth surface spalling theory have been also presented[3]. At present, there are two kinds of fault diagnosis methods for gear bearing rotor system. One is the model-based method, the other is the signal diagnosis method. Model-based method can determine the location of cracks based on the natural mode vectors of the system, which is based on the vibration signals. Signal diagnosis method judge the faults and identify the faults through the study of the time domain diagram, frequency spectrum, in which Fourier transform, Hilbert transform, frequency domain method are used.

Local wear faults of gear system occur only in individual teeth, so the vibration caused by the energy change in the entire gear vibration is very small, and often distribute in a wide frequency range. Wang et al. [4] used experimental method to collect vibration signals of normal gear and fault gear by sensors, and used wavelet analysis to process vibration signals. Parey et al. [5] set up a gear system fault model considering time-varying mesh stiffness and gear backlash. Using empirical mode method to diagnose tooth wear fault. Fakhfakh[6] diagnosed the faults of the gear system by spectrum and cepstrum analysis, and verified the correctness of the simulation analysis by the experiment. Dalpiaz [7] tested different degrees of tooth wear faults respectively, and used different signal analysis methods to analyze the vibration signals, the correctness and effectiveness of various methods were compared.

This paper will present a parameter identification method to identify multiple faults in gear system. Experimental study will be done to verify the reliability of theoretical model and algorithm.

2. Diagnosis model of faults in gear system

2.1. Diagnosis model of Crack Fault

Model-based method is used to diagnose crack fault in the gear system. By determining whether there are residual vibration equivalent forces on each node of rotor, the existence of cracks and location of the crack on the rotor can be effectively determined. Firstly, the vibration signal of the healthy rotor system are measured, then mass, stiffness and damping matrix are assembled and the dynamic equations of the system are established by the finite element method. Using the same method, the vibration state of rotor system with crack can be obtained. Finally, solving the residual vibration of the equivalent force by comparing the displacement, velocity and acceleration signals of the healthy and cracked rotor, the existence of cracks and location of the crack can be determined.

Assuming there is crack fault in a gear rotor system, dynamic characteristics of the whole system will be changed because of the existence of crack. Based on the finite element model of the system, the residual vibration model of the rotor system with crack can be written as:

$$\boldsymbol{M}\{\ddot{\boldsymbol{X}}\}+(\boldsymbol{C}+\boldsymbol{G})\{\dot{\boldsymbol{X}}\}+\boldsymbol{K}\{\boldsymbol{X}\}=\boldsymbol{P}(t)+\Delta\boldsymbol{P}(t) \tag{1}$$

Where, $\boldsymbol{M}$ is mass matrix of the system, $\boldsymbol{C}$ is damping matrix, $\boldsymbol{G}$ is gyroscopic matrix, $\boldsymbol{K}$ is stiffness matrix, $\boldsymbol{P}(t)$ is load matrix, $\Delta\boldsymbol{P}(t)$ is the residual vibration force matrix.

By comparing the vibration value between the healthy rotor system and the fault rotor system, the residual vibration value of acceleration, velocity and displacement can be obtained.

$$\Delta\ddot{\boldsymbol{X}}(t)=\ddot{\boldsymbol{X}}(t)-\ddot{\boldsymbol{X}}_o(t) \tag{2}$$

$$\Delta\dot{\boldsymbol{X}}(t)=\dot{\boldsymbol{X}}(t)-\dot{\boldsymbol{X}}_o(t) \tag{3}$$

$$\Delta\boldsymbol{X}(t)=\boldsymbol{X}(t)-\boldsymbol{X}_o(t) \tag{4}$$

Where, $\boldsymbol{X}_o(t)$ is displacement of healthy rotor, $\boldsymbol{X}(t)$ is displacement of the fault rotor.

In order to obtain the vibration state of all degrees of freedom of the gear rotor system, modal expansion method is used. By measuring the residual vibration of the nodes, all the remaining vibration values are extended, and the relationship between them can be established.

$$\Delta\boldsymbol{X}_M(t)=\boldsymbol{C}_t\Delta\boldsymbol{X}(t) \tag{5}$$

Where, $\Delta\boldsymbol{X}_M(t)$ is vibration value from the measured nodes, $\Delta\boldsymbol{X}(t)$ is vibration value of all nodes, $\boldsymbol{C}_t$ stands for a matrix formed from zero and one.

When the rotor is cracked, it is equivalent that an external torque applied to the rotor system. Therefore, the equivalent load is the direct basis for judging the existence of cracks. The model based method for the diagnosis of crack is to transform the crack into an equivalent load.

$$\Delta\boldsymbol{P}(t)=\boldsymbol{M}\Delta\ddot{\boldsymbol{X}}_M+(\boldsymbol{C}+\boldsymbol{G})\Delta\dot{\boldsymbol{X}}_M+\boldsymbol{K}\Delta\boldsymbol{X}_M \tag{6}$$

2.2. Diagnosis model of Tooth Wear Fault

In this paper, a new method to diagnosis the tooth wear fault is proposed. Firstly, using empirical mode method (EMD) to process the vibration signal with noise and choose appropriate intrinsic mode function (IMF), then using Hilbert transform (HHT) to obtain Hilbert spectrum and marginal spectrum. Finally, the frequency band information are analyzed from the marginal spectrum.

According to the definition of EMD, a signal $x(t)$ can be decomposed by the following steps.

$$x(t)=\sum_{i=1}^{n}c_i(t)+r_n(t) \tag{7}$$

Where, $x(t)$ is the original signal, $c_1(t)$ is intrinsic modal component, $r_n(t)$ is residual signal component.

Hilbert-Huang transform is a method of obtaining the best local signal by changing the frequency and amplitude of the sinusoidal curve.

$$H[c_i(t)]=\frac{1}{\pi}\int_{-\infty}^{+\infty}\frac{c_i(\tau)}{t-\tau}d\tau \tag{8}$$

Amplitude function and phase function of the signal can be solved through Hilbert-Huang transformation.

$$a_i(t)=\sqrt{c_i^2(t)+H^2\left[c_i(t)\right]},\ \phi_i(t)=\arctan\frac{H\left[c_i(t)\right]}{c_i(t)} \tag{9}$$

Where, $a_i(t)$ is amplitude function of $x(t)$, $c_i(t)$ is mode function of signal, can be obtained from EMD decomposition, $\phi_i(t)$ is phase function of the signal.

Then, the original signal can be expressed as:

$$x(t)=\mathrm{Re}\sum_{i=1}^{n}a_i(t)e^{i\phi_i(t)} \tag{10}$$

Hilbert energy spectrum is defined as:

$$H(\omega,t)=\mathrm{Re}\sum_{i=1}^{n}a_i(t)e^{i\int\omega_i(t)dt} \tag{11}$$

By Hilbert transformation, the marginal spectrum of the Hilbert transformation is defined. Marginal spectrum can be obtained by integrated along the time axis.

$$h(\omega)=\int_0^T H(\omega,t)dt \tag{12}$$

Where, T is time length of the original vibration signal analysis.

3. Coupling faults diagnosis method

Coupling faults of gear box include two types of the typical failures, crack rotor and gear wear. Model-based method is used to diagnose crack fault in the gear system. By solving the residual vibration of the equivalent force, the existence of cracks and location of the crack can be determined. Hilbert transform method is used to diagnose gear wearing fault. After analyzing the marginal spectrum of Hilbert transformation, it can be found that the change of amplitude of the meshing frequency and its harmonic components, which used to diagnosis the tooth wear fault.

4. Analysis and discussion

Figure 1 shows a gear rotor system test bench, the length of axis is 350mm, modulus of elasticity is 2×10^{11}Pa, Poisson's ratio is 0.3, which supported by four 7306 ball bearings. Torque of gear transmission is 200N·m. The tests of the gear rotor system with coupling faults of crack and wear have been done in this bench. The high-speed shaft and low-speed shaft have been divided into ten nodes, respectively. Eddy current sensors are installed in high speed shaft, which can measure displacement of four nodes. The rotating speed is adjusted to 3600rpm by a frequency converter.

In the test bench, there is a crack between the third node and fourth node in the high-speed shaft to simulate the crack fault. Teeth wear fault were simulated by installing some hard iron in the gear teeth. First, the first to fifth order modes of the high speed and low speed shafts in horizontal and vertical directions are calculated respectively. Then, the test with the health of the rotor has been done, the residual vibration displacement of each node are obtained according to the modal expansion method. Using the model of diagnosis method based on residual vibration calculation of each node, crack fault is diagnosed.

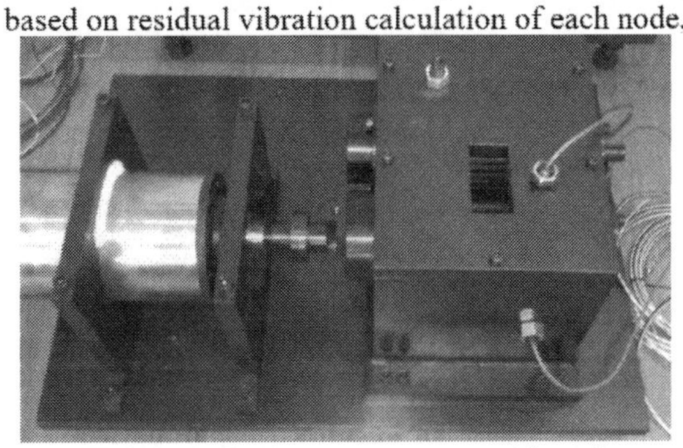

Figure 1. Gear rotor system test bench

IOP Conf. Series: Materials Science and Engineering **417** (2018) 012022 doi:10.1088/1757-899X/417/1/012022

Figure 2 shows vibration modes from first order to fifth order of high-speed shaft, low-speed shaft in gear rotor system. According to the modal expansion method, the residual vibration displacement of each node are obtained. Residual vibration velocity and residual vibration acceleration of all nodes are solved by differential method.

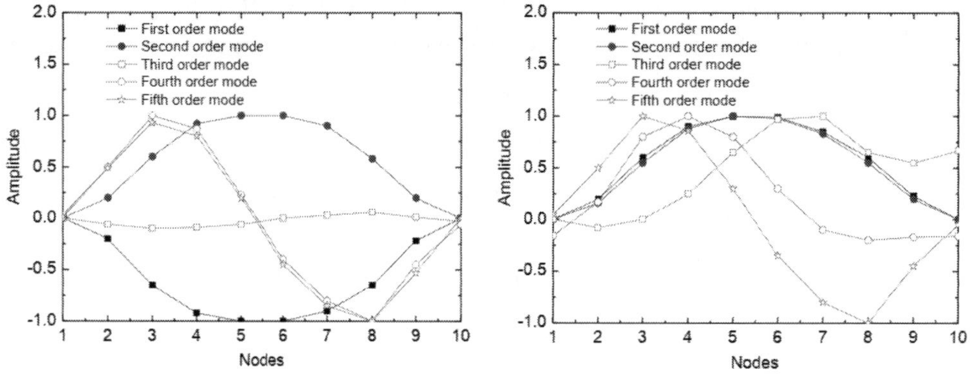

a)high-speed shaft b) low-speed shaft

Figure 2. Vibration modes of the gear-rotor system

The residual vibration forces are calculated and are shown as Figure 3, Figure 4. From Figure 3, it can be clearly seen that there is the maximum residual vibration force between the third node and fourth node in the high speed shaft. Figure 4 shows there is a small residual vibration force in the horizontal direction due to influence of high speed shaft, but there is no residual force in the vertical direction. Therefore, it can be inferred that there is a crack between the third node and fourth node in the high-speed shaft.

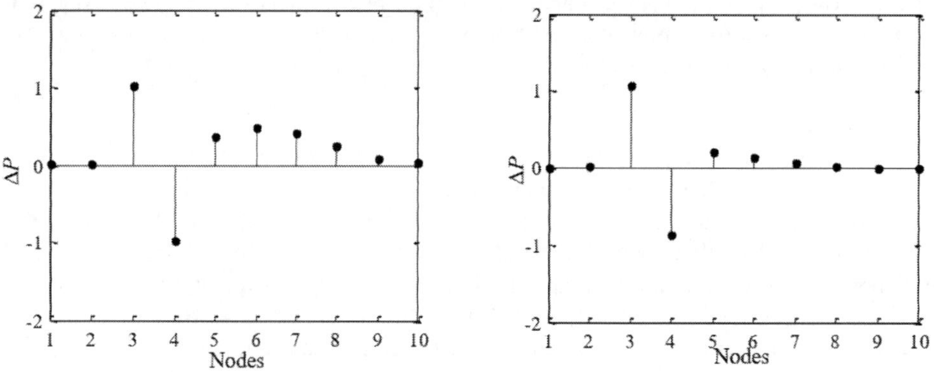

a)horizontal direction b) vertical direction

Figure 3. Residual vibration force of high-speed shaft

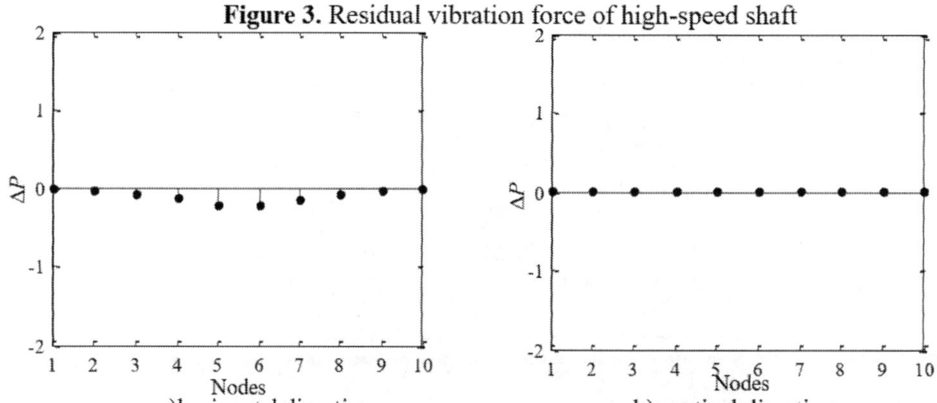

a)horizontal direction b) vertical direction

Figure 4. Residual vibration force of low-speed shaft

The second step is to diagnose the wear fault of the tooth surface. First, the decomposed modal functions and residues are obtained by EMD method. The marginal spectrum of each IMF component is given in Figure 5. It can be seen that the central frequency of the marginal spectrum of each IMF component is gradually reduced, which is in accordance with the characteristics of the EMD decomposition. Figure 6 gives the marginal spectrum after Hilbert transformation.

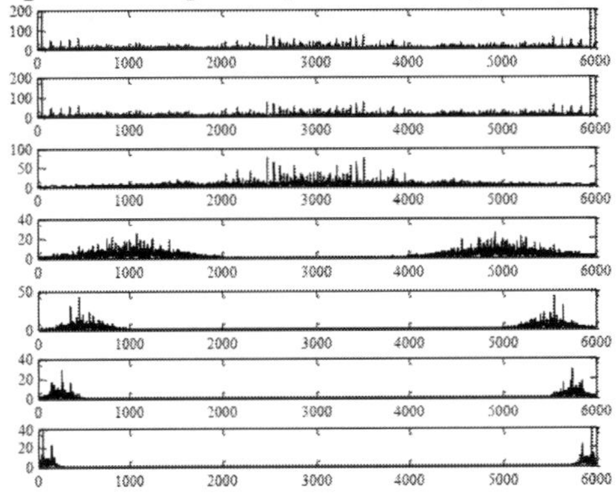

Figure 5. Marginal spectrum of IMF component

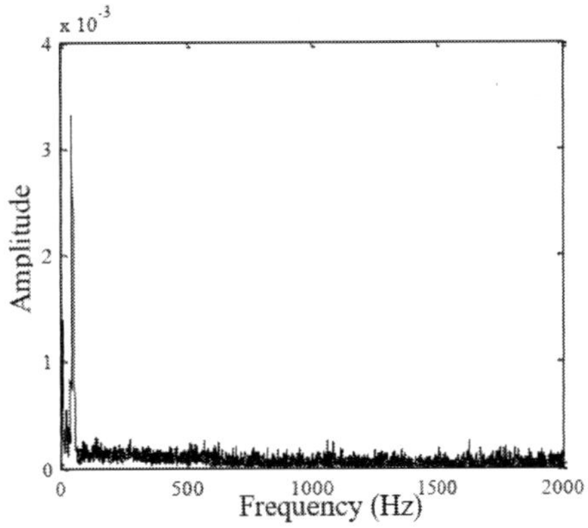

Figure 6. Marginal spectrum of gear wear fault signal

From Figure 5 and Figure 6, it can be seen clearly that the amplitude of rotating frequency of input shaft is much larger than other frequency components, and the amplitude of its meshing frequency is also very large. Because of the interference of noise signal, the rotating frequency the output shaft can hardly be observed. Therefore, it can be concluded that the gear wear fault occurs on the input shaft.

5. Conclusions

Aiming to the crack, gear wearing faults of gear rotor system, a multi fault identification algorithm for gear rotor system is proposed. The shaft crack fault and tooth wear gear bearing rotor system coupled by the sensors, vibration displacement, using model-based method to calculate the residual vibration based on the vibration displacement signals sampled by the sensors, the crack fault and the position of crack can be judged correctly. Using Hilbert-Huang transform to get the marginal spectrum, wear fault and the wear position are determined by the frequency spectrum. Experiments prove the effective of the methods.

Acknowledgments

This work was financially supported by the National Natural Science Foundation of China (No.51675323), Shanghai Natural Science Foundation (No.14ZR1416800), The key subject of Shanghai Polytechnic University (Material Science and Engineering, XXKZD1601).

References

[1] CaiWan, ChangJian. Strong nonlinearity analysis for gear-bearing system under nonlinear suspensionbifurcation and chaos[J]. Nonlinear Analysis: Real World Applications. 2010,11: 1760-1774.

[2] Peng Z, He Y, Chen Z, et al. Identification of theshaft orbit for rotating machines using wavelet modulus maxima[J]. Mechanical Systems and Signal Processing. 2002, 16(4):623-635.

[3] Miroslav Byrtus, Vladimír Zeman. On modeling and vibration of gear drives influenced by nonlinear couplings[J]. Mechanism and Machine Theory.2011, 46:375-397.

[4] Wilson Q, Wang. Fathy Ismail and M. FairdGolnaraghi. Assessment of gear damage monitoring techniques using vibration measurements. Mechanical Systems and Signal Processing. 2001, 15(5):905-922.

[5] A. Parey, N. Tandon. Impact velocity modeling and signal processing of spur gear vibration for the estimation of defect size，Mechanical Systems and Signal Processing. 2007, 21:234-243.

[6] Tahar Fakhfakh, Fakher Chaari. Numerical and Experimental Analysis of a Gear System with Teeth Defects. International Journal of Advanced Manufacture Technology. 2005, 25: 542-550.

[7] G.Dalpiaz, A.Rivola, R. Rubini. Effectiveness and Sensitivity of Vibration Processing Techniques for Local Fault Detection in Gear[J]. Mechanical Systems and Signal Processing. 2000,14(3):387-412.

IOP Conf. Series: Materials Science and Engineering **417** (2018) 012023 doi:10.1088/1757-899X/417/1/012023

Crowning Analysis of Involute Helical Pairs Shaved by CNC Gear Shaving Machine with Two Synchronous Axes

Ruei-Hung Hsu[1], Shih-Sheng Chen[2]

[1] Bachelor Program in Precision System Design, Feng Chia University, No. 100, Wenhwa Road, Seatwen, Taichung 407, Taiwan
[2] Department of Mechanical Engineering National Chung Hsing University, No. 250, Kuo-Kuang Road, Taichung 40227, Taiwan

e-mail: rhhsu@fcuoa.fcu.edu.tw

Abstract. Gear Shaving is one of the most efficient and economical ways to finish gears after gear hobbing or gear shaping. In tradition, the shaved gear can be crowning by shaving machine with the rocking mechanism. In this paper, we propose a mathematical model of a CNC shaving machine with two synchronous axes. The proposed mathematical model can be used to calculate the crowning tooth surface of the gear in parallel-shaving process.

1. Introduction

Gear shaving today is diffusely used in the automotive industry, truck and bus industry gears. The shaving cutter is meshed with the gear is similar to that in a crossed helical gear set. Gear shaving is a free cutting process that removes small amounts of metal from the working surface of gear teeth. The shaved tooth of the pinion and gear will be in line contact in an aligned gear drive. The new tendencies design for crowning gears are used to reduce the sensitivity of mating gears to misalignment and reduce the noise. In this paper, we propose a mathematical model to simulate the shaving process with crowning motion.

Basic mesh conditions for the involute helical gear set have been outlined in many textbooks [1]. In 1990, both Endoy [2] and Dugas [3] discussed important aspects of the shaving process, including the approximate life cycle and conditions for even contact. Moriwaki [4] and Moriwaki and Fujita [5] then proposed a stochastic model to predict the effect of shaving cutter performance on the finished tooth form. Hsu and Fong[6] proposed a mathematical model of pitch cylinder operation for the tooth profile of a gear finished by parallel shaving with an auxiliary crowning mechanism. Hsu[7] proposed a shaving method for double crowning that has no natural twist in the tooth flanks on the work gear surfaces is proposed, which uses a variable pressure angle shaving cutter in a parallel gear shaving process.

In this paper, we derive a mathematical models for the shaving cutter and coordinate systems for the CNC shaving machine and shaving gear. The parallel-shaving process is only need a CNC shaving machine with two synchronous axes.

2. Mathematical model of the parallel-shaving gears

The gear shaving process can be simplified as shown in Fig. 1. The coordinate systems and are rigidly connected to the shaving cutter and the gear, respectively. The coordinate system S_a is rigidly connected to the frame of the CNC shaving machine. The machine settings of the gear shaving machine include the traverse movement of the shaving cutter along the axis of the gear z_t, the crossed angle between the cutter's and the gear's axes γ, and the center distance between the gear and the shaving cutter E_o-a_1z_t-$a_2z_t^2$.The parameters ϕ_1 and $\phi_2(\phi_1)$ are the rotating angle of the shaving cutter and the gear, respectively.

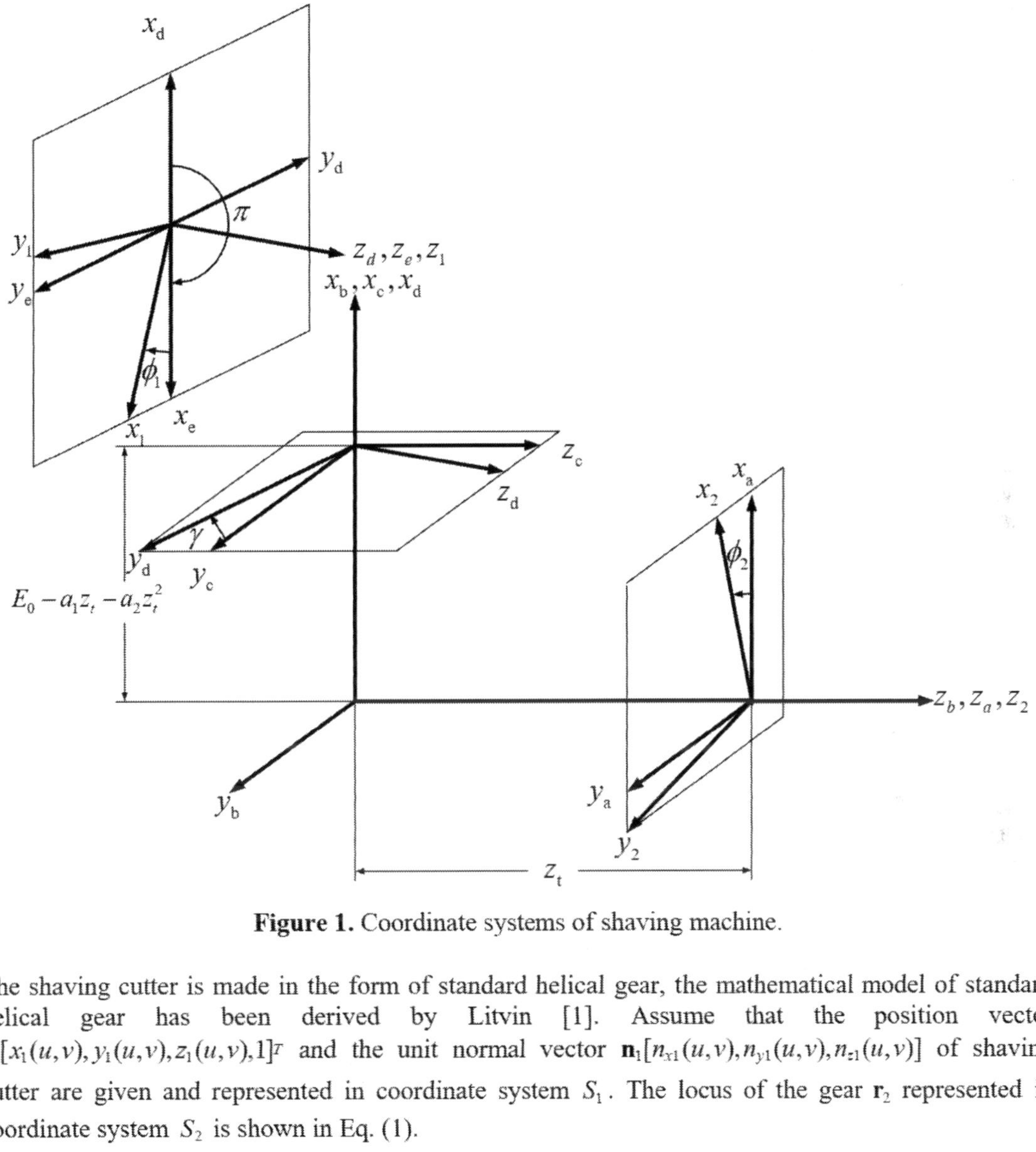

Figure 1. Coordinate systems of shaving machine.

The shaving cutter is made in the form of standard helical gear, the mathematical model of standard helical gear has been derived by Litvin [1]. Assume that the position vector $\mathbf{r}_1[x_1(u,v), y_1(u,v), z_1(u,v), 1]^T$ and the unit normal vector $\mathbf{n}_1[n_{x1}(u,v), n_{y1}(u,v), n_{z1}(u,v)]$ of shaving cutter are given and represented in coordinate system S_1. The locus of the gear $\mathbf{r}_2$ represented in coordinate system S_2 is shown in Eq. (1).

$$\mathbf{r}_2(u,v,\phi_1,\theta) = \mathbf{M}_{21}(\phi_1, z_t) \cdot \mathbf{r}_1(u,v) \quad (1)$$

where

$$\mathbf{M}_{21} = \mathbf{M}_{2c} \cdot \mathbf{M}_{c1} \quad (2)$$

$$\mathbf{M}_{2c}=\begin{bmatrix}\cos\phi_2 & \sin\phi_2 & 0 & sin\phi_2(E_o-az_t^2E_o\text{-}a_1z_t\text{-}a_2z_t^2)\\ -\sin\phi_2 & \cos\phi_2 & 0 & \cos\phi_2(E_o\text{-}a_1z_t\text{-}a_2z_t^2)\\ 0 & 0 & 1 & -z_t\\ 0 & 0 & 0 & 1\end{bmatrix} \tag{3}$$

$$\mathbf{M}_{c1}=\begin{bmatrix}-\cos\phi_1 & \sin\phi_1 & 0 & 0\\ -\cos\gamma\sin\phi_1 & -\cos\gamma\cos\phi_1 & \sin\gamma & 0\\ \sin\gamma\sin\phi_1 & \cos\phi_1\sin\gamma & \cos\gamma & 0\\ 0 & 0 & 0 & 1\end{bmatrix} \tag{4}$$

In the parallel-shaving process, the cutter is fed into the gear with table traverse in the direction of the gear longitudinal. If the gear needed to crown, the cutter is radial feed-in $a_2z_t^2$ along the center distance between the gear and the cutter. If the gear tooth surface requires taper modification, it needs to add a correction factor a_1z_t

Therefore, the locus of the gear $\mathbf{r}_2$ is simplified as shown in Eq. (5):

$$\mathbf{r}_2(u,\theta,\phi_1,z_t)=[x_2(u,\theta,\phi_1,z_t),y_2(u,\theta,\phi_1,z_t),z_2(u,\theta,\phi_1,z_t),1]^T \tag{5}$$

where

$$x_2=y_1\cos\phi_2\sin\phi_1+(E_o\text{-}a_1z_t\text{-}a_2z_t^2+z_1\sin\gamma-\mathrm{x}_1\cos\gamma\sin\phi_1)\sin\phi_2-\cos\phi_1(\mathrm{x}_1\cos\phi_2+\mathrm{y}_1\cos\gamma\sin\phi_2) \tag{6}$$

$$y_2=(E_o\text{-}a_1z_t\text{-}a_2z_t^2+z_1\sin\gamma-\cos\gamma(\mathrm{y}_1\cos\phi_1+\mathrm{x}_1\cos\phi_1))\cos\phi_2+\sin\phi_2(\mathrm{x}_1\cos\phi_1-\mathrm{y}_1\sin\phi_1) \tag{7}$$

$$z_2=-z_t+z_1\cos\gamma+\mathrm{y}_1\cos\phi_1\sin\gamma+\mathrm{x}_1\sin\phi_1\sin\gamma \tag{8}$$

The equation of meshing between the gear and the shaving cutter is shown in Eq. (9) and (10):

$$f_1(u,v,z_t,\phi_1)=\mathbf{n}_2\cdot\mathbf{v}_2^{(12)}=\frac{d\phi_1}{dt}\left(\frac{\partial}{\partial\phi_1}\left[x_2(u,\theta)\quad y_2(u,\theta)\quad z_2(u,\theta)\right]^T\right)=0 \tag{9}$$

$$f_2(u,v,z_t,\phi_1)=\mathbf{n}_2\cdot\mathbf{v}_2^{(12)}=\frac{dz_t}{dt}\left(\frac{\partial}{\partial z_t}\left[x_2(u,\theta)\quad y_2(u,\theta)\quad z_2(u,\theta)\right]^T\right)=0 \tag{10}$$

where

$$\mathbf{n}_2=\mathbf{L}_{21}\cdot\begin{bmatrix}n_{x1}(u,\theta)\\ n_{y1}(u,\theta)\\ n_{z1}(u,\theta)\end{bmatrix} \tag{11}$$

$\mathbf{L}_{21}$ is the up-submatrix of the order(3×3) of the matrix $\mathbf{M}_{21}$.

Gear shaving is a free cutting process that removes small amounts of metal from the working surface of gear teeth, therefore the operating pitch helical angle will be constant. According to literature the rotating angle between the shaving cutter and the gear is as shown in following equation:

$$\phi_1=\frac{N_2}{N_1}\phi_2\pm\frac{\tan\beta_{o2}}{r_{o2}}z_t \quad \text{(+ for R.H. , – for L.H.)} \tag{12}$$

Where N_1 and N_2 are teeth number of the shaving cutter and the gear. The parallel-shaving shaved gear be solved by using Eqs. (5), (9), (10) and (12) simultaneously.

3. Numerical examples

The basic gear data is shown in Table 1. The basic shaving cutter setup on the CNC shaving machine, we adopt the basic meshing condition for a crossed helical gear set derived by Litvin [1]. The simulation crowning amount of the shaved gear in the pith circle as shown in Fig. 2 and Fig. 3. The vertical axis is the measured face width of the gear and the unit is mm, and the horizontal axis is the crowning amount of the shaved gear and the unit is μm in Fig. 2 and Fig. 3.

Table 1. Basic data for gear, shaving cutter and machine settings.

Work gear data	
Number of teeth	3
Normal module	2.65 mm
Normal circular-tooth thickness	5.174 mm
Normal pressure angle	20
Helix angle	10° L.H.
Outer diameter	102.717 mm
Form diameter	92.43865 mm
Face width	30 mm
Shaving cutter data	
Number of teeth	73
Helix angle	22° R.H.
Normal circular-tooth thickness	2.502 mm
Machine settings	
Operating center distance (E_o)	151.846 mm
Operating crossed angle (γ_o)	11.931°

Table 2. Coefficient value of tooth surface modification in the axial direction

	a_1 (for taper)	a_2 (for crown)
Case1	0	1.0×10^{-4}
Case2	5.0×10^{-4}	1.0×10^{-4}

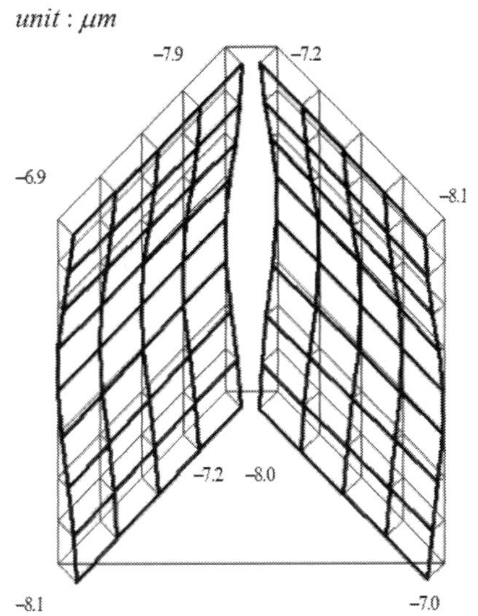

Figure 2. The topography of simulation of the crowned tooth form for Case1

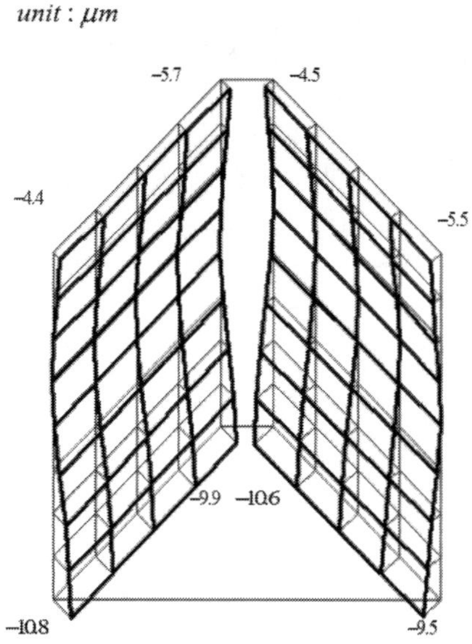

Figure 3. The topography of simulation of the crowned plus tapered tooth form for Case2

Coefficients a1 and a2 in Table 2 represent the first-order and second-order of polynomial function for center distance between shaving cutter and the gear, to achieve the effect of taper and crown modification. The results show that the maximum amount of crowning in Figure 2 is close to 8 μm. Figure 3 shows that the amount of crowing is changed due to the effect of tapering. According to the numeral examples, the proposed crowning and tapering method can shave a crowned or tapered helical gear by CNC gear shaving machine with two synchronous axes.

4. Conclusions

In this paper, we propose a mathematical model of tooth profile modification by using the two-axis synchronous motion of the gear shaving machine. The use of this processing method can replace the traditional shaving mechanism to do crowning tooth processing, simplifying the machine structure, and can make the tooth shaving process more stable.

Acknowledgments

The authors are grateful to the Ministry of Science and Technology of the R.O.C for the financial support, under the projects No: MOST106-2221-E-035 -050 -.

References

[1] Litvin F L 1994 *Gear Geometry and Applied Theory* (PTR Prentice Hall) pp 412–468
[2] Dugas J P 1990 *The Process of Gear Shaving Gear Design* (SAE Manufacturing and Inspection Manual) AE 15 pp 327 334
[3] Endoy R 1990 *Gear Hobbing, Shaping and Shaving – A Guide to Cycle Time Estimation and Process Planning* (SME Publications Development Department) pp 63–74
[4] Moriwaki I and M Fujita 1994 *Effect of Cutter Performance on Finished Tooth Form in Gear Shaving* (Journal of Mechanical Design, Transactions of the ASME) pp 701–705
[5] Moriwaki I 1990 *Numerical Analysis of Tooth Forms of Shaved Gears* (JSME International Journal Series III) pp 608–613
[6] Hsu R and Fong Z 2010 *An Auxiliary Crowning Analysis of Parallel Gear Shaving* (Mechanism and Machine Theory) pp 1298–1313
[7] Hsu R and Wu Y 2017 *Manufacturing helical gears with double-crowning and twist-free tooth flanks using a variable pressure angle shaving cutter* (Proceedings of the Institution of Mechanical Engineers, Part B: Journal of Engineering Manufacture) pp 1–10

Gradual Change Reliability Sensitivity Design for Spindle System on Fully Stochastic Process

Xin-gang WANG, Bao-yan WANG, Xiaohui CHEN, Dongxiao HOU

School of Mechanical Engineering and Automation, Northeastern University, Qinhuangdao 066004, P. R. China

xgwang@neuq.edu.cn; wangbaoyan2005@163.com;

Abstract. Gradual change reliability design of mechanical components is considered by semi-stochastic process model of loading with change of time and fully stochastic process model of known degradation function of strength at present, yet it can't obtain reasonable reliability design by existing information of components for these models. Thus the influence of existing information and time-dependent parameters of components on reliability should be taken into account in order to access reliability correctly. By studying effect of loading action and time-dependent character of strength, strength of components is taken as a process of independent increments and autocorrelation coefficient of strength is also calculated, meanwhile, we present a method for computing time-dependent reliability. Combining the reliability design theory with sensitivity analysis method, a numerical method for time-dependent reliability sensitivity design of components based on fully stochastic process is proposed, the variation rules of reliability sensitivity of parameters are given at any moment and the effects of design parameters on reliability of components are also studied. The method presented in this paper provides the theoretical basis for structural design and life prediction of mechanical components.

1. Introduction

Most of the characteristics data of mechanical products vary with time gradually, making reliability of the product show the characteristic of gradual or gradual change change, for example, the decrease of mechanical strength caused by fatigue, wear and corrosion. The change of the product's characteristics is in fact a process of gradual change over time. We must have it processed as a stochastic process, of course, the product's reliability also inevitable is a function of time, and now its focus is transforming from the invariant-time reliability to the gradual change reliability of the whole life structure process composed by design, operation, maintenance and degradation. At present, a lot of research on mechanical gradual change reliability have been did by many scholars, yet most of which is to set up the mathematical model of pure theory, don't put the current state of components and the given gradual change information into the theoretical model, so certain error is brought into the reliability design. Although gradual change characteristics of load are now very in-depth research, gradual change characteristics of strength still need further research[1-3]. Mainly because there are ample causes for the strength degeneration of components, coupled with the change of physical properties of the material itself, it is hard to show the non-stationary stochastic process of strength degeneration by a certain mathematical model. In this paper, certain work was done about this problem, in which we fully used the measured value of components at the present moment to make correction about the hypothetical stochastic process model of strength[4]. Here, what is assumed is the model on stochastic process of strength when the parts to be failure, and then use the measured values to amend. Thus we can get the model on stochastic process of strength for existing components in the future using period.

2. Time-dependnt performance of stress

For mechanical structure and components in existence, the fact that strength degenerate and random loads change due to the influence of external work environment and internal factors cannot be neglected, and strength and loads is related with time. Thus, it is necessary to simulate the structural strength and the effect of random loads during the course of the stochastic process. The expression of state function is given as,

$$g(t)=r(t)-\sigma(t) \tag{1}$$

where r(t) is the stochastic process of strength degeneration, and (t) is the stochastic process of load effect. Formula for the time-dependent reliability is represented as follows

$$R(t)=\Phi(\beta(t))=P\{r(t)>\sigma(t) \quad t\in[0,T]\} \tag{2}$$

Because in the calculation of the structural reliability index, we adopted the first-order second-moment analysis method which considers probability distribution types of basic random variable, the stochastic process of load effect must be converted to random variable, which is maximum load, in the design service period or required service period in future. Although load changing with time randomly, its expected value of maximum distribution will not decrease with the passage of time. In order to identify the effect on components from the equivalent maximum load of this random load, the random load is required to be equally discreted as n parts from the moment ti to a certain moment ti+Δ t in actual measurement. The components should not be failure, when components were not failure under the action of the maximum load Smax of n parts. It can be represented as follows,

$$P(t)=P\{r>\sigma(S_{\max})\}=P\{r>\sigma(S_1),r>\sigma(S_2),\cdots,r>\sigma(S_n)\} \tag{3}$$

Therefore, the reliability after the effect of n times of random load is equivalent to the corresponding reliability of maximum load value of n load samples. The maximum load in the load samples can be defined as the equivalent load of n times' load effect. According to order statistic theory, the maximum load Smax actually is the maximum order statistic variable Sn which is defined by the load sample (S_1, S_2,..., S_n).

$$F_\tau(x)=P(\max_{1\le i\le n} S_i\le x)=P(S_1\le x)P(S_2\le x)\cdots P(S_n\le x)=\prod_{i=1}^{n}P(S_i\le x)=[F_\tau(x)]^n \tag{4}$$

Thus the probability model of stochastic process of load is set up.

The formula of reliability under the action of the stochastic process of load is given as

$$R(t)=P\{r>\max\sigma(t) \quad t\in[t_i,t_i+\Delta t]\} \tag{5}$$

3. Time-dependnt performance of stress

At present, there are two kinds of gradual change model studying strength degeneration of mechanical components basically, the first model, directly converting strength to random variable of each period, substituting section random variable at one moment for stochastic process, which cannot reflect the correlation and other random characteristics among every random variable of strength at each moment; the second model, to show the strength degeneration process by some certain function, to transform the non-stationary stochastic process to stationary stochastic process. Due to the extreme complexity of the strength degeneration process, it is difficult to use certain mathematical model to express its degeneration function, and some information of the given components can't be used fully.

Because the great uncertainty of correlation among each strength at any time exists, in this paper, using independent increment principle of stochastic process, we view correlation among each strength as independent increment of stochastic process, and make corrections to the original stochastic process model of strength by the measured value at current time, thus the stochastic process model of strength degeneration for components in future using time is formed. Then suppose stochastic process of strength degeneration to be $\{r(t), t\in[t_i,t_i+\Delta t]\}$, the strength $r(t_i)$ is a random variable at the present moment t_i, its mean value is $E[r(t_i)]$, variance is $D[r(t_i)]$. The stochastic process of strength degeneration in design is $\{r_0(t), t \in [t_i, t_i + \Delta t]\}$, and its mean value function is $E[r_0(t)]$, variance function is $D[r_0(t)]$. $E[r(t_i)]$ is equal to $E[r_0(t_i)]$, $D[r(t_i)]$ is equal to $D[r_0(t_i)]$ at current moment t_i in the meaning of statistics and theory. Then the stochastic process of strength is defined as

$$r(t) = r(t_i) + [r_0(t) - r_0(t_i)] \tag{6}$$

The mean value function and the variance function are given as follows

$$E[r(t)] = E[r(t_i)] + E[r_0(t)] - E[r_0(t_i)] \tag{7}$$

$$D[r(t)] = D[r(t_i)] + D[r_0(t)] + D[r_0(t_i)] - 2\operatorname{cov}[r_0(t), r_0(t_i)] \tag{8}$$

where $cov[r_0(t), r_0(t_i)]$ is the covariance function of strength degeneration. If $\{r(t), t \in [t_i, t_i + \Delta t]\}$ is independent increment process, the autocorrelation coefficient of strength is

$$\rho_r[r(t_i), r(t_i + \Delta t)] = \frac{\operatorname{cov}[r(t_i), r(t_i + \Delta t)]}{\sqrt{D[r(t_i)]D[r(t_i + \Delta t)]}} \tag{9}$$

According to formula (2), the structural reliability of components is expressed by function equation as follows,

$$R(t) = P\{g(t) > 0, t \in [0, T]\} \tag{12}$$

The formula (11) is to show that when $r(t)$ of every time t in design service period is greater than $\sigma(t)$, the component structure is in the reliable state. The probability of failure is

$$F(t) = 1 - R(t) = P\{r(t) < \sigma(t), t \in [0, T]\} \tag{13}$$

Considering measured value of components at current time t_i, reliability in the mechanical parts' future service period $[t_i, t_i+\Delta t]$ can be given as

$$R(t) = P\{r(t_i) - \sigma(t_i) > 0, t_i \in [t_i, t_i + \Delta t]\} \tag{14}$$

where $r(t_i)$ is the strength of any time t_i for components; (t_i) is the effect of random load of any time t_i. According to formula (14), a component is equal to a series system made up by n samples in the whole design service period. From the reliability definition of the series system, if to make the whole system effective, every subsystem is effective too. And the time-dependent reliability is

$$R(t) = P\left\{\bigcap_{i=1}^{n} g(t_i) > 0, t_i \in [t_i, t_i + \Delta t]\right\} \tag{15}$$

On the basis of the independent increment process, correlation coefficient of the state function is

$$\rho_g[g(t_i), g(t_j)] = \frac{\operatorname{cov}[g(t_i), g(t_j)]}{\sqrt{D[r(t_i)]D[r(t_j)]}} = \frac{\rho_r[r(t_i), r(t_j)]\sqrt{D[r(t_i)]D[r(t_j)]}}{\sqrt{\{D[r(t_i)] + D[\sigma(t_i)]\}\{D[r(t_j)] + D[\sigma(t_j)]\}}} \tag{16}$$

$(i \neq j, i=1, 2, \ldots, n; \ j=1, 2, \ldots, n)$

4. Design on gradual change reliability sensitivity

4.1. Mechanical Model of a Land Axle

Spindle force distributing diagrammatic is shown in figure 1.

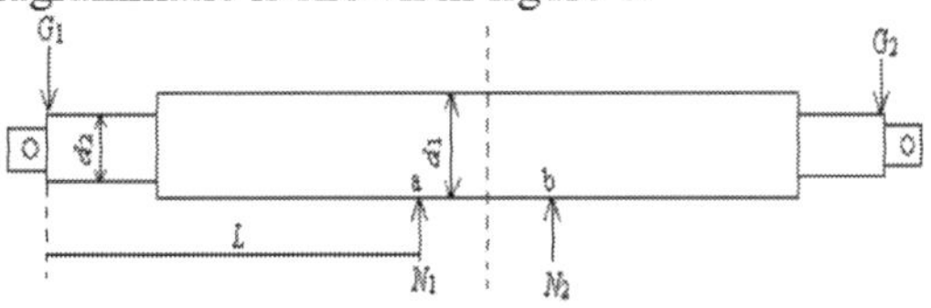

Figure 1. Force analysis of the spindle.

After the analysis of the spindle, it can be known that the closer spindle to the symmetric line. the greater bending moment produced by gravity, and as a result

$$G_1 = G_2 = \frac{G}{2},\ N_1 = N_2 = \frac{G}{2} \tag{17}$$

$$M_{\max} = M_a = M_b = G_1 L \tag{18}$$

4.2. Computation of the model

After being detected a spindle of a certain type of spindle system, which have worked 500 hours, the mean value function and variance function of its strength are respectively $E[r(t)]$ and $D[r(t)]$, $E[r(t)]=2.6e^{-0.00003t}$,$D[r(t)]=5.3\exp(-2\times10^{-9}t^2)$. The first two moments of diameters are d_1 and d_2 of the stepped shaft, d_1=(10.24, 0.087)mm, d_2=(7.92, 0.06)mm and L=(128, 0.18)mm. The measured load G with the first two moments at the current moment is G=(310.88×10^3 N, 3.5×10^3 N)Nmm, and the maximum equivalent load effect of load stochastic process obeys extremum distribution. Suppose the strength of the land axle obeys exponential distribution at the moment t_i, to calculate the reliability and gradual change reliability sensitivity of d_1, d_2 and L within next 5000 hours.

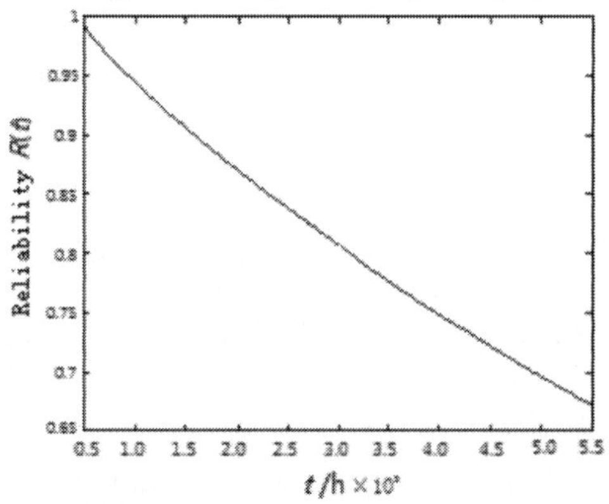

Figure 2. Curve of time-dependent reliability of land axle

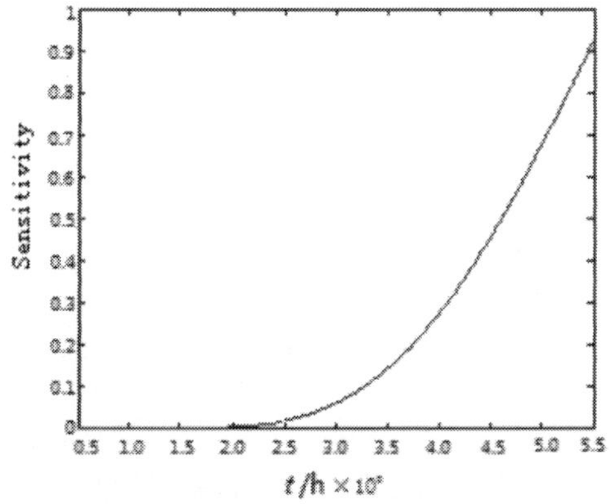

Figure 3. Curve of time-dependent reliability of land axle

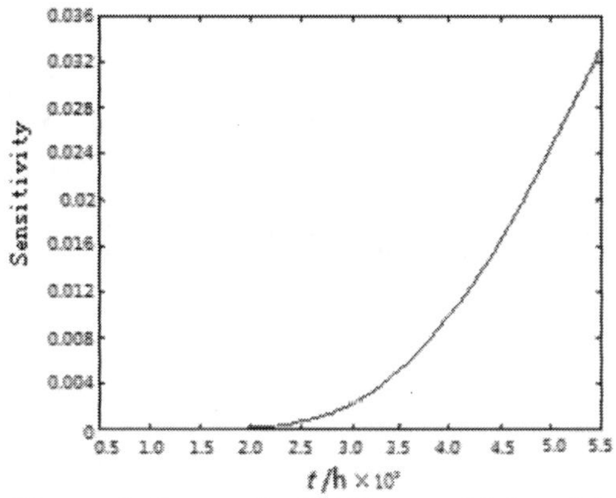

Figure 4. Curve of reliability sensitivity of d_2

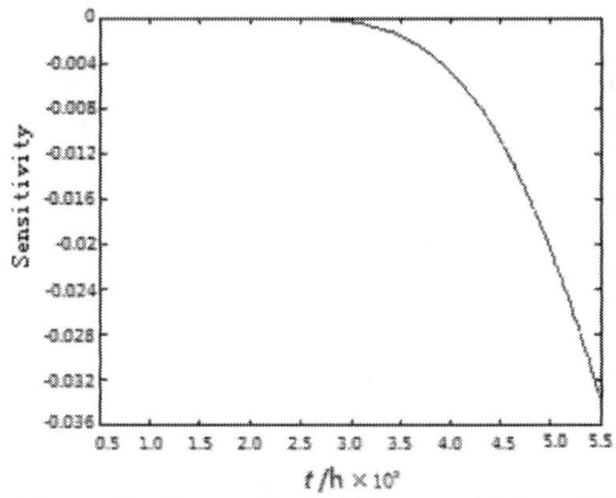

Figure 5. Curve of reliability sensitivity of L

From figure 2, it is accord with the actual working condition that structural reliability R of the spindle system is gradually reduced with the increase of service time. Figure 3 to 5 show changes of sensitivity of parameters d_1, d_2 and L, it can be seen that reliability R with respect to sensitivity of d_1 is larger than that of d_2 and L at any time, thus d_1 is the most sensitive. Mean value of d_1 and d_2 is the larger, and the spindle system is more reliable. Sensitivity of L is negative value, which shows the spindle system is unreliable, that is improving failure rate, with the increase of mean value of L.

5. Conclusions

(1) Based on practical engineering and the actual effect of the load, derived from the theory of the stress-strength interference and independent increment process model of strength, the computing method of gradual change reliability for mechanical components which is on the basis of the fully stochastic process is established, and the change rule of reliability with time is also given in this paper.
(2) On the basis of gradual change reliability consideration for a spindle system, a numerical method of gradual change reliability sensitivity is proposed. It reflects the sensitive degree of each structural parameter to reliability of the whole mechanical structure within different use period.
(3) Owing to the consideration of measured information and time response of strength and load for components, it is closer to the reliability problem in practical engineering, meanwhile, the complexity for computing higher order integrals of series system is avoided.

Acknowledgment

The research described in this paper was financially supported by the National Natural Science Foundation of China (Grant No. 51475086) and Project Supported by the Natural Science Foundation of Hebei province of China (Grant No. E2015501073) and the Fundamental Research Funds for the Central Universities (Grant No. N152304004; N162312001).

References

[1] SALVATORE Benfratello, LIVIA Cirone, FRANCESCO Giambanco. A multicriterion design of steel frames with shakedown constraints[J]. Computers and Structures, 1 (84),269-282(2006)
[2] Zhang, J.Z., The Random Time-Dependent Reliability Analysis Method of Existing Structure[J]. Journal of Nanchang College of Water Conservancy and Hydroelectric Powe ydroelectric Power,19(4),5-10(2000)
[3] Wang, Z. Xie, L.Y., Li, B. Time-dependent reliability model of component under random load[J]. Chinese Journal of Mechanical Engineering,43(12),20-25(2007)
[4] Du, B. Fully stochastic analysismethod for structural time-independent reliability[J]. ENGINEERING SCIENCES,12(3),108-112(2010)

IOP Conf. Series: Materials Science and Engineering **417** (2018) 012025 doi:10.1088/1757-899X/417/1/012025

Preparation Process of Sliding Bearing of Precision Motorized Spindle

Wang Jianlei[1, a], Wang Yunlong[1], Zhang Chen[1], Wang Dongping[1], Jia Qian[2], Cui Yahui[1], Yuan Xiaoyang[2]

[1] School of Mechanical and Precision Instrument Engineering, Xi'an University of Technology, Xi'an, PR China 710048;
[2] Key Laboratory of Education Ministry for Modem Design and Rotor-bearing System, Xi'an Jiao tong University, Xi'an 710049, China

[a] E-mail:jlwang@xaut.edu.cn

Abstract. Aiming to realizing the high efficiency and high precision of manufacture, the requirements of the bearing manufacturing process were pointed out for the application of high-precision motorized spindles. A standardized bearing manufacturing process was proposed. Furthermore, the manufacturing process flow was verified with three steps of manufacture of sliding bearing, including 3D modelling, simulation processing, bearing processing. Stiffness and vibration of the sliding bearing manufactured were tested on motorized spindle bearing testbed. The result showed that the sliding bearing manufactured has the characteristics of high rigidity and low vibration, and meet the requirements of precision motorized spindle. Process design method in this paper provides an important reference for high-precision bearing manufacturing.

1. Introduction

The direction of machine tool technology researches is high efficiency and high precision at present. For example, milling of aero-engine impeller should ensure design geometric accuracy of blade profile to meet aerodynamic requirement, which provide better maneuverability and bigger take off power for modern warplanes [1]. Again for instance, large-caliber optical lenses need more efficient manufacturing process to shorten the process of exact-grinding and polishing from several weeks to one week [2]. Motorized spindle is the power source for achieving high efficiency and high speed for machine tool. But for the motorized spindle, high efficiency and high precision is a pair of contradictions [3]. For the motorized spindle which shaft is support by rolling bearing, excessive speed will lead to decrease of spindle rotation accuracy, the main reason of which is heat of the coils caused change of fit clearance between rotor and bearing. The root of the problem is rolling bearing lack damping characteristics [4]. Replacing rolling bearing with sliding bearing can resolve the defect of rolling bearing. Although support stiffness of rolling bearing is double larger than sliding bearing with the same size, the rotor supported by sliding bearings can get higher speed, which is 3-4 times higher than the rotor supported by rolling bearing [5][6].

Sort by the way of lubrication, sliding bearings include liquid bearing, gas bearing and electromagnetic bearing. But, the support stiffness of gas lubricated bearing is so low although it can use in the higher speed working condition, so it is unfit use in a motorized spindle work for milling [7]. While electromagnetic bearing need electromagnetic induction to control the clearance with rotor, which will produce coupling action against spindle motor coils to make bearing operation unstable [8].

So, it is a feasible scheme to use liquid lubricated bearing. There are several materials commonly used for liquid lubricated bearing, like steel, tin bronze and Babbitt metal. Babbitt bearing is a kind of bimetallic bearing, the Babbitt of bearing surface form a whole with metallic matrix [9].

The purpose of this paper is design a manufacturing process of sliding bearing using the scientific method, to provides an important reference for high-precision bearing manufacturing.

2. Manufacturing process design of sliding bearing of precision motorized spindle

2.1. General process of sliding bearing manufacturing.

The manufacturing of sliding bearings is generally based on manuals and experiences. The main problems faced in process design are: (1) How to meet the required manufacturing accuracy, including dimensions and tolerances; (2) How to meet the technical requirements for bearing linings, including adhesion and defect control; (3) How to obtain higher reliability and lifetime. General process of bearing manufacturing is shown in the Figure 1.

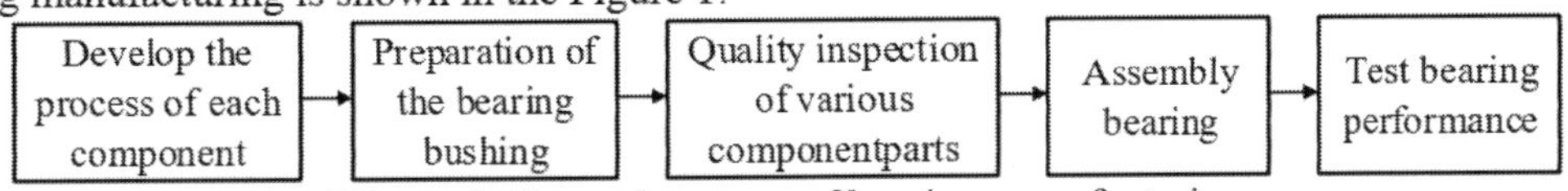

Figure 1. General process of bearing manufacturing

2.2. Design principle of sliding bearing manufacturing process.

Aimed to the requirements of high efficiency, high precision, high stiffness of manufacture, the requirements of the bearing manufacturing process were pointed out for the special application of high-precision motorized spindles. Design purpose of manufacturing process is the selection of bearing manufacturing program, processing equipment and quality inspection equipment. The following ways are usually used to estimate the manufacturing quality of sliding bearing:(1) Measured dimension error of bearing. Dimension error is the difference between work piece dimension and drawing dimension, work piece dimension can be measured by vernier calliper or micrometer. While for high precision bearing which has special requirements, it can be measured by three coordinates measuring instrument. (2) Measured surface error of bearing. Surface error of bearing mainly include internal lubricated surfaces and external mounting surface, measured index are surface runout and surface roughness. (3) Measured defect and bonding strength of Babbitt layer. Internal defect of Babbitt can be detected by penetrant test or ultrasonic testing; bonding strength can be measured with universal tensile testing machine. (4) Measured static-dynamic characteristic of bearing. Static and dynamic characteristics of bearing were measured by bench test which simulated the work conditions of bearing. (5) Measured life of bearing. Lift test of bearing is carried out on bench test too. According to enlarging load applied or increasing test speed, simulation of long-time operation of bearing was realized to estimate life.

2.3. Design of sliding bearing manufacturing process

This paper considers the bearing manufacturing evaluation criteria and process design principles, based on a large number of basic research carried out by the studying team on the motorized spindle system, proposes a standardized sliding bearing manufacturing process. Refine and concretize the design method of the sliding bearing manufacturing process, and integrate design decentralization of traditional manufacturing processes, thereby improving the efficiency and quality of the bearing manufacturing. Machining process flow chart of sliding bearing for high precision motorized spindle was shown in Figure 2. Processes include manufacture and detection, the core links were manufacture of bearing shape and preparation of Babbitt layer. Advanced CNC machine tool, reasonable machining parameters and high digital centrifugal casting machine will be the guarantee to achieve high quality manufacture

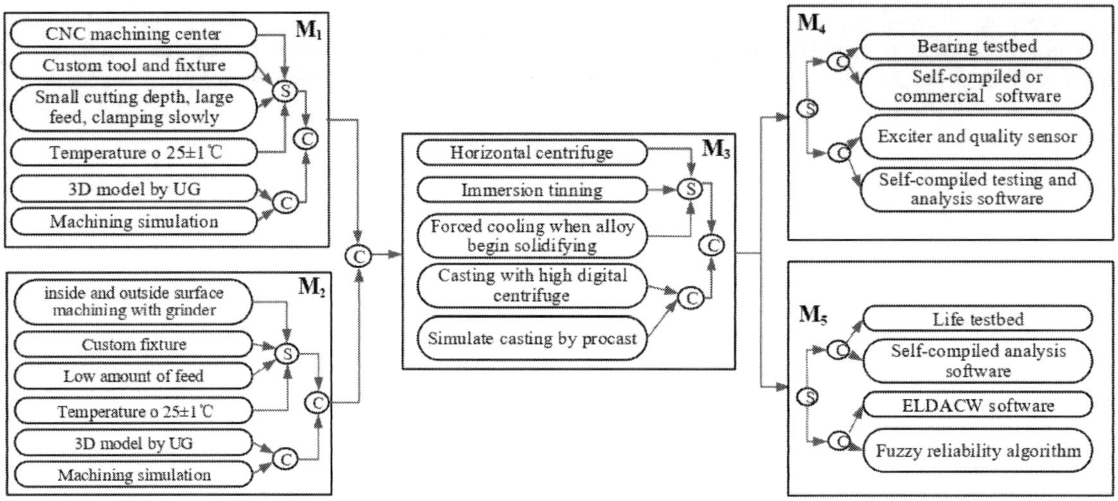

Figure 2. Machining process flow chart of sliding bearing of precision motorized spindle

3. Manufacturing example and property tests of the bearing

3.1. Manufacturing example of the bearing

The bearing to be manufactured is a hybrid bearing, and which was designed for a kind of high precision motorized spindle. Structure diagram of the bearing is shown in the Figure.3. And the values of main design parameters are listed in the Table 1.

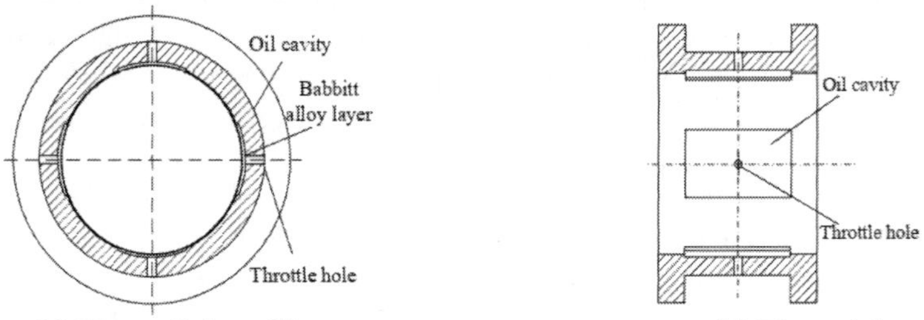

(a) The radial profile (b) The axial profile

Figure 3. Structure diagram of the bearing to be manufactured

Table 1. The values of main design parameters of the bearing

Parameter	Value	Parameter	Value
Radial motion error ε_r /	0.3	Design speed n / r/min	6000
Axial motion error ε_a /	0.3	Inside diameter r / mm	85
Radial stiffness K_r /	1000	Outside diameter R /	100
Axial stiffness K_a / N/μm	1000	Width B / mm	90

The manufacturing process of bearing was strictly in accordance with the figure.2. Several core links of the manufacturing process are described as follows. First of all, a 3D model of the bearing to be manufactured was built by UG software. The model is shown in the Figure.4, from which the appearance of the bearing after manufacture can be seen clearly. Then, machining simulation of the bearing was carried out by VERICUT software according to the 3D model. Machine tool in the machining simulation is a milling-grind machine (DMG HSC75), which is the same as that in the actual processing of the bearing. The manufacturing process sequence of the bearing is rough Machining-Babbitt casting-finish machining. The simulation results show that all the machine processes of the bearing could be carried out on the machine tool smoothly, indicating there has no interference between each machinery process. The total time of machining processes was 576 min. screenshots of the machining simulation are shown in the Figure.5.

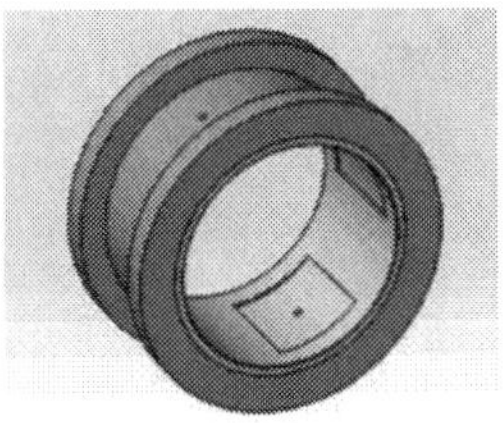

Figure 4. 3D model of the bearing to be manufactured

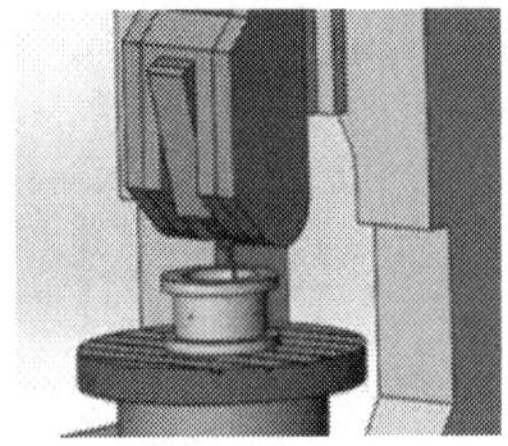

Figure 5. Machining simulation of bearing

The equipment used for preparation of the Babbitt layer of bearing was a digital centrifuge. Digital centrifuge and the bearing after finish machining were shown in Figure.6.

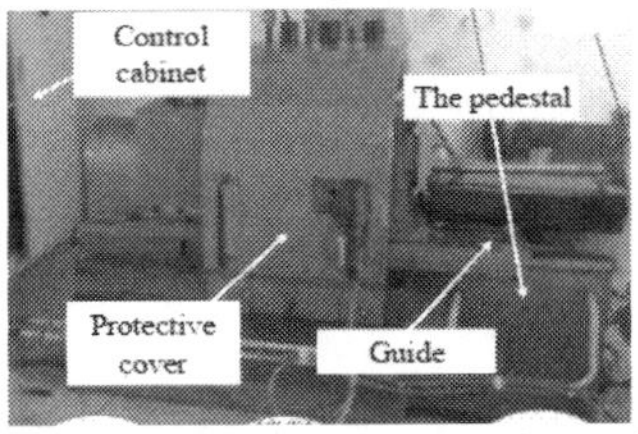

(a) Digital centrifuge for preparation of the Babbitt layer

(b) The bearing after finish machining

Figure 6. Digital centrifuge and the bearing after finish machining

3.2. Property test of bearing

Schematic diagram and photo of testbed were shown in Figure.7. Rotor of testbed was derived by coil directly, the same as actual motorized spindle.

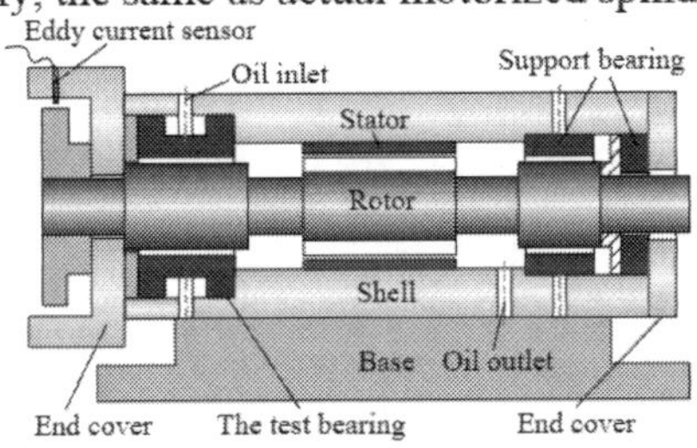

(a) The schematic diagram of testbed

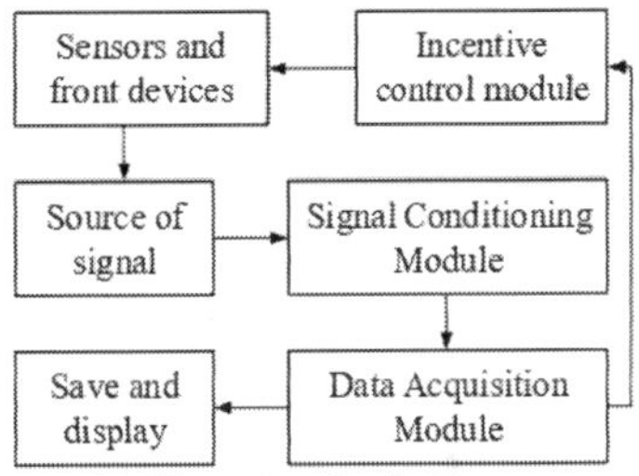

(b) Test Analysis System

Figure 7. The schematic diagram of testbed and test analysis system

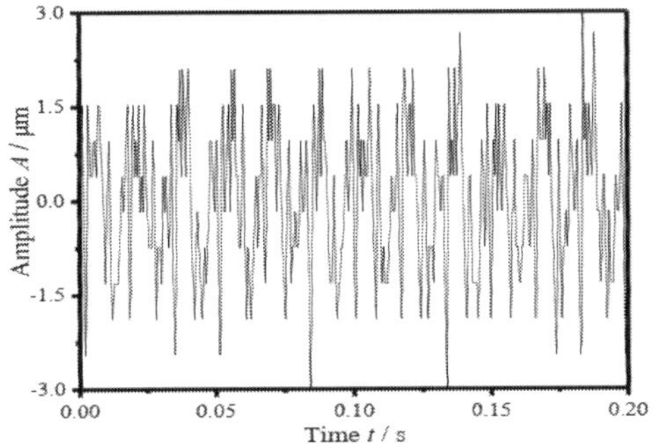

(a) Horizontal vibration time-domain waveform

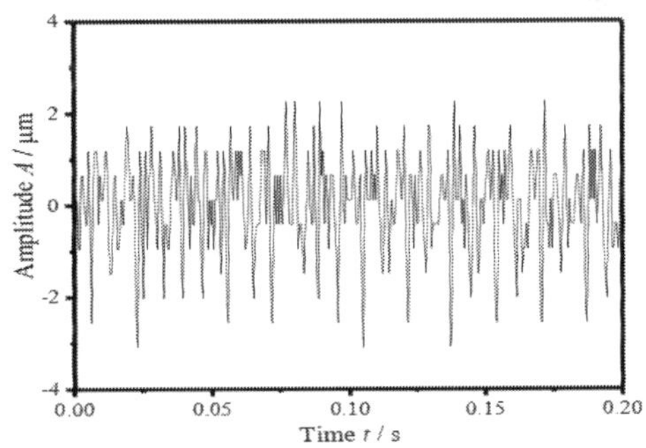

(b) Vertical vibration time-domain waveform

Figure 8. The result of test data of rotor under speed 4000 r/min

Test data were measured under different speed ranging from 500 r/min to 4000 r/min, the data was measured every additional 500 r/min. The test period of amplitude was 0.20s. Maximum amplitude under each speed was considered as corresponding amplitude. Fig. 9 showed the test data under speed 4000 r/min.

The amplitude and stiffness of rotor under each test speed were shown in Fig. 10. The maximum horizontal amplitude and vertical amplitude of rotor are 3.2 μm and 3.0 μm respectively. When speed is 500 r/min, the stiffness of bearing is minimum, i.e. 3.2×107 N/m. It presented that the sliding bearing manufactured design method meet the requirements of precision motorized spindle and grinding of thin-walled Aluminium alloys components and optical lenses, signifying reliability of process design method.

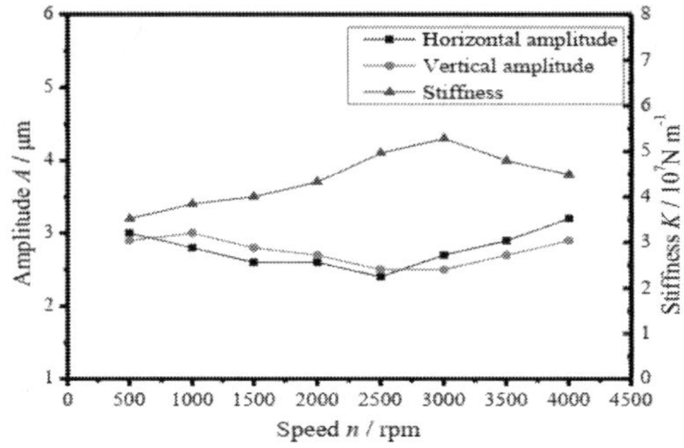

Figure 9. The amplitude and stiffness of rotor under each test speed

4. Conclusions

The requirements of the bearing manufacturing process were pointed out for the special application of high-precision motorized spindles. A standardized bearing manufacturing process was proposed. According to the proposed process, a bearing was manufactured, and the performance of the bearing was evaluated. The evaluation results showed that the bearing has the characteristics of high rigidity and low vibration, and it meets the requirements for high-precision spindles. Process design method in this paper provides an important reference for high-precision bearing manufacturing.

Acknowledgments

This article is funded from Shaanxi Provincial Department of Education Special Scientific Research Project (No. 15JK1528) and Xi'an University of Technology Innovation Project (No. 2014CX006)

References

[1] Tao Wang, et.al. 2016 Application of low-melting alloy in the fixture for machining aeronautical thin-walled component. *The International Journal of Advanced Manufacturing Technology*., 87(09):2797 - 2807 .

[2] Jiang, C., Wang, C., Li, H. 2016. Experimental investigation of brittle material removal fraction on an optical glass surface during ultrasound-assisted grinding. *International Journal of Advanced Manufacturing Technology*, 86(1-4), 419-426.

[3] Q Jia, J Zha, CX Zhang, et al. 2015 Precision prediction model and experimental verification of hydrostatic bearing-rotor system of ultra-precision machine tools. *ASME 2015 International Design Engineering Technical Conferences and Computers and Information in Engineering Conference*, IDETC/CIE, Boston, Massachusetts, USA.

[4] Zhou Chang, Qian Jia, Xing Yuan, et al. 2017 Main failure mode of oil-air lubricated rolling bearing installed in high speed machining. *Tribology International*.,112:68-74.

[5] Guoyuan Zhang, et.al. 2010The tribological behavior of Ni–Cu–Ag-based PVD coatings for hybrid bearings under different lubrication conditions. *Tribology International*.(43):197-201.

[6] C. Chen, J. Yang, D. Han, et al. 2014 Research on speed's influence on hydrostatic bearing's stiffness. *Vibroengineering Procedia* ,04 :113-118.

[7] Mohd F.B., Mohd A.A., Hilmi A., et.al. 2013 Frictional behavior of bearing material under gas lubricated conditions. *Procedia Engineering*.68:688-693.

[8] Putz Matthias, Blau Peter, Koriath Hans-Joachim, et.al. 2017 Adaptive spindle damping system with active electromagnetic bearing. *Procedia Manufacturing*.08:557-562.

[9] Dongya Zhang, John K.L.Ho, et.al. 2015 Tribological properties of Tin-based Babbitt bearing alloy with polyurethane coating under dry and starved lubrication conditions. *Tribology International*.,90:22-31.

IOP Conf. Series: Materials Science and Engineering **417** (2018) 012026 doi:10.1088/1757-899X/417/1/012026

Determination of Optimal Exchanged Grinding Wheel Diameter when Internally Grinding Alloy Tool Steel 9CrSi

Le Xuan Hung[1], Vu Ngoc Pi[1]*, Luu Anh Tung[1], Hoang Xuan Tu[1], Gong Jun[2] and BanhTien Long[3]

[1.] Thai Nguyen University of Technology,Thai Nguyen 23000, Vietnam
[2.] Lanzhou University of Technology, Lanzhou, 730050, China
[3.] Ha Noi University of Science and Technology, Ha Noi 100000, Vietnam

Corresponding author's email: vungocpi@tnut.edu.vn

Abstract. This paper introduces a study on optimal determination of exchanged grinding wheel diameter in internal grinding. In the study, the influences of grinding process parameters including the wheel life, the total dressing depth, the radial grinding wheel wear per dress and the initial grinding wheel diameter on the exchanged grinding wheel diameter were investigated. In addition, the influence of cost components including the machine tool hourly rate and the grinding wheel cost were taken into account. For evaluating the effects of these factors on the optimal exchanged grinding wheel diameter, an "experiment" was designed and a computer program was built for performing the "experiment". Based on the results of the experiment, a model for calculating the optimal exchanged grinding wheel diameter was proposed

1. Introduction

Grinding is a machining process which uses high-speed abrasive wheels, pads, and belts [1]. It is mentioned that this type of machining accounts for about 20-25% of the total expenditures on machining operations in industries [2]. Consequently, optimization of grinding process as well as of internal grinding process have been subjected to many studies.

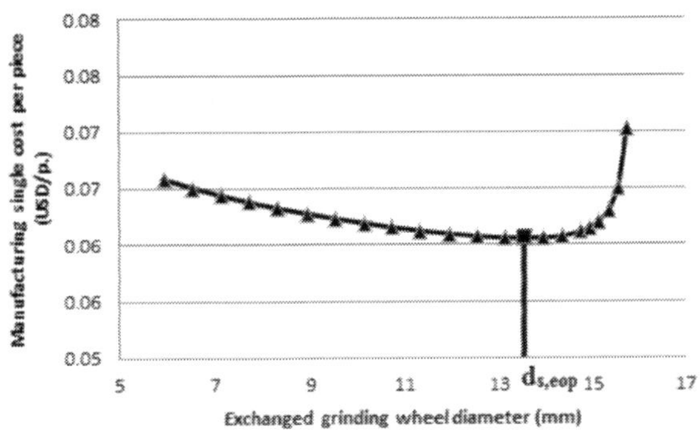

Figure 1. Manufacturing single cost versus exchanged grinding wheel diameter

*Corresponding author: Vu Ngoc Pi, Ass. Prof., Ph.D., E-mail: vungocpi@tnut.edu.vn.

So far, numerous studies have been focusing on optimization for external cylindrical grinding [3, 4, 5, and 6] or for surface grinding [7, 8 and 9]. For internal grinding process, there have been done several studies on determination of the wheel life [10], on online-optimization of grinding process and dressing parameters in order to reduce the grinding time [11], and on adaptive process control for increasing the efficiency of the grinding process [12]. Recently, there has been a cost optimization study on internal grinding process [13] for finding optimum exchanged grinding wheel diameter. It was found that there is an optimum exchanged grinding wheel diameter at which the grinding cost is minimal (Figure 1). Also, it was presented that grinding with optimal exchanged diameter significantly reduce not only the grinding cost but also the grinding time. However, in this study, the influences of grinding process parameters on optimal exchanged diameter still not evaluated in scientific way.
Hence, this paper presents a study on calculation of optimal exchanged grinding wheel diameter when internally grinding alloy tool steel 9CrSi. In the study, the effects of grinding process parameters as well as the effect of cost components on the optimum exchanged diameter were analyzed in detail. To evaluate the effects of these factors on the optimal exchanged grinding wheel diameter, an "experiment" was designed and a computer program was built for achievement the "experiment". From the results of the "experiment", the influences of process parameters and cost elements on the optimal exchanged grinding wheel diameter were analyzed scientifically. Moreover, a model for determining the optimal exchanged diameter was presented.

2. Cost analysis

In internal grinding process, the manufacturing single cost per piece $C_{\sin}$ can be determined as follows [13]:

$$C_{\sin} = t_s \cdot C_{mt,h} + C_{gw,p} \tag{1}$$

Wherein,

$C_{mt,h}$ - Machine tool hourly rate (USD/h) including wages, overhead, and cost of maintenance etc.;

$C_{gw,p}$ - Grinding wheel cost per part (USD/part); $C_{gw,p}$ is calculated by:

$$C_{gw,p} = C_{gw} / n_{p,w} \tag{2}$$

Where, C_{gw} is the cost of an internal grinding wheel (USD/piece); $n_{p,w}$ is total number of parts ground by a grinding wheel and it can be written [1]:

$$n_{p,w} = \left(d_{s,0} - d_{s,e}\right) \cdot n_{p,d} / \left[2\left(\delta_{rs} + a_{ed,ges}\right)\right] \tag{3}$$

In which, $d_{s,0}$ is the initial grinding wheel diameter (mm); $d_{s,e}$ is exchanged grinding wheel diameter (mm); δ_{rs} is the radial grinding wheel wear per dress (mm/dress); $a_{ed,ges}$ is total depth of dressing cut (mm); $n_{p,d}$ is number of parts per dress and is given by:

$$n_{p,d} = t_w / t_c \tag{4}$$

Where, t_w is wheel life (h) and t_c is grinding time (h);The grinding time can be determined as:

$$t_c = l_w \cdot a_{e,tot} / \left(v_{fa} \cdot f_r\right) \tag{5}$$

In which, $a_{e,tot}$ is total depth of cut (mm), l_w is length of part (mm), v_{fa} is axial feed speed (mm/min) and f_r is radial wheel feed (mm/double stroke).

The axial feed speed v_{fa} is calculated by [14]: $v_{fa} = 22.88 \cdot D_{gw}^{0.9865} \cdot d_w^{0.0821} \cdot tg^{-2.9833} \cdot n_w^{1.2471}$ (when grinding alloy tool steel 9CrSi).

In the above formulas, D_{gw} is grinding wheel diameter; d_w is workpiece diameter; tg is tolerance grade; n_w is workpiece speed; As the workpiece is alloy tool steel 9CrSi, from the tabulated data in [15], n_w can be determined by the following regression model [14]: $n_w = 1255.8 \cdot d_w^{-0.3491}$.

f_r –radial wheel feed (mm/double stroke); f_r is determined by [14]:

$$f_r = f_{r,tab} \cdot c_1 \cdot c_2 \cdot c_3 \cdot c_4 \tag{6}$$

In which, $f_{r,tab}$ is tabled radial wheel feed (mm/double stroke); From the tabulated data in [15], the following regression equation was found for determining $f_{r,tab}$ [14]:

$$f_{r,tab} = 30.2944 \cdot a_{e,tot}^{0.567} \cdot v_{fa}^{-0.9693} \cdot d_w^{-0.1269} \tag{7}$$

Wherein, $a_{e,tot}$ is the total depth of cut (mm).

c_1 - coefficient which depends on workpiece material and tolerance grade tg; As the workpiece material is alloy tool steel 9CrSi, c_1 is calculated by [14]: $c_1 = 0.0857 \cdot tg^{1.2767}$;

c_2 - coefficient which depends on grinding wheel diameter d_s. Based on data in [15], c_2 can be calculated by the following regression equation (with $R^2 = 0.9977$):

$$c_2 = 0.5657 \cdot d_s^{0.153} \tag{8}$$

c_3 - coefficient which depends on measurement type; $c_3 = 1$ if a micrometer is used for measurement and $c_3 = 1.4$ if a snap gauge is used [15];

c_4 - coefficient which depends on the character of workpiece's hole and the ratio of the length of workpiece (l_w) to its diameter (d_w). Based on the data in [15], the following formulas were found for determination of coefficient c_4: $c_4 = 1.0642 \cdot (l_w / d_w)^{-0.5079}$ when grinding continuous cylindrical hole (with $R^2 = 0.9637$); $c_4 = 1.375 \cdot (l_w / d_w)^{-0.5058}$ when grinding non-continuous cylindrical hole (with $R^2 = 0.955$); $c_4 = 0.8514 \cdot (l_w / d_w)^{-0.5079}$ when grinding cylindrical hole with a curved shoulder (with $R^2 = 0.9637$);

t_s - Manufacturing time includes auxiliary time (h); in internal grinding process, the manufacturing time can be expressed as:

$$t_s = t_c + t_{lu} + t_{sp} + t_{d,p} + t_{cw,p} \tag{9}$$

Where, t_{lu} - time for loading and unloading workpiece (h); t_{sp} - spark-out time (h); $t_{d,p}$ - dressing time per piece (h):

$$t_{d,p} = t_d / n_{p,d} \tag{10}$$

In which t_d is dressing time (h). Substituting (4) into (10) we have:

$$t_{d,p} = t_d \cdot t_g / t_w \tag{11}$$

$t_{cw,p}$ is time for changing a grinding wheel per workpiece (h); $t_{cw,p}$ can be calculated as:

$$t_{cw,p} = t_{cw} / n_{p,w} \tag{12}$$

Where, t_{cw} is time for changing a grinding wheel (h).

Substituting (3) into (12) we have:

$$t_{cw,p} = 2t_{cw}\left(\delta_{rs} + a_{ed,ges}\right) / \left[n_{p,d}\left(d_{s,0} - d_{s,e}\right)\right] \tag{13}$$

t_c - grinding time (h); it can be calculated as [15]:

$$t_c = \frac{l_w \cdot a_{e,tot}}{v_{fa} \cdot v_{fr}} \tag{14}$$

3. Experiment work

Table 1. Grinding parameters for "the experiment"

Factor	Code	Unit	Low	High
Op. exchanged grinding wheel diameter	D_0	mm	5	30
Total depth of dressing cut	t_{sd}	Mm	0.01	0.03
Wheel life	T_d	min.	10	30
Radial grinding wheel per dress	D_{max}	Mm	0.01	0.03
Machine tool hourly rate	C_m	USD/h	1.5	10
Grinding wheel cost	C_d	USD/p.	0.2	2

To explore the influence of factors on the optimum exchanged grinding wheel diameter an "experiment" was designed and performed by a computer program. In this "experiment", 6 process parameters were selected as the input factors for the exploring. They are the initial grinding wheel diameter D_0, the total depth of dressing cut t_{sd}, the wheel life T_d, the radial grinding wheel per dress D_{max} and the machine tool hourly rate C_m. Also, a 2-level full factorial design was chosen for the design of the "experiment". Therefore, the design was prepared with $2^6 = 64$ numbers of experiments. Fig. 2 shows the creation of factorial design with the factors were described in the Table 1. To accomplish the experiment, a computer program was built. Table 2 shows the input parameters and the output results of the computer program (the optimum exchanged grinding wheel diameter D_{op}) with many levels.

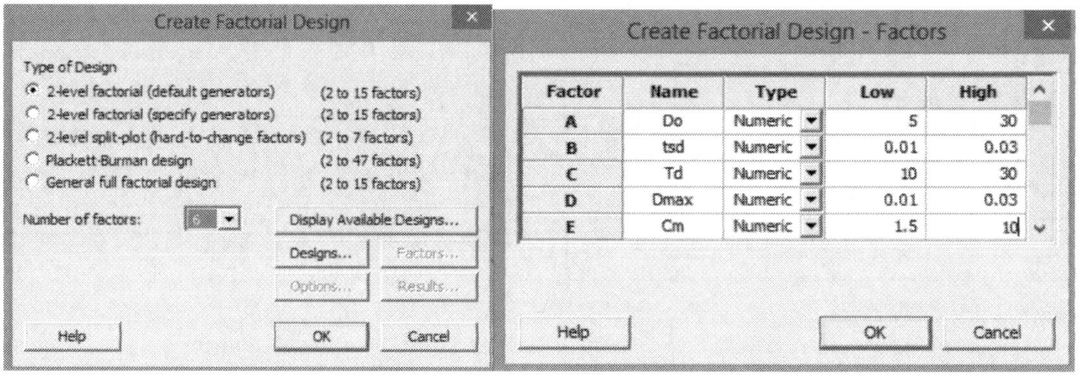

Figure 2. Creation of factorial design

4. Results and discussions

Figure 3 presents the graph of the main effect of each factor. The graph is used for evaluating the influences of factors on the response and for comparing the relative strength of the influence. From this graph, it is clear that the optimum exchanged grinding wheel diameter D_{op} depends strongly on the initial grinding wheel diameter. Furthermore, it is affected by the machine hourly rate C_m and the grinding wheel cost C_d. Besides, the other factors including the total depth of dressing cut t_{sd}, the wheel life T_d and the radial grinding wheel per dress D_{max} do not affect the optimum diameter D_{op} because the line is nearly parallel to the mean value of all responses.

Fig. 4 shows the Pareto chart of the standardized effects from the largest effect to the smallest effect. From the chart, the bars that represent factors A (the initial grinding wheel diameter), F (grinding wheel cost), E (the machine hourly rate), AF, AE, and EF cross the reference line. Consequently, these factors are statistically significant at the 0.05 level with the response model.

Figure 5 reports the Normal Plot of the standardized effects. From this Figure, it is seen that, the initial grinding wheel diameter (factor A) and the grinding wheel cost (factor F) are the significant effects factors. Also, all the effects which lie along the line (including B, C, D and their interactions) are negligible. In addition, the initial grinding wheel diameter (factor A) has a positive standardized effect. When it changes from the low level to the high level of the factor, the optimum exchanged diameter increases. Besides, the grinding cost has a negative standardized effect. When it grows the optimum exchanged diameter decreases.

Figure 6 describes the estimated effects and coefficients for D_{op} after ignoring insignificant effects. It was found that factors which have a significant effect on a response have P-values lower than 0.05 are initial grinding wheel diameter D_0, machine hourly rate C_m, grinding wheel cost C_d and the interactions between D_0 and C_m, D_0 and C_d, C_m and C_d (Figure 6). Therefore, the following equation can be used for describing the relation between the optimum diameter and significant effect factors:

$$D_{op} = 13.264 + 10.24D_0 + 1.214C_{m,h} - 1..437C_d + 0.803D_0C_{m,h} - 0.949D_0C_d + 0.534C_{m,h}C_d \quad (15)$$

Table 2. Experimental Plans and Output Response

StdOrder	RunOrder	CenterPt	Blocks	Do (mm)	tsd (mm)	Td (min.)	Dmax (mm/dress)	Cm,h (USD/h)	Cd (USD/p)	**Dop (mm)**
12	1	1	1	30	0.03	10	0.03	1.5	0.2	**24.7**
51	2	1	1	5	0.03	10	0.01	10	2	**2.94**
36	3	1	1	30	0.03	10	0.01	1.5	2	**17.74**
61	4	1	1	5	0.01	30	0.03	10	2	**3.06**
19	5	1	1	5	0.03	10	0.01	10	0.2	**3.83**
7	6	1	1	5	0.03	30	0.01	1.5	0.2	**3.17**
26	7	1	1	30	0.01	10	0.03	10	0.2	**27.08**
46	8	1	1	30	0.01	30	0.03	1.5	2	**18.43**
45	9	1	1	5	0.01	30	0.03	1.5	2	**2**
47	10	1	1	5	0.03	30	0.03	1.5	2	**2**
63	11	1	1	5	0.03	30	0.03	10	2	**2.94**
18	12	1	1	30	0.01	10	0.01	10	0.2	**27.08**
9	13	1	1	5	0.01	10	0.03	1.5	0.2	**3.28**
43	14	1	1	5	0.03	10	0.03	1.5	2	**2**
14	15	1	1	30	0.01	30	0.03	1.5	0.2	**25.08**
11	16	1	1	5	0.03	10	0.03	1.5	0.2	**3.17**
21	17	1	1	5	0.01	30	0.01	10	0.2	**3.91**
…										
40	61	1	1	30	0.03	30	0.01	1.5	2	**17.74**
25	62	1	1	5	0.01	10	0.03	10	0.2	**3.91**
32	63	1	1	30	0.03	30	0.03	10	0.2	**26.85**
55	64	1	1	5	0.03	30	0.01	10	2	**2.94**

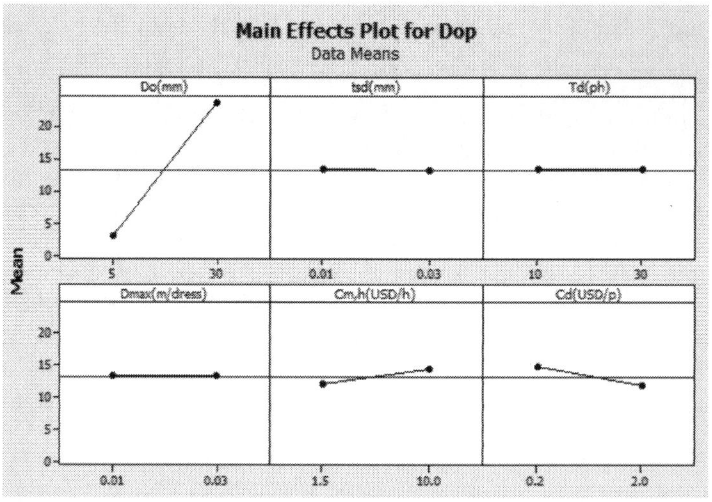

Figure 3. Main effects plot for optimum exchanged grinding wheel diameter

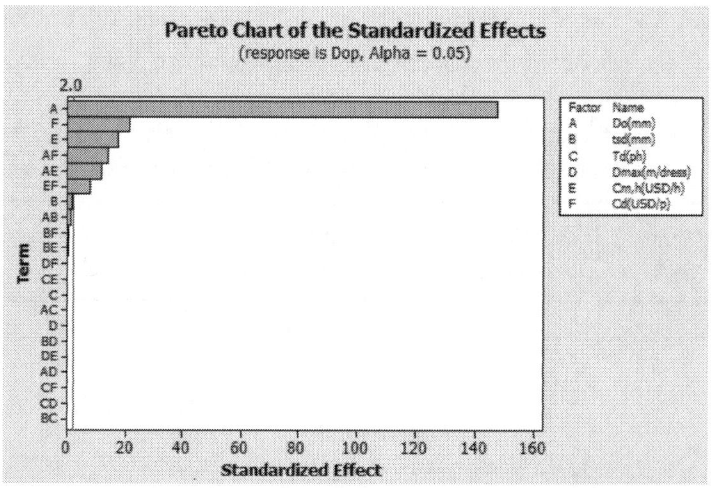

Figure 4. Pareto Chart of the Standardized Effects

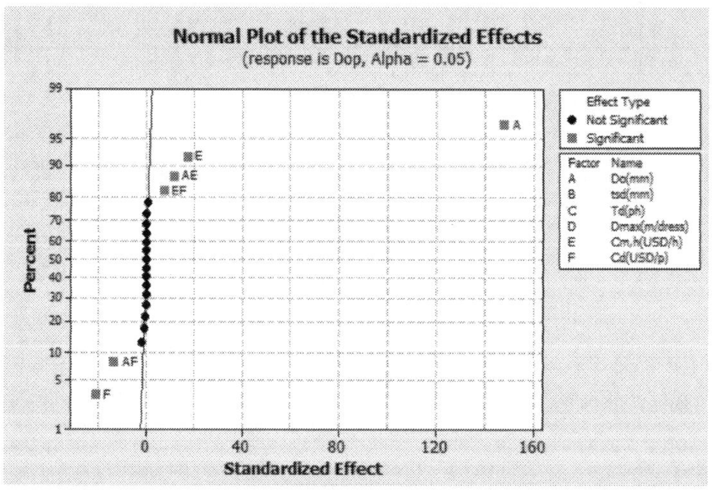

Figure 5. Normal Plot for D_{op}

Estimated Effects and Coefficients for Dop (coded units)

Term	Effect	Coef	SE Coef	T	P
Constant		13.264	0.06299	210.58	0.000
Do(mm)	20.481	10.241	0.06299	162.57	0.000
Cm,h(USD/h)	2.429	1.214	0.06299	19.28	0.000
Cd(USD/p)	-2.946	-1.473	0.06299	-23.39	0.000
Do(mm)*Cm,h(USD/h)	1.606	0.803	0.06299	12.75	0.000
Do(mm)*Cd(USD/p)	-1.899	-0.949	0.06299	-15.07	0.000
Cm,h(USD/h)*Cd(USD/p)	1.069	0.534	0.06299	8.48	0.000

S = 0.503925 PRESS = 18.2481
R-Sq = 99.80% R-Sq(pred) = 99.74% R-Sq(adj) = 99.77%

Figure 6. Estimated Effects and Coefficients for D_{op}

5. Conclusions

A study on optimal determination of exchanged grinding wheel diameter when internally grinding alloy tool steel 9CrSi was carried out. In the study, the cost analysis for the grinding process was analyzed. In addition, the effects of grinding process parameters and cost components on the optimal exchanged diameter were investigated scientifically based on an "experiment" designed and performed by a computer program. From the results of the "experiment", a model for determining the optimal exchanged diameter was proposed. As the model is an explicit equation, the optimal exchanged diameter for internal grinding process can be calculated in a simple way.

Acknowledgements

The work described in this paper was supported by Thai Nguyen University of Technology for the project N_0 T2017-B05.

References

[1] W. Brian Rowe, *Principle of Modern Grinding Technology,* William Andrew, 2009.

[2] Malkin S., Guo C., *Grinding Technology: Theory and Applications of Machining with Abrasives,* Industrial Press, 2008.

[3] G. F. Li, L. S. Wang, L. B. Yang, *Multi-parameter optimization and control of the cylindrical grinding process,*Journal of Materials Processing Technology, Volume 129, Issues 1-3, 11 October 2002, Pages 232-236.

[4] R. Gupta, K. S. Shishodia, G. S. Sekhon, *Optimization of grinding process parameters using enumeration method,* Journal of Materials Processing Technology, Volume 112, Issue 1, 3 May 2001, P. 63-67.

[5] Y. C. Fu, H. J. Xu, J. H. Xu, *Optimization design of grinding wheel topography for high efficiency grinding,* Journal of Materials Processing Technology, Volume 129, Issues 1-3, 11 October 2002, Pages 118-122.

[6] Vu Ngoc Pi, Phan Quang The, Vu Hong Khiem, Nguyen Ngoc Huong, *Cost optimization of external cylindrical grinding,* Applied Mechanics and Materials, Vols. 312 (2013), pp. 982-989.

[7] X.M. Wen, A.A.O. Tay, A.Y.C. Nee,*Micro-computer-based optimization of the surface grinding process,* Journal of Materials Processing Technology, Volume 29, Issues 1-3, January 1992, P. 75-90

[8] G. Warnecke, C. Barth, *Optimization of the Dynamic Behavior of Grinding Wheels for Grinding of Hard and Brittle Materials Using the Finite Element Method,* CIRP Annals - Manufacturing Technology, Volume 48, Issue 1, 1999, Pages 261-264.

[9] Vu Ngoc Pi, Luu Anh Tung, Le Xuan Hung, Nguyen Van Ngoc, *Experimental Determination of Optimum Exchanged Diameter in Surface Grinding Process,* Journal of Environmental Science and Engineering A (2017), pp. 85-89.

[10] I. Inasaki, *Monitoring and Optimization of Internal Grinding Process,* CIRP Annals - Manufacturing Technology, Volume 40, Issue 1, 1991, Pages 359-362.

[11] G. Xiao, S. Malkin, *On-Line Optimization for Internal Plunge Grinding*, CIRP Annals – Manufacturing Technology, Volume 45, Issue 1, 1996, Pages 287-292

[12] H.K. Tönshoff, A. Walter, *Self-tuning fuzzy-controller for process control in internal grinding,*Fuzzy Sets and Systems, Volume 63, Issue 3, 10 May 1994, Pages 359-373.

[13] Vu Ngoc Pi, Le Xuan Hung, Luu Anh Tung, Banh Tien Long, *Cost Optimization of Internal Grinding,* Journal of Material Science and Engineering B 6 (11-12) (2016), pp. 291-296.

[14] Banh Tien Long, Vu Ngoc Pi, Le Xuan Hung, Luu Anh Tung, *Building cutting regime formulas for internal grinding (In Vietnamese)*, Journal of Science and Technology, Thai Nguyen University, No. 9 (2016), Pages15-18.

[15] L.M. Kozuro, A.A. Panov, E.I. Remizovski, P.S. Tristosepdov, *Handbook of Grinding (in Russian),* Publish Housing of High-education, Minsk, 1981.

IOP Conf. Series: Materials Science and Engineering **417** (2018) 012027 doi:10.1088/1757-899X/417/1/012027

The Effect of Product Quality on the Pricing of New and Remanufactured Short Life-cycle Product

Shu-San Gan*, Didik Wahjudi, Yopi Yusuf Tanoto

Mechanical Engineering Department, Petra Christian University, Surabaya, Indonesia

gshusan@petra.ac.id

Abstract. Remanufacturing is one of the recovery process that has become significant among many attempts to mitigate the landfill exhaustion, especially from mountain of wastes that come from short life-cycle products disposal. However, remanufactured product are often perceived to have lower quality compared to the new one. There are misconception about remanufactured product and lack of knowledge about its characteristics. On the other hand, several studies show that price and product quality have positive relationship. This paper investigates the effect of product's perceived quality on the pricing decision, to maximize the profit of the retailer and the manufacturer. We develop pricing decision model for new and remanufactured short life-cycle product in a closed-loop supply chain consists of a manufacturer and a retailer, where the manufacturer is a Stackleberg leader. We find that lower product's perceived quality would decrease the retail and wholesale prices of new and remanufactured products, but does not affect the new product's sales volume significantly. Also, the speed of change of demand influences the optimum total profit.

1. Introduction

Due to the rapid development in technology and research innovation, products' life-cycle has become shorter, especially for technology-based product such as electronics products. The period between launching a product and the introduction of newer model, newer design, or addition of new features has become shorter, and this has convinced customers to buy new product even though the previous one is still perfectly functioning. Usually, in the introduction phase, the product's demand would increase significantly, but when newer product or model is introduced, it would decrease rapidly. The demand characteristics of short life-cycle product is totally different from durable product. Therefore it is important to develop demand function that could capture the dynamics.

Recently, there are numerous attempt to study closed-loop supply chain, which is a study that is not only considering forward chain but also the reverse chain. In the reverse chain, a used product is collected and sent for a recovery process and then put back to the market with higher value than its discarded stage. This approach could extend product's useful life and slower the disposal rate. There are several recovery processes i.e. repair, refurbishing, and remanufacturing [1]; and remanufacturing is considered to be the ultimate form of recycling [2]. Remanufacturing is a process of transforming used product into "like-new" condition, where the value added during manufacturing is recaptured [3]. Several studies show that one of the critical factors to ensure successful remanufacturing is durability of the product. A short life-cycle product is also suitable for remanufacturing, and doing so could be beneficial for the environment and yet maintaining profitability [4].

Pricing decision is one of the important tasks in attempting to gain economic benefit from remanufacturing practices. Atasu et al. [5] point out that remanufacturing could become an effective marketing strategy, where manufacturer perform price discrimination to protect its market share. Also, a proper pricing strategy developed based on the market composition could avoid cannibalization effect i.e. remanufactured product cannibalizes the sales of new product [6]. Similarly, Souza [7, 8] shows that adding remanufactured product to the market alongside with the new product could expand the market, but on the other hand it could result in cannibalization, hence it is very critical to correctly decide the prices. Pricing decision model for remanufactured product has been widely studied. Regardless of remanufacturing product's key feature which is a product as good as the new one, remanufactured product is often perceived to have lower quality compared to the new product [9, 10]. Several pricing models accommodate this situation by differentiating the price of new and remanufactured products [6, 11, 12, 13, 14]. Gaur et al. [14] identify main drivers for purchase intentions through grounded theory-based interview, and indicate that price, brand, product quality, and service quality are the underlying factors behind customers' decision to purchase remanufactured products over new ones. van Weelden et al. [15] conducting in-depth interviews with consumers of remanufactured and new mobile phones, identify that misconception of remanufactured products, lack of awareness, lack of availability, and lack of the thrill of newness as the barriers for remanufactured mobile phones to be considered in the consideration phase. Wahjudi et al. [16] shows that product knowledge and purchase attitude have positive correlation with purchase intention. In this paper we would investigate the effect of product's perceived quality on the pricing of new and remanufactured short life-cycle products.

2. Optimization Modelling

We consider a closed-loop supply chain that consists of two members of the supply chain, i.e. retailer and manufacturer, under the selling horizon. Initially, manufacturer performs the production of new products, and customers can buy them via a retailer. After a particular period of time, customers start to sell end-of-use products to the manufacturer. The used product is then remanufactured and sent to the retailer. Now, customer have the option to purchase a new or remanufactured product. Therefore, during the selling horizon, both retailer and manufacturer sells new and remanufactured products with different time boundaries as in [12].

The pricing decision model is developed under a Stackelberg power structure, where manufacturer acts as the leader. We use the whole selling periods as the planning horizon. The model is developed based on backward induction. Retailer finds her optimum retail prices dependent on the wholesale prices. The manufacturer is then decide the wholesale prices to maximize her profit, after knowing retailer's move. In this model, we consider a single item short life-cycle product, which is typically prone to obsolescence in function and/or desirability. This type of product would have brief introduction, growth, and maturity phases, before declining. Product demand is constructed to cover the time frames according to the given time boundaries, and to allow price-dependent relation. Remanufactured short life-cycle product is often considered to be inferior to its new counterpart and customers' willingness to pay is typically lower for remanufactured short life-cycle product [17]. Therefore, a remanufactured product usually is more sensitive to price changes, or is having higher price sensitivity index.

The governing demand functions, for new and remanufactured product, are formulated as in [12]:

$$d_n(t) = \begin{cases} d_{n1}(t) = \frac{U}{1+ke^{-\lambda Ut}};\ 0 \le t \le \mu \text{ where } k = \frac{U}{d_0 - 1} \\ d_{n2}(t) = \frac{U}{\lambda U(t-\mu)+\delta};\ \mu \le t \le t_3; \text{where } \delta = 1 + ke^{-\lambda U\mu} \end{cases} \quad (1)$$

$$d_r(t) = \begin{cases} d_{r1}(t) = \frac{V}{1+he^{-\eta V(t-t_1)}};\ t_1 \le t \le t_3\ ; \text{where } h = \frac{V}{d_{ro}-1} \\ d_{r2}(t) = \frac{V}{\eta V(t-t_3)+\varepsilon}\ ;\ t_3 \le t \le T\ ; \text{where } \varepsilon = 1 + he^{-\eta V(t_3-t_1)} \end{cases} \quad (2)$$

$d_n(t)$ is the demand pattern for new product and $d_r(t)$ is for remanufactured products. Total demand volumes during the selling horizon can be determined by integrating the demand functions with respect to time.

$$D_n = \int_0^{\mu} \frac{U}{1+ke^{-\lambda Ut}} dt + \int_{\mu}^{t_3} \frac{U}{\lambda U(t-\mu)+\delta} dt = \frac{1}{\lambda}\ln\left(\frac{\delta}{(1+k)e^{-\lambda U\mu}}\right) + \frac{1}{\lambda}\ln\left(\frac{\lambda U(t_3-\mu)+\delta}{\delta}\right) \quad (3)$$

$$D_r = \int_{t_1}^{t_3} \frac{V}{1+he^{-\eta V(t-t_1)}} dt + \int_{t_3}^{T} \frac{V}{\eta V(t-t_3)+\varepsilon} dt = \frac{1}{\eta}\ln\left(\frac{\varepsilon}{(1+h)e^{-\eta V(t_3-t_1)}}\right) + \frac{1}{\eta}\ln\left(\frac{\eta V(T-t_3)+\varepsilon}{\varepsilon}\right) \quad (4)$$

We introduce a product quality factor (g), and "remanufactured product's quality coefficient" (b) to the pricing model to study its impact to the overall pricing decision. Product quality factor is considered inversely proportional to the customers' perceived quality. It means, the lower the product's perceived quality, the higher the factor value. We use product quality factor as a price coefficient in the demand function, with P_m as the maximum price, therefore

$$\text{Demand of new product} = D_n\left(1 - g\frac{P_n}{P_m}\right) \quad (5)$$

$$\text{Demand of remanufactured product} = D_r\left(1 - bg\frac{P_r}{P_n}\right) \quad (6)$$

Under a symmetrical information setting, both retailer and manufacturer share the demand information. Retailer finds her optimum retail prices (P_n for new product, and P_r for remanufactured product) that maximize her profit. Based on retailer's optimum prices, manufacturer is then find the wholesale prices (P_{nw}, P_{rw}) that maximize manufacturer's profit. Considering that product has short life-cycle, we assume remanufacturing process is only applied to used product that is originated from new product, hence remanufacturing is only applied one time during the whole product's life.

Since this study is focusing on the effect of product's perceived quality to the pricing decision, we do not make an attempt to show detailed derivation of production and operational costs, and instead treat those costs as given parameters, which consist of unit manufacturing cost for new product (c_n), and unit manufacturing cost (c_r). Unit manufacturing cost includes raw material, manufacturing cost, etc. Unit remanufacturing cost includes cores' acquisition, collecting cost, remanufacturing cost, etc.

2.1. Retailer's Optimization

Retailer's optimization model is constructed to find retail prices that maximize retailer's profit.

$$\max_{P_n,P_r} \Pi_R = D_n\left(1 - g\frac{P_n}{P_m}\right)\cdot(P_n - P_{nw}) + D_r\left(1 - bg\frac{P_r}{P_n}\right)\cdot(P_r - P_{rw}) \quad (7)$$

Π_R is concave with respect to retail prices, therefore optimum prices can be found by solving the first derivative conditions, represented by

$$\frac{2D_n g}{P_m} P_n^{*3} - \left(\frac{D_n}{P_m}(P_m + gP_{nw}) + \frac{D_r}{4bg}\right) P_n^{*2} + \frac{D_r}{4} bgP_{rw}^{\;2} = 0 \quad (8)$$

$$P_r^* = \frac{1}{2}\left(\frac{P_n^*}{bg} + P_{rw}\right) \quad (9)$$

2.2. Manufacturer's Optimization

$$\max_{P_{nw},P_{rw}} \Pi_M = D_n\left(1 - g\frac{P_n^*}{P_m}\right)\cdot(P_{nw} - c_n) + D_r\left(1 - bg\frac{P_r^*}{P_n^*}\right)\cdot(P_{rw} - c_r) \quad (10)$$

subject to (8), (9), and

$$0 \le P_{nw} \le P_n \le P_m\,,\; 0 \le P_{rw} \le P_r \le P_n,$$
$$P_{rw} \le P_{nw},\, P_{rw} \ge c_r,\, P_{nw} \ge c_n$$

Due to the optimization problem's complexity, we use a computational approach and use Matlab to find the optimum prices.

3. Results and Discussions

We use a numerical example to show the optimization results as well as the effect of product quality. The parameters are adopted from [12]. Demand capacity parameters for new product are $U =$ 1,000, $D_0 = 90$, and demand capacity parameters for remanufactured product are $V = 500, D_{r0} = 50$. The speed of change in demands are $\lambda = 0.05, \eta = 0.05$. Selling horizon is divided into four time frames where t_1=1, μ=2, t_3=3, and T=4. The unit manufacturing cost for new product c_n=2,500, unit remanufacturing cost c_r=1,800, maximum price is P_m=12,000. The product quality factor and remanufactured product's quality coefficient will be varied to study their effects on the pricing decision. The decision variables are P_n, P_r and P_{nw}, P_{rw} which represent the retail price of new product, retail price of remanufactured product, wholesale price of new product, and wholesale price of remanufactured product.

3.1. The effect of product's perceived quality

In order to observe the effect of product's perceived quality to the price, product quality factor (g) and remanufactured product's quality coefficient (b) values are varied as follows: g=[1,1.1,1.2,1.3], b=[1,1.1,1.2]. The optimization results can be seen in Table 1.

Table 1. Numerical example's result – product's perceived quality

b	g	Pn*	Pr*	Pnw*	Prw*	Profit M	Profit R	Total Profit	sales 1	sales 2
1	1	10,010.05	8,283.40	7,160.40	6,556.75	2,405,395.59	1,263,036.99	3,668,432.58	336.71	175.79
1	1.1	9,112.54	6,938.78	6,638.91	5,593.44	2,011,823.96	1,049,809.35	3,061,633.31	334.39	165.50
1	1.2	8,369.89	5,917.85	6,201.97	4,860.80	1,698,048.66	880,820.62	2,578,869.28	330.99	154.45
1	1.3	7,745.09	5,124.07	5,830.49	4,290.38	1,443,561.69	744,586.82	2,188,148.51	326.80	142.61
1.1	1	9,966.70	7,553.90	7,174.29	6,047.16	2,327,957.35	1,216,073.10	3,544,030.44	344.05	169.47
1.1	1.1	9,076.21	6,336.90	6,650.05	5,172.81	1,949,223.29	1,011,791.62	2,961,014.92	341.15	158.16
1.1	1.2	8,338.80	5,412.45	6,211.00	4,507.63	1,646,954.87	849,788.35	2,496,743.22	337.30	145.97
1.1	1.3	7,718.00	4,693.40	5,837.84	3,989.59	1,401,688.11	719,176.58	2,120,864.69	332.76	132.89
1.2	1	9,930.44	6,949.53	7,185.58	5,623.69	2,265,131.33	1,177,681.12	3,442,812.45	350.18	163.28
1.2	1.1	9,045.73	5,838.00	6,659.06	4,823.18	1,898,708.75	980,903.15	2,879,611.90	346.82	150.92
1.2	1.2	8,312.63	4,993.34	6,218.24	4,214.03	1,606,056.41	824,792.39	2,430,848.80	342.62	137.58
1.2	1.3	7,695.10	4,336.13	5,843.66	3,739.52	1,368,556.27	698,954.22	2,067,510.49	337.80	123.26

The results show that the higher the product quality factor (which means the lower product's perceived quality in the market), the lower profit of both retailer and manufacturer. The effect is significantly recognized in the lower prices, both new and remanufactured, wholesale and retail prices. Interestingly, the new product's sales quantity is least affected. It is also observed that despite the shared prices decrease, the percentage of relative decrease is bigger for remanufactured products prices compared to new products'. The average relative decreases are 8.17%, 6.64%, 14.66%, 12.94% for new retail price, new wholesale price, remanufactured retail price, and remanufactured wholesale price. On the other hand, the relative decrease of profit is almost the same for manufacturer and retailer, i.e. 15.55% and 16.06% respectively.

The effect of remanufactured product's quality coefficient can be seen in Table 1 as well. The remanufactured prices are the ones most affected, with relative decreases 8.40% and 7.39% for retail and wholesale prices, on average. As for new product retail price, the relative decrease is relatively small, which is 0.40%. However, new product wholesale price is unexpectedly increasing even though the relative increase is very small i.e. 0.18%. Also, the sales of new product is increasing with average relative increase as much as 1.98%.

Table 2. Sensitivity analysis for demand's speed of change

lmb	nu	g	Pn*	Pr*	Pnw*	Prw*	Profit M	Profit R	Total Profit	sales 1	sales 2
0.01	0.01	1	9,947.48	7,544.21	7,178.86	6,045.26	2,267,507.28	1,182,401.60	3,449,908.87	341.92	157.29
0.01	0.01	1.1	9,061.04	6,330.04	6,653.34	5,171.62	1,901,415.72	985,390.07	2,886,805.79	338.64	146.79
0.01	0.01	1.2	8,326.74	5,407.52	6,213.35	4,506.90	1,608,782.46	828,910.09	2,437,692.55	334.48	135.48
0.01	0.01	1.3	7,708.40	4,689.83	5,839.50	3,989.16	1,370,988.95	702,562.22	2,073,551.17	329.68	123.34
0.05	0.05	1	9,966.70	7,553.90	7,174.29	6,047.16	2,327,957.35	1,216,073.10	3,544,030.44	344.05	169.47
0.05	0.05	1.1	9,076.21	6,336.90	6,650.05	5,172.81	1,949,223.29	1,011,791.62	2,961,014.92	341.15	158.16
0.05	0.05	1.2	8,338.80	5,412.45	6,211.00	4,507.63	1,646,954.87	849,788.35	2,496,743.22	337.30	145.97
0.05	0.05	1.3	7,718.00	4,693.40	5,837.84	3,989.59	1,401,688.11	719,176.58	2,120,864.69	332.76	132.89
0.1	0.1	1	9,967.19	7,554.14	7,174.18	6,047.21	2,318,950.58	1,211,421.47	3,530,372.04	342.54	169.02
0.1	0.1	1.1	9,076.59	6,337.08	6,649.97	5,172.84	1,941,609.01	1,007,880.25	2,949,489.26	339.67	157.73
0.1	0.1	1.2	8,339.11	5,412.58	6,210.94	4,507.64	1,640,463.78	846,469.56	2,486,933.34	335.84	145.58
0.1	0.1	1.3	7,718.24	4,693.49	5,837.80	3,989.60	1,396,117.35	716,340.54	2,112,457.89	331.33	132.54
0.2	0.2	1	9,966.87	7,553.98	7,174.25	6,047.17	2,309,829.83	1,206,621.89	3,516,451.72	341.31	168.22
0.2	0.2	1.1	9,076.34	6,336.96	6,650.02	5,172.82	1,934,019.82	1,003,914.03	2,937,933.85	338.44	156.99
0.2	0.2	1.2	8,338.91	5,412.49	6,210.98	4,507.63	1,634,089.16	843,160.54	2,477,249.70	334.62	144.89
0.2	0.2	1.3	7,550.52	4,508.70	5,736.54	3,856.41	1,326,646.52	679,581.67	2,006,228.19	328.66	127.85

3.2. Sensitivity analysis for demand's speed of change

In order to study the effect of demand's speed of change, λ and η, we varied $\lambda, \eta = [0.01, 0.05, 0.1, 0.2]$. Table 2 shows the optimization results. The highest profit is attained at 0.05. This result shows that increasing speed of change in demand does not always increasing the profit. The lower the product's perceived quality, which means the higher product quality factor, the sales quantity is decreasing as well as the profits. While the first three values (0.05, 0.01, 0.1) show decreasing rate along with lower product's perceived quality, the last value, 0.2, shows an exception. The percentage of decrease is 8.93% when product quality factor is increasing from 1 to 1.2, then 8.12% from 1.1 to 1.2, and 9.45% from 1.2 to 1.3. It appears that for higher speed of change the effect of product quality becomes more significant.

4. Conclusion

This paper studies the effect of product quality on the pricing decision of new and remanufactured short life-cycle products in a closed-loop supply chain. We develop a pricing decision model involving product quality factor (which is inversely proportional to the product's perceived quality), and the optimization is conducted under Stackelberg pricing game with manufacturer as the leader. The results show that lower product's perceived quality could increase price sensitivity and hence decrease the sales quantity and the profits. The speed of change in demand influence pricing decision, when it reach a higher value, the effect of product quantity becomes more significant.

References

[1] Thierry M, Salomon M, Nunen J A E E van and Wassenhove L N Van 1995 Strategic issues in product recovery management *Calif. Manage. Rev.* 37 114–35

[2] Steinhilper R 1998 *Remanufacturing - The Ultimate Form of Recycling* (Fraunhofer IRB Verlag)

[3] Lund R T and Hauser W M 2010 Remanufacturing – An American Perspective *5th International Conference on Responsive Manufacturing - Green Manufacturing (ICRM 2010), Ningbo, China, 11-13 January 2010.* (Institution of Engineering and Technology) pp 1–6

[4] Gan S-S, Pujawan I N and Suparno 2014 Remanufacturing of Short Life-cycle Products *Oper. Supply Chain Manag.* 7 13–22

[5] Atasu A, Sarvary M and Wassenhove L N Van 2008 Remanufacturing as a marketing strategy *Manage. Sci.* 54 1731–46

[6] Atasu A, Guide V D R J and Wassenhove L N Van 2010 So what if remanufacturing cannibalizes my new product sales? *Calif. Manage. Rev.* 52 56–76

[7] Souza G C 2013 Closed-Loop Supply Chains : A Critical Review and Future Research *Decis. Sci.* 44 7–38

[8] Souza G C 2009 Remanufacturing in Closed-Loop Supply Chains *Prod. Invent. Manag. J.* 45 56–66

[9] Agrawal V V., Atasu A and Van Ittersum K 2015 *Remanufacturing, third-party competition, and consumers' perceived value of new products*

[10] Wu C 2013 OEM product design in a price competition with remanufactured product *Omega* 41 287–98

[11] Wu C 2012 Product-design and pricing strategies with remanufacturing *Eur. J. Oper. Res.* 222 204–15

[12] Gan S-S, Pujawan I N, Suparno and Widodo B 2015 Pricing decision model for new and remanufactured short-life cycle products with time-dependent demand *Oper. Res. Perspect.* 2 1–12

[13] Gan S-S, Pujawan I N, Suparno and Widodo B 2017 Pricing decision for new and remanufactured product in a closed-loop supply chain with separate sales-channel *Int. J. Prod. Econ.* 190 120–32

[14] Jighyasu Gaur, Mehdi Amini, Pratyush Banerjee, Ritu Gupta 2015 Drivers of consumer purchase intentions for remanufactured products *Qual. Mark. Res. An Int. J.* 18 30–47

[15] Van Weelden E, Mugge R and Bakker C 2016 Paving the way towards circular consumption: Exploring consumer acceptance of refurbished mobile phones in the Dutch market *J. Clean. Prod.* 113 743–54

[16] Wahjudi D, Gan S-S, Anggono J and Tanoto Y Y 2017 Factors Affecting Purchase Intention of Remanufactured Short Life-Cycle Products *International Conference on Management and Enterpreneurship*

[17] Guide V D R J and Li J 2010 The Potential for Cannibalization of New Products Sales by Remanufactured Products *Decis. Sci.* 41 547–72

IOP Conf. Series: Materials Science and Engineering **417** (2018) 012028 doi:10.1088/1757-899X/417/1/012028

Test Technology in General Assembly of Large Satellites

Rongxin Yan*, Yong Wang, Wei Sun, Mian Chen, Zaihua Yang

Beijing Institute of Spacecraft Environment Engineering, No.104, Youyi Road,Haidian, Beijing,China,100094

*corresponding author's e-mail: pigsheepdog@126.com

Abstract. The special test technology of the assembly, integration and test (AIT) of large satellites was studied, which accurately determine the leakage, alignment and mass property of large satellites, and ensure the reliable operation of large satellites in orbit. The local and total leakage testing technique was developed which based on quadrupole mass spectrometer and helium mass spectrometer. To meet large satellite quality and high precision, the satellite mass property testing equipment of large load bearing based on three coordinate conversion machine has been set up. A laser tracker uniting theodolite measure system with high positional precision and high angle measurement accuracy has been developed. Several large satellites have been tested in the process of AIT with these testing equipment. The quality of the large satellite has been effectively guaranteed.

1. Introduction

In the next few years, China will develop many large satellites, and these large satellites in the AIT. In the process, each satellite must be at least 4 times leakage testing, one time mass property measurement, 4 times alignment testing. This requires that the satellite chemical propulsion system leakage should be tested by helium mass spectrometer used helium as search gas, and then the satellite electric propulsion system leakage should be able to conduct direct leakage detection using work gas Xenon. In mass property measurement, satellite quality reached more than 5000 kg. In terms of installation alignment, the envelope of satellite structure size exceeds 10 meters, satellite angle measurement accuracy up angle of second order of magnitude. Therefore, it is necessary to develop the special test equipment and technology of the AIT of large satellites, to accurately determine the leakage, alignment and mass property of large satellites, and ensure the reliable operation of large satellites in orbit.

2. Large satellite leakage testing

2.1. Large satellite local leakage testing

Chemical propulsion system of satellite leakage testing mainly adopts the helium mass spectrometer atmosphere accumulation method [1]. Satellite structure leakage are tested by the method of pressure change leak detection [2,3]. Mass spectrometer system leak detection are used in pipelines [4]. The total leakage detection technology in vacuum container is used to a higher sensitivity of spacecraft leakage [5,6]. Helium mass spectrometer is also used to test the total leak rate of the NASA space station and the Russian spacecraft leak detection test.

For large satellite electric propulsion system requirements of a local leakage detection, the test system was developed based on quadrupole mass spectrometer and helium mass spectrometer, as shown in figure 1. The test system is mainly composed of quadrupole mass spectrometer with a sniffer,

IOP Conf. Series: Materials Science and Engineering **417** (2018) 012028 doi:10.1088/1757-899X/417/1/012028

xenon positive pressure standard leak. A xenon calibration device with known leakage rate is developed to calibrate test system from 10^{-6} Pa·m^3 /s to 10^{-8} Pa·m^3/s, and synthetic standard uncertainty is better than 5%. With the development of the test system, on the basis of linear comparison test, through the sniffer suck the local position of large satellite which is most likely to leak, quadrupole mass spectrometer or helium mass spectrometer leak detector can give leakage signal I_2.With the same testing conditions, background signal I_0 and xenon positive pressure standard leak signal I_1 are also tested, therefore, the large satellite local leakage is gotten as the Eq. 1 shown.

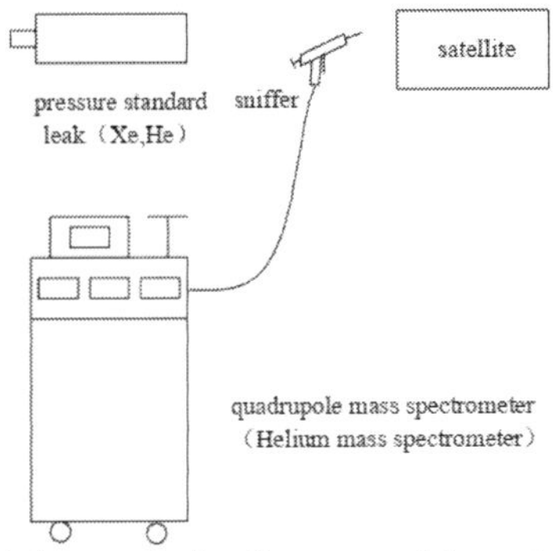

Figure. 1 the diagram of the satellite local detector leakage test system

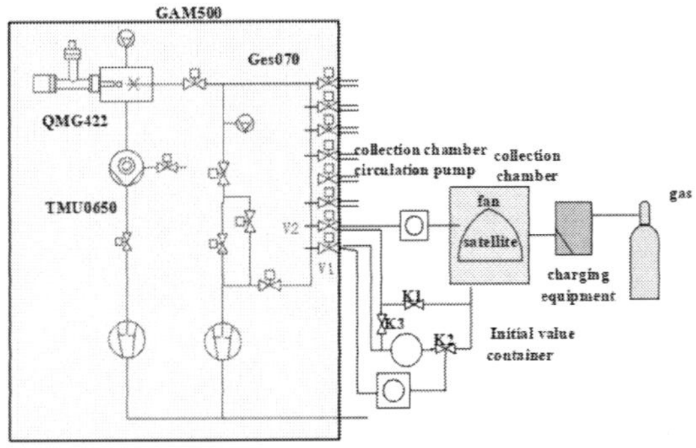

Figure.2 the principle of satellite multi-system leak by mass spectrometer

$$Q = \frac{I_2 - I_0}{I_1 - I_0} Q_0 \tag{1}$$

2.2. Satellite total leakage testing with multi-system leak detector by mass spectrometer.
Nowadays, helium mass spectrometer technology is widely used in the leak detection of spacecraft. This technology can effectively solve leak problem of single sealing system. As the development of spacecraft, multi-system leak detection is needed. Based on the old detector, ESA developed a piezoelectricity crystal pulse sampling mass spectrometer leak detector [7], which contrast the ion flow of search gas to the ion flow of compared gas. This method reduces effect of instability of vacuum system and mass spectrometer, the sensibility and reliability are increased. On the same sensibility level, the cost decreases clearly. Sodal [8] researched on the mass spectrometer using a pulsed gas sampling system, and acquired satisfying result. After many experiments of detectable gas analysis machine, and based on the GAM500 on-line mass produced by Germany IPI, we developed alternate circle sampling system and get over the instability of long-time measure [9]. The multi-system leak detector by mass spectrometer has been developed and used in the large satellite successfully.

*2.2.1. Composition of total leakage testing equipment.*The equipment is composed by on-line mass spectrometer named as GAM500, gas cycle sampling system, collection chamber, charging equipment and so on, the principle chart is shown on figure 2. GAM500 on-line mass spectrometer is combined with QMA422 mass spectrometer, Ges070 sampling system, TMU0650 molecular pump vacuum system and 8 sampling valve, which can acquire more sampling points.

Gas cycle sampling system is combined with two route: one route includes sampling valve, collection chamber cycle pump, stop valve, the other route includes initial value container, sampling valve, stop valve and initial value container cycle pump. A fan cycle system is installed in the collection chamber which blows chamber air mixture equally in 5 minutes. The airproof capability of collection chamber is: after charging 2kPa pure Nitrogen, the pressure reduces 150 Pa in 24 hours.

Charging equipment is combined with compressor, gas cylinder, standard container, mechanical pump, control meter, control valve and so on. The charging equipment can pump the satellite system, charge and recycle the search gas.

2.2.2. Test principle. Charge specific pressure search gas (there are some kinds of inert gases such as He, Ne, Kr, ,Xe,CF_4) into different test systems in satellite, seal the inter-face and push into the collection chamber, operate the fan in the collection chamber, reduce the background search gas in the chamber. Close the chamber door and seal the chamber. Startup two cycle pump, open K3, close K1 and K2 valves to make the initial value container air the same as chamber air. Close K3, open K1 and K2 valves, as the GAM500 operates well, cycle test collection chamber and the initial value container have the same detectable gas ion flow. Stay the satellite in the chamber for t seconds to make the leak search gas accumulate and the density increases. Open the fan cycle system and initial value container cycle pump. The gas in the initial value container and the gas in the collection chamber flow into the mass spectrometer alternately, and we achieve corresponding ion flow u_{i10} and u_{i1}. After testing several steady sampling cycle, inject standard gas W_i ($P_{ib} \cdot V_b$) into collection chamber, and then open the fan cycle system to make the standard gas, search gas and air mixed uniformly. The ion flow in initial value container gas and chamber gas are u_{i20} and u_{i2} respectively. Because the test time is short and we test alternately, the stability of pumping speed, and the stability of QMS can be ensured, and the test ion flow keep stable and can be repeated well. The ratio between standard gas W_i ($P_{ib} \cdot V_b$) and chamber atmosphere gas is about 1%. After injecting standard gas, the collection chamber pressure changes a little. The total leakage of different satellite system can be acquired as following equation:

$$Q_i = \frac{W_i(u_{i1} - u_{i10})}{t[(u_{i2} - u_{i20}) - (u_{i1} - u_{i10})]} \tag{2}$$

In the equation, W_i ($P_{ib} \cdot V_b$) is the standard gas (Pa·m^3), $Wi = P_{ib} \cdot V_b$, V_b is the volume of standard gas flow container, P_{ib} is the pressure of search gas of injecting collection chamber.

In the atmosphere, the minimal He concentration can be test is 13×10^{-9}, CF_4 concentration is 10×10^{-9}, Ne concentration is 3×10^{-6}, SF_6 concentration is 0.3×10^{-9}, Kr and Xe concentration is 10×10^{-9}. For the collection chamber with 100 m^3 volume, in 24 hours, the minimal He leak rate can be test is 1.5×10^{-6} Pam^3/s, CF_4 leak rate is 1.1×10^{-6} Pam^3/s, Ne leak rate is 3.5×10^{-5} Pam^3/s, SF_6 leak rate is 3×10^{-8} Pam^3/s, Kr and Xe leak rate are 1.1×10^{-6} Pam^3/s.

3. Mass property testing

Mass Properties of an object, including mass, center of gravity, moment of inertia, and product of inertia are determined by satellite material, shape, and structure. They are important parameters that reflect the satellite's intrinsic mechanical characters. Mass and center of gravity are measured by three point support equipment. Moment of inertia, and product of inertia are tested by the torsional pendulum technology [10]. But those measurement can only carry out single axis direction, the horizontal axis and vertical axis direction need different adapter tools, lifting and rotating satellite repeatedly, the security hidden danger of satellite, and the test efficiency is very low.

Since 2009, we have successfully developed three coordinate conversion instrument with capacity of 3 tons. According to the need of large satellite quality, high precision, as shown in figure 3, we developed satellite mass property equipment based on three coordinate conversion machine of large load bearing. The equipment is divided into two parts. One part is three coordinate conversion and other part is test platform. The test platform is three point support equipment or torsional pendulum equipment. The mass, center of gravity, and moment of inertia for larger satellite are tested just one lifting. The longitudinal center measurement principle of its satellite is shown in Eq.3:

$$\mathrm{Xc} = {\left(\mathrm{Y}_c' - \mathrm{Yc} \cdot \cos\alpha\right)}\Big/{\sin\alpha} - \mathrm{R} - \mathrm{H} \tag{3}$$

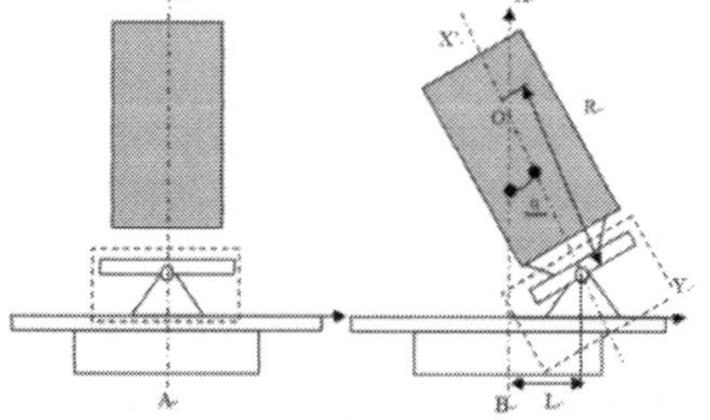

Figure. 3 The large satellite mass property equipment

Using three coordinate conversion equipment, in vertical A state, α = 0, get the direction cosine (0, 1), I_X can be measured. Using three coordinate conversion , change satellite install direction α, , and test out the several groups of direction cosine respectively corresponding to the moment of inertia of I_1, I_2, I_3, I_4, can use the Eq.4, test out along the different moment of inertia of axis of rotation of the satellite.

$$I_y = \frac{1}{2\sin^2\alpha}(I_1 + I_2 - 2I_x) + I_x, \qquad I_z = \frac{1}{2\sin^2\alpha}(I_3 + I_4 - 2I_x) + I_x \qquad I_{xy} = \frac{I_1 - I_2}{2\sin 2\alpha} \qquad I_{xz} = \frac{I_3 - I_4}{2\sin 2\alpha} \tag{4}$$

Large satellite mass property test system maximum load reach 5000 kg, 50000 Nm moment of inertia. One lifting can detect different axis center of gravity, moment of inertia, synthetic standard uncertainty is better than 0.1%.

4. Alignment of satellite assemble

Satellite installation alignment is for relative to the satellite coordinate system to the installed instrument precision position and normal vector measurement products generally used the principle of theodolite intersection test [11], however, the space dimension of the large satellite has reached dozens of meters, requires both direction test to high precision, also requires a positional accuracy is high. With the laser tracker measurement location high precision, widely used in aerospace assembly manufacture, we developed a laser tracker uniting theodolite measurement system [12], Which effective use the characteristics of the theodolite with high Angle measurement and laser tracker the advantages of high precision of observation points. A set of conversion standard benchmark was made. The measurement system can simultaneously test satellite by a theodolite and the laser tracker. Fig 4 is theodolite measurement system. The position can be tested according to Eq.5:

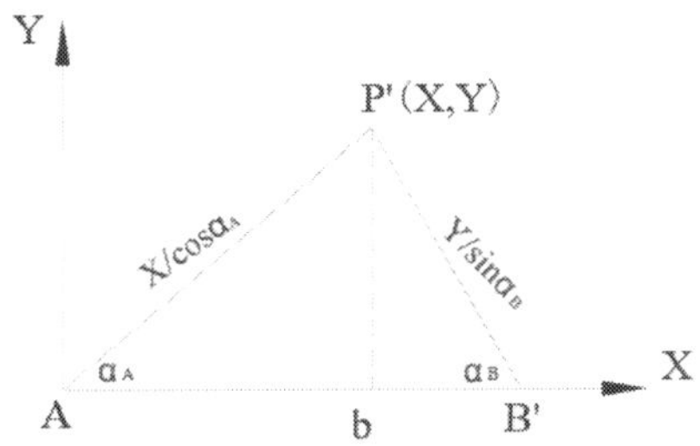

Figure. 4 Principle of theodolite measurement system

$$X = \frac{\sin\alpha_B \cos\alpha_A}{\sin(\alpha_B + \alpha_A)} b, \qquad Y = \frac{\sin\alpha_A \sin\alpha_B}{\sin(\alpha_B + \alpha_A)} b \qquad Z = \frac{1}{2}\left[\frac{\sin\alpha_A \cot\gamma_{BP} + \sin\alpha_B \cot\gamma_{AP}}{\sin(\alpha_B + \alpha_A)} b + h\right] \tag{5}$$

5. Summary

The large satellite of leakage, mass property, alignment testing and testing calibration problems have been effectively solved. The testing systems have been set up and the testing reliability is effectively improved. Several large satellites have been tested in the process of AIT. The quality of the larger satellite has been effectively guaranteed.

Large satellite calibration device of leakage test system has less than the 5% synthetic standard uncertainty, range of the total leakage rate from 1×10^{-3} Pam^3/s to 1×10^{-5} Pam^3/s. The local leakage testing range is from1×10^{-3} Pam^3 / s to 1×10^{-8} Pam^3/s and synthetic standard uncertainty of calibration is better than 20%. The test system of large satellite mass property can be used to test more than 5000kg satellite. The moment of inertia can reach 50000Nm, the precision of the testing system is t better than 0.1%. Large satellite alignment testing system has been developed. The calibration device synthesis standard uncertainty is better than that of 1 um, angle synthesis standard uncertainty is better than that of 3 ". The position synthesis standard uncertainty of alignment testing system is about 10 um, and angle testing synthetic standard uncertainty is better than that of 12 ".

Acknowledgments
This work were supported by the National Science Foundation of China (U1537109) and technological base (JSJL2015203B019)

References

[1] Wangyong, SunLichen, Yan Rongxin,Sun,lizhi. A new method for testing linear performance of helium mass spectrometer leak detactor[J].Chinese Space Science and Technology,2011,31(2)pp39-42

[2] Yan Rongxin, Liu ping, FengQi, Hong Xiaopeng,ShiLixia. Factor of leak detection for pressure decay[J]. Chinese Space Science and Technology , 2006,26(4)pp37-41

[3] Sunwei et al. Study of the space station on-oribit leak detection based on the differential pressure gas sensor[C]. iThings 2013.8pp 1718-1721

[4] Yan Rongxin, Liu ping, FengQi, Hong Xiaopeng. The study of leak detection for spacecraft with quadrupole mass spectrometer[J]. Chinese Space Science and Technology 2001,21(4)p44

[5] Yan Rongxin, Hong Xiaopeng, Zhong Liang,Shilia. The discuss of the reliability and sensitivity for spacecraft helium mass spectrometer leak testing in vacuum chamber[J]. Vacuum(Chinese) 2011,48(5)

[6] Donghui Meng, Lichen Sun, Rongxin Yan, Rongpin Shao, Xinfa Yu. Effects of cryopump on vacuum helium leak detection system[J]. Vacuum, 2017 ,143,316-319

[7] Sanger S. Leak detector for spacecraft testing with high accuracy. AIAA 81 p2463

[8] Sodal I E, Hanna D A. Improved mass spectrometer measurements using a pulsed gas sampling system[J]. J. Vac. Sci. Technol. A 1997, 15 (1) p196

[9]Yan Rongxin. Research on multi-detector gas analysis machine[J]. Chinese Space Science and Technology. 1998, 18(6)p 49

[10]Gao Shenbing, The technique of satellite manufacture[M], China Astronautic Publishing House 1998,pp345-360

[11]Liu Jianxin. The influencing factors to the equipment installation precision and the technique for optimizing the spacecraft structure[J], Spacecraft Environment Engineering. 2015, 32 (6)pp 337-343

[12]Ren Chunzhen,Jia Shijin,Liu Haomiao, Sun Gang, Ji Yu. Laser tracker used in multi-radius test of microwave radar for rendezvous and docking, Spacecraft Environment Engineering[J].2013,30(6) pp550-555

IOP Conf. Series: Materials Science and Engineering **417** (2018) 012029 doi:10.1088/1757-899X/417/1/012029

A Liquid Spring with High-static-low-dynamic Stiffness

Zhanyong Li and Qian Chen

State Key Laboratory of Mechanics and Control of Mechanical Structures, Nanjing University of Aeronautics and Astronautics, Nanjing 210016, China

zhy.li@nuaa.edu.cn

Abstract. The conventional multi-rate spring with high-static-low-dynamic (HSLD) stiffness may not apply to massive payload low frequency vibration isolation due to the poor lateral stability of the spring with pre-compression. The new type of multi-rate spring refers to as a kind of liquid spring consists of a hydraulically driven spring and several pre-compressed slave springs. The payload is supported by the piston of the liquid spring, and the excitation force from the payload is transferred to the pistons of slave springs by hydraulic pressure. The liquid spring shows an HSLD stiffness property. It is found that the proper design of pistons' cross-section area and the number of slave springs of the liquid spring can achieve an ideal piecewise linear stiffness property. Moreover, the slave spring owns better lateral stability under the condition that both the conventional multi-rate spring and the liquid spring have the same stiffness property. The lateral stability of a pre-compressed linear spring and the stiffness model of the liquid spring have been investigated.

1. Introduction

An isolator with HSLD stiffness may overcome the dilemma that low natural frequency and small static displacement cannot be achieved at the same time for a linear spring. Many passive nonlinear isolators with HSLD stiffness have been developed recently. Carrella[1,2] applied oblique springs and attracting magnets as two types of negative stiffness elements, and parallel combined them with vertical spring to achieve two typical isolators with HSLD stiffness. Huang[3,4] utilized Euler-bulked beams as the negative stiffness element. Zhou[5,6] invented cam–roller–spring mechanisms as the negative element. Those additional mechanisms providing negative stiffness are proved benefit from the vibration isolation performance, while those complex mechanisms may need more spaces. Furthermore, the nonlinear stiffness properties of those nonlinear isolators may produce harmful nonlinear dynamic phenomena even the payload under a small excitation force.

The conventional multi-rate spring can be designed to possess a piecewise linear HSLD stiffness. As a result, the vibration isolation system applying the conventional multi-rate spring can be regarded as a linear system when the excitation force has not reached the critical value. However, the conventional multi-rate spring may not apply to massive payload low frequency vibration isolation due to the poor lateral stability of the spring with pre-compression.

The liquid spring mentioned in this paper is also a multi-rate spring characteristic of piecewise linear HSLD stiffness. And the liquid spring also utilizing pre-compressed spring, while the lateral stability of the pre-compressed linear spring can be designed to be better than the conventional counterpart. In other words, the liquid spring is potentially applying to massive payload low frequency vibration isolation. Moreover, the liquid spring can achieve rich structures by taking advantage of Pascal's principle and the fluid connectivity.

2. Multi-rate spring with HSLD stiffness

2.1. The conventional multi-rate spring

The schematic diagram of the conventional multi-rate spring[7] with HSLD stiffness is shown in Figure 1 (a). A linear spring with high stiffness K_0 is series connection with a pre-compressed linear spring with low stiffness K_1. When the external force F is smaller than the pre-compressed force F_1, the conventional multi-rate spring shows a high static stiffness. Otherwise, the conventional multi-rate spring presents a series stiffness of these two linear spring. Therefore, the conventional multi-rate spring shows a piecewise linear stiffness property. According to Figure 1 (a), the force-displacement relationship is given by

$$F = \begin{cases} K_0 \cdot x & F \le F_1 \\ \dfrac{K_0 \cdot K_1}{K_0 + K_1} \cdot x & F_1 < F \end{cases} \tag{1}$$

Define the working stiffness as

$$K_w = \frac{K_0 \cdot K_1}{K_0 + K_1} \tag{2}$$

Then the stiffness property of the conventional multi-rate spring is shaped like Figure 1 (b).

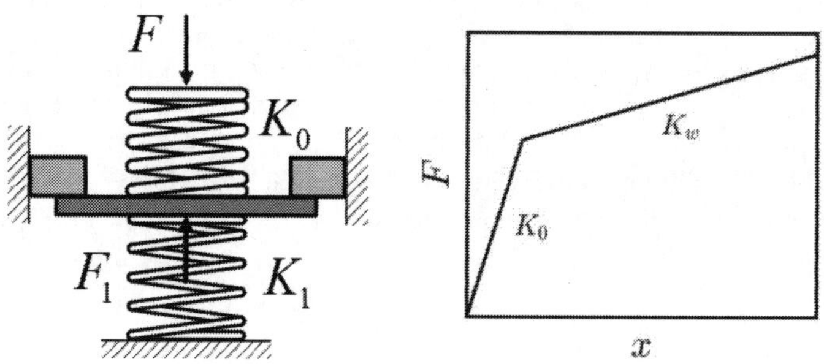

(a) (b)

Figure 1. Conventional multi-rate spring, (a) schematic diagram, (b) stiffness property.

2.2. The liquid spring with one slave spring

The schematic diagram of the liquid spring with one slave spring is plotted in Figure 2. The primary piston with diameter r_1 is subject to the external force F. The slave piston with diameter r_2 is subject to a pre-compressed force F_2 provided by the linear slave spring with stiffness K_2. The primary piston and the slave piston are communicated by the liquid. Before the slave spring operating, the stiffness of the liquid spring can be designed to be equal to K_0 by proper arrangement of the type of the liquid, the volume of the liquid and the size of the primary piston according to hydrostatics. While when the external force can push the slave piston moving, the movement of the primary piston not only contributed by the compressibility of the liquid but also contributed by the movement of the slave piston. Therefore, the liquid spring presents a softer spring. Theoretically, the force-displacement relationship of the liquid spring can be derived as

$$F = \begin{cases} K_0 \cdot x & F \le F_2 \cdot \mu^2 \\ \dfrac{K_0 \cdot K_2 \cdot \mu^4}{K_0 + K_2 \cdot \mu^4} \cdot x & F_2 \cdot \mu^2 < F \end{cases} \tag{3}$$

where $\mu = r_1 / r_2$ is the diameter ratio between the primary piston and the slave piston.

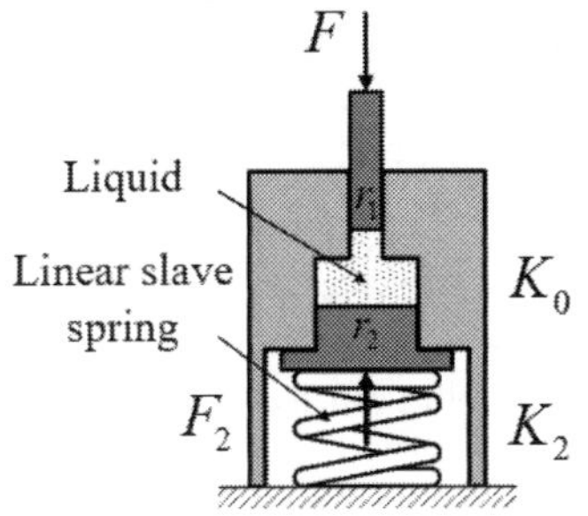

Figure 2. Schematic diagram of the liquid spring with one slave spring.

To obtain the same stiffness property shown in Figure 1 (b) as the conventional multi-rate spring, the configuration of the liquid spring should follow the bellow relationship with the configuration of the conventional multi-rate spring according to Equation (1) and Equation (3).

$$K_2 \cdot \mu^4 = K_1 \tag{4}$$

$$F_2 \cdot \mu^2 = F_1 \tag{5}$$

The Equation (4) indicates that a linear slave spring with any stiffness can be applied to the liquid spring to achieve the same stiffness property as the conventional multi-rate spring by setting a proper size of the slave piston. Moreover, a corresponding pre-compressed force of the slave spring is acquired according to Equation (5). Then the lateral stability of each slave spring can be determined. As a result, it is significant that taking the lateral stabilities of a series of pre-compressed slave springs and the lateral stability of the pre-compressed spring to have a compare. Especially, it is noticed that when the diameter of the slave piston is equal to the diameter of the primary piston (i.e., $\mu = 1$), the lateral stability of the slave spring is equal to the lateral stability of the pre-compressed linear spring in the conventional multi-rate spring.

Concerning whether the lateral stabilities of those slave springs is better than the conventional counterpart. The definition of the lateral stability of a pre-compressed linear spring is proposed in this paper. Taking no consideration of the installation of the pre-compressed linear spring, the lateral stability of a pre-compressed linear spring is only based on the stiffness and the deformation of the spring. It is well known that, as for a specific spring, the smaller deformation, the better lateral stability. Moreover, when springs under the same deformation, the spring with higher stiffness shows better lateral stability. As a result, the lateral stability of a pre-compressed linear spring is defined as

$$S = \frac{K}{\Delta l} \tag{6}$$

where Δl is the deformation of the spring, and K is the stiffness of the spring. It indicates the higher the value, the better the lateral stability.

Then the lateral stability of the pre-compressed spring in the conventional multi-rate spring is

$$S_C = \frac{K_1}{F_1 / K_1} = \frac{K_1^2}{F_1} \tag{7}$$

Moreover, the lateral stability of the slave spring in the liquid spring is

$$S_L = \frac{K_2}{F_2 / K_2} = \frac{K_2^2}{F_2} \tag{8}$$

Substituting Equation (4) and Equation (5) into Equation (8) derive

$$S_L = \frac{1}{\mu^6} \frac{K_1^2}{F_1} \tag{9}$$

To design a productive liquid spring with one slave spring, the lateral stability of the slave spring in the liquid spring should be better than the lateral stability of the pre-compressed spring in the conventional multi-rate spring, i.e., $S_C < S_L$. Then the design criteria of the liquid spring with one slave spring can be reduced as

$$\mu < 1 \tag{10}$$

Concerning to the liquid spring with one slave spring, the diameter of the slave piston should be greater than the diameter of the primary piston to ensure the slave spring has better lateral stability than the conventional counterpart. While the consequence is that the stiffness and the pre-compressed force of the slave spring are higher than the conventional counterpart according to Equation (4) and Equation (5). Theoretically, any high stiffness spring can be applied in the liquid spring to achieve an ideal low working stiffness.

2.3. The liquid spring with more than one slave spring

As for the liquid spring with more than one slave spring shown in Figure 3 (a), it can be seen that the configuration of each slave spring element may be different. The stiffness property of the liquid spring may be quite complicated on account of the various configuration of every slave spring element. While, concerning the influence of the number of slave springs on the liquid spring's stiffness property, it assumes that all slave spring elements are with the same configuration as Figure 3 (b). Correspondingly, the force-displacement relationship of the liquid spring with more than one slave spring can be derived as

$$F = \begin{cases} K_0 \cdot x & F \le F_2 \cdot \mu^2 \\ \dfrac{K_0 \cdot K_2 \cdot \mu^4 / n}{K_0 + K_2 \cdot \mu^4 / n} \cdot x & F_2 \cdot \mu^2 < F \end{cases} \tag{11}$$

To obtain the same stiffness property as the conventional multi-rate spring, the configuration of the liquid spring with more than one slave spring should follow the bellow relationship with the configuration of the conventional multi-rate spring according to Equation (1) and Equation (11).

$$\frac{K_2 \cdot \mu^4}{n} = K_1 \tag{12}$$

$$F_2 \cdot \mu^2 = F_1 \tag{13}$$

Then the lateral stability of the slave spring in the liquid spring with more than one slave spring is

$$S_L = \frac{n^2}{\mu^6} \frac{K_1^2}{F_1} \tag{14}$$

As a result, the design criterion of the liquid spring with more than one slave spring can be written as

$$\mu^3 < n \tag{15}$$

where n is the number of slave springs.

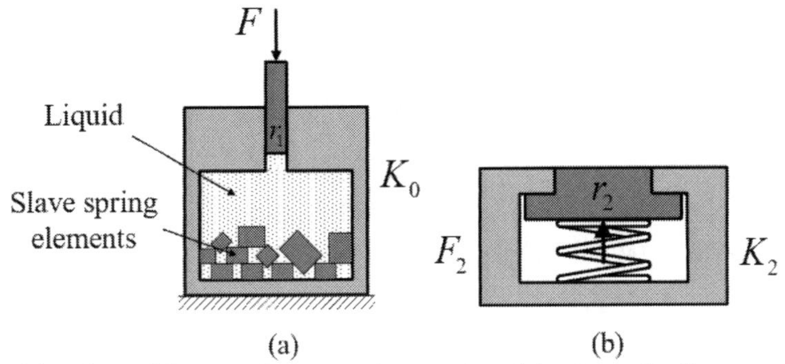

Figure 3. The liquid spring with more than one slave spring, (a) schematic diagram, (b) configuration of a slave spring element.

The Equation (12) and the Equation (15) demonstrate that any stiffness of the slave spring, even the stiffness of the slave spring is equal to the conventional counterpart, can be applied to the liquid spring to obtain the desired stiffness property while considering the lateral stability of the slave spring. Furthermore, the diameter ratio can be greater than one due to the number of slave springs is a positive integer. In another word, the slave piston can be designed as small as possible to achieve the same stiffness property so long as the number of slave springs can obey the Equation (12).

3. Conclusion

The liquid spring presented in this paper is a new type of multi-rate spring and possesses the characteristic of piecewise linear HSLD stiffness. And the lateral stability of the pre-compressed slave spring in the liquid spring can be designed to be better than the conventional counterpart. The stiffness property of the liquid spring is devisable by adjusting the size of pistons and the number of slave springs. A linear spring with arbitrary magnitudes stiffness is able to apply to the liquid spring to achieve the ideal piecewise linear stiffness property. The liquid spring is especially applicable for massive payload low frequency vibration isolation.

Acknowledgments

The authors would like to appreciate Dr. Fangshuo Li and Dr. Muchun Yu for their helpful advice.

References

[1] Carrella A, Brennan M J and Waters T P 2007 Static analysis of a passive vibration isolator with quasi-zero-stiffness characteristic *J. Sound. Vib.* 301 (2007) 678–689
[2] Carrella A, Brennan M J, Waters T P and Shin K 2008 On the design of a high-static-low-dynamic stiffness isolator using linear mechanical springs and magnets *J. Sound. Vib.* 315 (2008) 712–720
[3] Huang X C, Liu X T, Sun J Y, Zhang Z Y and Hua H X 2014 Vibration isolation characteristics of a nonlinear isolator using Euler buckled beam as negative stiffness corrector: A theoretical and experimental study *J. Sound. Vib.* 333 (2014) 1132–1148
[4] Huang X C, Su Z W and Hua H X 2018 Application of a dynamic vibration absorber with negative stiffness for control of a marine shafting system *Ocean. Eng.* 155 (2018) 131-143
[5] Zhou J X, Wang X L, Xu D L and Bishop S 2015 Nonlinear dynamic characteristics of a quasi-zero stiffness vibration isolator with cam–roller–spring mechanisms *J. Sound. Vib.* 346 (2015) 53–69
[6] Wang K, Zhou J X and Xu D L 2017 Sensitivity analysis of parametric errors on the performance of a torsion quasi-zero-stiffness vibration isolator *Int. J. Mech. Sci.* 134 (2017) 336-346
[7] Nicholas P C 1961 *Spring Design and Application* (New York: McGraw-Hill) P 172

IOP Conf. Series: Materials Science and Engineering **417** (2018) 012030 doi:10.1088/1757-899X/417/1/012030

Preferential Brownian Motion Induced by a Fluid with Inhomogeneous Temperature

Qucheng Li, Deming Nie*

Institute of Fluid Mechanics, China Jiliang University, Hangzhou, China

*Correponding author: nieinhz@cjlu.edu.cn

Abstract. In the present work the fluctuating lattice Boltzmann method was used to numerically study the Brownian motion of particles in a fluid with inhomogeneous temperature. It has been shown that the Brownian particles are preferential to move into the cold fluid area due to small thermal fluctuations. In addition, the boundary between hot fluid and cold fluid prevents the particles going outside once they enter into the cold area. As a consequence, the Brownian particles can be captured or collected by the cold fluid area if the temperature of cold fluid is low enough.

1. Introduction

Brownian motion, also called Brownian movement, is the random motion of particles suspended in a fluid arising from those particles being struck by individual molecules of the fluid, which was named for the Scottish botanist Robert Brown. Brownian motion can be seen in many industrial applications. For instance, recent developments demonstrated that Brownian motion can help to measure the local viscoelastic response of soft materials [1] or the topography of a surrounding polymer network [2]. The trajectories of a Brownian particle can be used to characterize the mechanical properties of molecular motors [3]. In particular, experimental study [4, 5] showed that the nano-fluids exhibit enhanced thermal conductivity and convective heat transfer coefficients compared to those of base fluids, which is believed to be connected with the Brownian motion of nanoparticles [6].

In the past a variety of studies have been devoted to study the Brownian motion of particles in a viscous fluid because of its importance in industrial applications. Lin et al. [7] investigated the constrained Brownian motion of a sphere between two walls through an experimental work. The same problem was also studied by Benesch and Yiacoumi [8] who used the method of reflection. Iwashita and Yamamoto [9] examined the effect of fluid inertia on the short-time motion of Brownian particles by using the direct numerical simulations. Recently, Radiom et al. [10] conducted an experiment to measure the hydrodynamic interactions between two Brownian spheres at low Reynolds numbers, to check the effect of fluid inertia. As is known, the Brownian motion is very sensitive to the temperature. Higher temperature leads to more-rapid Brownian motion. Brownian motion is isotropic if the temperature of fluid is homogeneous everywhere, which leads to a totally random movement of particles in the fluid. In other words, if a number of particles subject to Brownian motion are present in a given fluid and there is no preferred direction for the random oscillations, then over a period of time the particles will tend to be spread evenly throughout the medium. However, this may not be the truth if the temperature of fluid is inhomogeneous. For example, the Brownian motion may become preferential if there are hot fluid and cold fluid at the same time because the thermal fluctuations are determined by the temperature. However, as far as we know this issue has not been reported, which motivates the present work. The fluctuating lattice Boltzmann method [11] is adopted in this work to study the preferential Brownian motion in a fluid with inhomogeneous temperature.

2. Numerical Method

The fluctuating lattice Boltzmann method is used to solve the fluid flow [11]. The discrete lattice Boltzmann equations are described by,

$$f_i(\boldsymbol{x}+\boldsymbol{e}_i\Delta t, t+\Delta t)-f_i(\boldsymbol{x},t)=-\frac{1}{\tau}\left[f_i(\boldsymbol{x},t)-f_i^{(eq)}(\boldsymbol{x},t)\right]+f_i^{(B)}(\boldsymbol{x},t) \tag{1}$$

where f_i $(\boldsymbol{x}, t)$ is the distribution function on the i-direction discrete velocity $\boldsymbol{e}_i$, $f_i^{(eq)}(\boldsymbol{x}, t)$ is the equilibrium distribution function, Δt is the time step, $f_i^{(B)}(\boldsymbol{x}, t)$ is a stochastic term representing the thermal fluctuations, which is related to the fluctuating stress in the Navier-Stokes equations [11].

The macro-variables density ρ and momentum $\rho\boldsymbol{u}$ are obtained from,

$$\rho=\sum_i f_i, \quad \rho\boldsymbol{u}=\sum_i f_i\boldsymbol{e}_i \tag{2}$$

The equilibrium distribution function is chosen as,

$$f_i^{(eq)}(\boldsymbol{x},t)=w_i\rho\left[1+\frac{\boldsymbol{e}_i\cdot\boldsymbol{u}}{c_s^2}+\frac{(\boldsymbol{e}_i\cdot\boldsymbol{u})^2}{2c_s^4}-\frac{\boldsymbol{u}^2}{2c_s^2}\right] \tag{3}$$

where $c_s^2=c^2/3$ is the speed of sound.

As shown by Nie and Lin [11], the stochastic term is related to the fluctuating stress in the following way,

$$\sigma_{\alpha\beta}^{(B)}=-\tau\sum_i f_i^{(B)}e_{i\alpha}e_{i\beta} \tag{4}$$

According to the fluctuation-dissipation theorem, $\sigma_{\alpha\beta}^{(B)}$ has the following property,

$$\begin{gathered}\left\langle\sigma_{\alpha\beta}^{(B)}\right\rangle=0\\ \left\langle\sigma_{\alpha\beta}^{(B)}(x_1,t_1)\sigma_{\gamma\delta}^{(B)}(x_2,t_2)\right\rangle=2k_BT\mu\left(\delta_{\alpha\gamma}\delta_{\beta\delta}+\delta_{\alpha\delta}\delta_{\beta\gamma}-\frac{2}{3}\delta_{\alpha\beta}\delta_{\gamma\delta}\right)\delta(x_1-x_2)\delta(t_1-t_2)\end{gathered} \tag{5}$$

where <> denotes averaging over an ensemble, k_B is the Boltzmann constant, T is the temperature of fluid, μ is the dynamic viscosity of the fluid. The fluctuating stress is sampled from a Gaussian distribution with zero mean and a given variance of $2k_BT\mu$.

In this work we assume the stochastic term $f_i^{(B)}(\boldsymbol{x}, t)$ to be the following form to make sure of the conservation of mass and momentum [11],

$$\begin{aligned}&f_0^{(B)}=0\\ &f_1^{(B)}=f_3^{(B)}=\frac{1}{2\tau}\sigma_{yy}^{(B)}\\ &f_2^{(B)}=f_4^{(B)}=\frac{1}{2\tau}\sigma_{xx}^{(B)}\\ &f_5^{(B)}=f_7^{(B)}=-\frac{1}{4\tau}\left[\sigma_{xx}^{(B)}+\sigma_{yy}^{(B)}+\sigma_{xy}^{(B)}\right]\\ &f_6^{(B)}=f_8^{(B)}=-\frac{1}{4\tau}\left[\sigma_{xx}^{(B)}+\sigma_{yy}^{(B)}-\sigma_{xy}^{(B)}\right]\end{aligned} \tag{6}$$

3. Validation

The Brownian motion of 81 particles in a periodic domain was simulated to validate the present method. In the simulations only the hydrodynamic force was considered. The periodic domain is set to be 240 × 240. The density of the fluid is fixed at $\rho = 1$ and the non-dimensional relaxation time $\tau =$ 0.65, which leads to the viscosity of fluid $\nu = (2\tau\text{-}1)/6 = 0.05$. The radius of particle is $a = 3.5$. The solid/fluid density ratio is fixed at $\rho_s/\rho = 11$ in the simulations. In order to determine the magnitude of the fluid fluctuation, the temperature of fluid is chosen as $k_B T = 10^{-4}$.

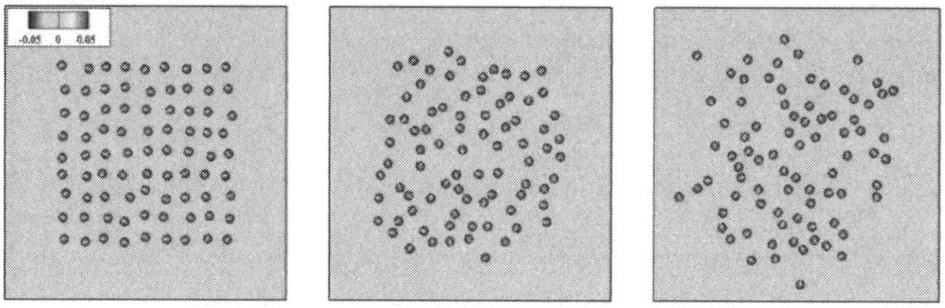

(a) (b) (c)

Figure 1. Instantaneous fluctuations of fluid pressure at different times: (a) t=1×10^4, (b) t=1×10^5, (c) t=2×10^5, (d) t=4×10^6. The white arrow on each particle is used to visually track its rotation with horizontal orientation at the beginning (the same as below).

The instantaneous fluctuations of fluid pressure at different times are shown in Fig. 1, along with the Brownian particles, which clearly illustrate the thermal fluctuations. This is the origin of the Brownian motion of particles, resulting from the essence of the present fluctuating lattice Boltzmann method, which is different from the Langevin dynamics. As shown in Fig. 1, the particles tend to spread out with time as they undergo Brownian motion, displaying the classical motion of Brownian diffusion. In addition, it is worth stating here that the rotation of particles can be realized in the present simulations, as shown in Fig. 1.

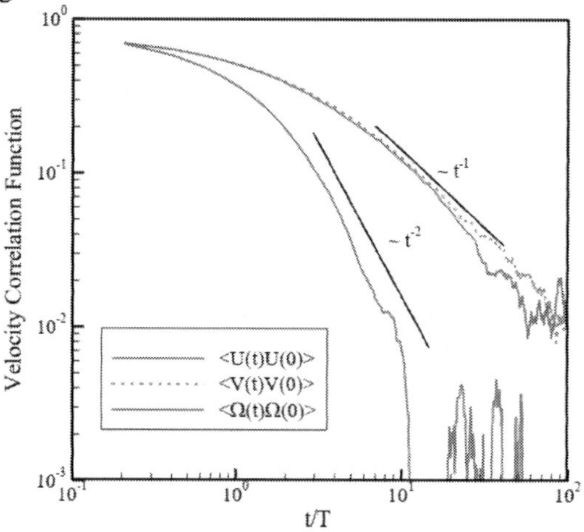

Figure 2. The translational/rotational velocity correlation functions of Brownian motion.

The long time tails can be used to validate the effectiveness of numerical method. Fig. 2 shows the translational and rotational velocity autocorrelation functions (VACFs) of particles. In the figure U and V refer to the translational velocity, and Ω refers to the rotational velocity. All the results are normalized by their initial values, i.e. the values at $t = 0$. According to the theoretical predictions, the translational and rotational VACFs of particles undergoing Brownian motion have power-law decays over long times that are t^{-1} and t^{-2}, respectively. As shown in Fig. 2, the similar long time tails are observed for the particles in the present simulations, which is consistent with the theoretical prediction.

4. Numerical Results

First of all, the Brownian motion of a circular particle in a fluid with inhomogeneous temperature is numerically investigated. The particle is freely moving in the domain of $L \times L$, which is filled with the hot fluid with temperature T_h. There is a circular area with radius R inside which the fluid is cold. The temperature is denoted as T_c. For simplicity the no-slip boundary conditions are imposed on all four fixed walls of the domain. In all simulations, the parameters are fixed at $a = 3.5$, $R = 3.5a$, $T_h = 1.0\times10^{-3}$ and $\rho_s/\rho = 11$. The particle is initially placed in the center of the domain.

Fig. 3 shows the instantaneous flow (the magnitude of fluid velocity $|\boldsymbol{u}|$ normalized by ν/a) of particle Brownian motion at different times. The computational domain is 64 × 64. Two kinds of cold temperature are taken into account, i.e. $T_c = 1.0\times10^{-3}$ (top) and 5.0×10^{-5}(bottom). For the results of $T_c = 1.0\times10^{-3}$, the root mean square (RMS) of fluid velocity is homogeneous everywhere because there is no temperature difference in the fluid. However, things are different for the other results. By displaying the RMS of fluid velocity a cold temperature zone is observed in the central area, as can be seen in Fig. 3. For the case of $T_c = 1.0\times10^{-3}$, the particle is randomly moving in the whole domain. However, the particle will exhibit preferential Brownian motion when there is cold fluid. In other words, the particle prefers to move into cold fluid area. It is found that the particle will stay in the cold area if T_c is low enough, which is seen for the cases of $T_c = 5.0\times10^{-5}$. This makes it possible that the Brownian can be captured by the cold temperature area somewhere in the fluid.

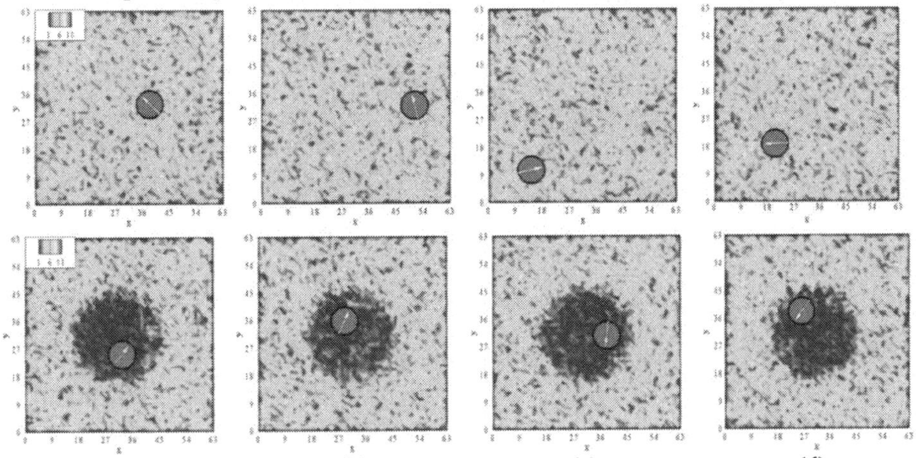

Figure 3. Instantaneous flow (the magnitude of fluid velocity $|\boldsymbol{u}|$ normalized by ν/a) of Brownian motion at different times: (a) t=1×10^4^, (b) t=1×10^5^, (c) t=4×10^5^, (d) t=1×10^6^. The values of T_c are 1×10^-3^ (top) and 5×10^-5^ (bottom).

In what follows another two cases are presented to further illustrate the influence of fluid temperature. The first case is depicted in Fig. 4, showing the instantaneous flow of Brownian motion of 16 particles at different times. The computational domain is 240 × 240. The left half of domain is set to be cold area with temperature $T_c = 5\times10^{-5}$. At the beginning of simulations, the particles are placed homogeneously over the computational domain, as one can see in Fig. 4(a). The cold fluid results in very small thermal fluctuations in the left part of domain, as a consequence, most of particles quickly go into the cold area and do not leave, which can be seen in Fig. 4. The second case is illustrated in Fig. 5, which shows the preferential Brownian motion of 36 particles in an inhomogeneous fluid. For this case the cold area is located in the center of domain, instead. Other parameters are the same to those used in the case of Fig. 3. Generally speaking, similar behavior is observed from Fig. 5. The cold area captures almost all particles when they are randomly moving in the fluid. However, it will take a much longer time for the rest of particles to be captured. The reason behind this is clear. There is not much room for these particles in the central area. When approaching this area they are repelled due to the hydrodynamic interactions as well as particle collisions.

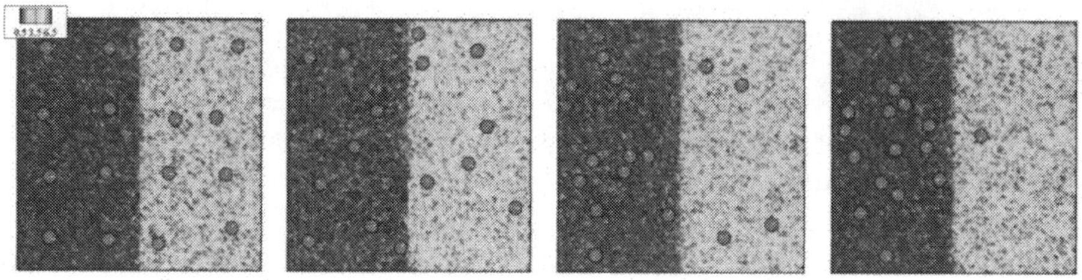

(a) (b) (c) (d)

Figure 4. Instantaneous flow of Brownian motion of 16 particles for $T_c = 5\times10^{-5}$ at different times: (a) t=1×10^4, (b) t=1×10^5, (c) t=5×10^5, (d) t=2×10^6.

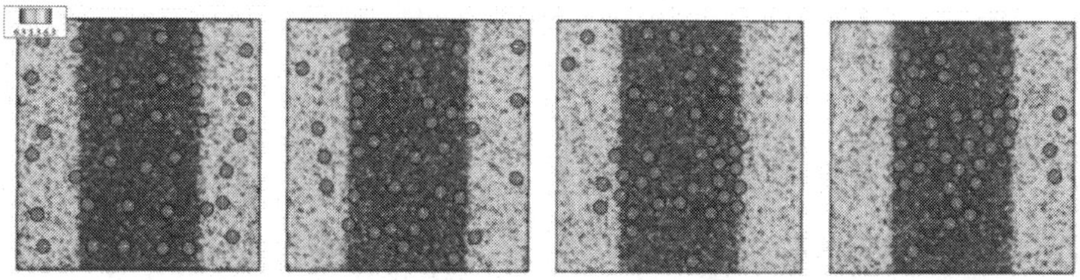

(a) (b) (c) (d)

Figure 5. Instantaneous flow of Brownian motion of 36 particles for $T_c = 5\times10^{-5}$ at different times: (a) t=1×10^4, (b) t=1×10^5, (c) t=5×10^5, (d) t=1.8×10^6.

In the final, it should be stated that the viscosity is kept constant for every single simulation in this work. However, it is believed that there exist some fluids whose viscosities are not sensitive to the temperature. Therefore, the conclusion of this work still stands for these fluids.

5. Summary

In this work the fluctuating-lattice Boltzmann method was adopted to numerically investigate the preferential Brownian motion of particles in a fluid with inhomogeneous temperature. Results show that the particles are preferential to randomly move in the cold fluid area. Most important of all, the particles go into the cold area and stay there eventually if the temperature of the cold area is low enough. In other words, the cold fluid can capture or collect the Brownian particles.

References

[1] MacKintosh F and Schmidt C 1999 *Curr. Opin. Colloid Interface Sci.* **4** 300.
[2] Tischer C, Altmann S, Fisinger S, Hörber J, Stelzer E and Florin E 2001*Appl. Phys. Lett.* **79** 3878.
[3] Jeney S, Stelzer E, Grubmüller H and Florin E 2004 *Chem. Phys. Chem.* **5** 1150.
[4] Keblinski P, Phillpot S, Choi S and Eastman J 2002 *Int. J. Heat Mass Transfer* **45** 855.
[5] Lee S, Choi S, Li S and Eastman J 1999 *J. Heat Transfer* **121** 280.
[6] Prasher R, Bhattacharya P and Phelan P 2005 *Phys. Rev. Lett.* **94** 025901.
[7] Lin B, Yu J and Rice S 2000 *Phys. Rev. E* **62** 3909.
[8] Benesch T and Yiacoumi S 2003 *Phys. Rev. E* **68** 021401
[9] Iwashita T and Yamamoto R *Phys. Rev. E* **79** 021401.
[10] Radiom M, Robbins B, Paul M and Ducker W 2015 *Phys. Fluids* **27** 022002.
[11] Nie D and Lin J 2009 *Particuology* **7** 501.

IOP Conf. Series: Materials Science and Engineering **417** (2018) 012031 doi:10.1088/1757-899X/417/1/012031

Research Study on Mechanics Mechanism of Roadbed Grouting for Shallow Buried Tunnel

Shangyang YANG[1,2],Longyun ZHANG[2*]

[1]Department of Physics and Mathematics, Shandong Jiaotong University, Jinan 250023, China

[2]Geotechnical & Structural Engineering Research Center, Shandong University, Jinan, 250061, China

77959140@qq.com

Abstract: The traffic congestion has become a prominent problem in many cities under the rapid development of urbanization at present in China. In order to solve it, more and more cities have accelerated the development of the subway traffic. In consequence of that, there will be a large number of tunnel construction technical problems to be solved. The strata deformation caused by the tunnel construction, especially in the dense urban area, has always been a subject of great concern. Especially in shallow tunnel construction, the surface will form a settlement trough due to the excavation of the stratum. The construction sink may inflict the road pavement damage, underground pipeline damage, as well as the damage of buildings. The grouting process is one of the most effective methods to solve the construction settlement. The grouting process has been widely used in tunnel construction because it may contribute to the reinforcement of the rock and the soil in situ and will change their inherent physical and mechanical properties. In this paper, combined with the construction of one shallow buried tunnel in a city, the method of roadbed grouting has been adopted to make the reinforcement for the roadbed. This paper studies the precautions which are needed when the soft soil roadbed does not interrupt the traffic. The study may provide some certain guiding significance for the construction of other city underground road on soft soil foundation [1-4].

1. Introduction

Yushun Road tunnel which is mentioned in this paper is 2335 meters long, with a depth of 6~10 meters and a road width of 32 meters, heavy traffic, dense underground pipelines, furthermore, the key line is about 8m from the tunnel vault. There is a large number of buildings on both sides of the tunnel, some of them only 2 meters far from the tunnel border. The geology is complex along the tunnel. For example, soil and stone are distributed at intervals, mainly in clay soil, with loess intrusion. Rocks are mainly composed of limestone with karst development and large changes in bedrock surface. The Quaternary system is under filling field which is composed of collapsible loess, clay and gravel soil. Loess having general intensity and medium collapsible is plasticity, and the local was hard. Clay in the absence of local affected largely by the thickness is plastic completely, and has a low compressibility. From top to bottom, the rock mass in the site is strongly weathered gabbro, moderately-weathered limestone, moderately-weathered marl, and the bedrock bearing capacity is generally normal. The integrity of the rock mass is between 0.13 and 0.36, and the rock quality grades are Ⅲ-Ⅳ.

IOP Conf. Series: Materials Science and Engineering **417** (2018) 012031 doi:10.1088/1757-899X/417/1/012031

2. Grouting theory

According to Darcy's law

$$q = K_g iA \tag{1}$$

When $r = r_0, h = H$; and when $r = r_1, h = h_0$,
Therefore, it is derived from the upper type

$$H - h_0 = h_1 = \frac{q\beta}{2\pi aK}\ln\frac{r_1}{r_0} \tag{2}$$

$$q = \frac{2\pi aKh_1}{\beta\ln\frac{r_1}{r_0}} \tag{3}$$

It is known as $Q = \pi r_1^2 an$, $Q = qt$, and then

$$r_1 = \sqrt{\frac{2Kh_1 t}{n\beta\ln\left(r_1/r_0\right)}} \tag{4}$$

$$t = \frac{n\beta r_1^2\ln(r_1/r_0)}{2Kh_1} \tag{5}$$

In the above formula,

K_g - The permeability coefficient of strata (cm/s) ;
Q - Slurry flow (cm^3);
K - The permeability coefficient of sand (cm/s) ;
q - The amount of grouting per unit time (cm^3/s);
r_1 - Radius of slurry diffusion (cm) ;
a - Length of lumnar grouting (cm);
h_1 - Hydraulic head of water (cm);
β - The ratio of the viscosity of the slurry to the water;
A - The area of penetration (cm^2);
h_0 - Hydraulic head of water (cm);

Limit pressure of penetration grounting

When grouting in the formation of the semi-infinite space, if the grouting pressure exceeds a certain limit p_u, slurry flow will be transformed by penetration into fracture grouting mode. It is only when the grouting pressure is less than p_u that the slurry is permeable in the formation. p_u can be expressed as:

$$p_u = \frac{2(1-\nu)(\sigma_c + 2K_0\gamma H)}{\frac{2+(1-2\nu)}{(\ln r_1 - \ln r_0)}} \tag{6}$$

Grouting pressure generally increases rapidly after grouting, and the unstable slurry will emerge near the grouting hole. If the effect on the soil of the permeability stress of the slurry is omitted, it can be simplified as:

$$p_u = \sigma_c + 2K_0\gamma H(1-\nu) \tag{7}$$

σ_c - The compressive strength of soil (MPa);
γ - Weight density of soil (KN/m^3);
H - Depth of grouting holes (m);
K_0 - Coefficient of earth pressure at rest
ν - Poisson' ratio.

Calculation of grouting quantity

The type of grouting, the porosity and the crack rate of the rock and the filling degree of the slurry should be taken into account in the calculation of the grouting quantity. The quality of the penetration grouting depends on the pore filling degree of the volume soil within the penetration radius. The higher the filling rate is, the better the grouting effect. However, the slurry will not be able to infuse unlimited, and there should be an optimal amount of grouting. In the condition of no splitting, the diameter of the slurry is very limited as well as its amount. The commonly used formula of the slurry volume in penetration grouting is shown as follows,

$$Q = \pi r^2 hn\alpha(1+\beta) \tag{8}$$

r - Penetration radius;
h - Thickness of the grouting;
n - Porosity of the soil;
α - Effective perfusion coefficient;
$(1+\beta)$ - Loss coefficient, 1.1-2.0 is advisable.

Grouting at the site

Figure 1. Site of roadbed grouting

This project has designed a total of 204 drilling holes in the ground and one small Angle in the hole. Among them, and the total drilling penetration was 1994m. The roadbed grouting is shown in Figure 1.

In order to guide the construction of grouting, a timely monitoring measurement is carried out for the settlement of the arch of the tunnel. The daily subsidence statistics and changes of the monitoring points of each range vault are shown in Figure 2.

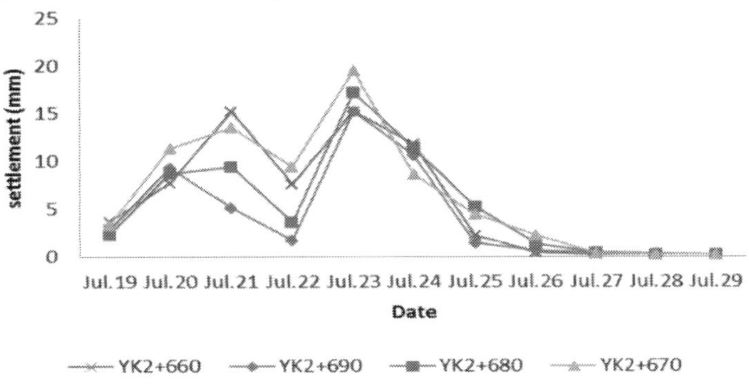

Figure 2. The daily settlement convergence curve of each monitoring point at the vault

Figure 3. Grouting effect

By analyzing the vault subsidence and the relationship of the grouting operations, vault subsidence change is bigger which is affected by the process of grouting for grouting pressure and filling into the stratigraphic gap size weight pressure influence. However, after the grouting stops for a period of time, the settlement amount gradually becomes smaller and the arch becomes stable. The reason why the settlement decreases and the vault tends to stabilize In addition to eliminating the influence caused by the grouting pressure disturbance, the early strength characteristics of the slurry consolidation also play a role, that is, the strength of the injected slurry rapidly increases, so that the consolidated formation Self-stabilizing ability gradually strengthened. Grouting effect is shown in Figure 3.

3. Conclusions

In geological survey and detailed, accurate and reasonable tunnel in soft soil foundation design and construction measures properly, under the premise of monitoring information feedback in time, new urban road underground engineering safety can be guaranteed. Adopting the grouting method to reinforce the soil of the existing composite roadbed, as well as putting forward the grouting pressure control parameters and the grouting amount control parameters, I have strengthened the reinforcement of the existing composite roadbed which can provide a layer of force for the high carrying capacity for the road traffic.

The monitoring results show that the settlement of the existing composite subgrade has been gradually reduced and the settlement of the subgrade is gradually convergent with the formation of reinforcement and strengthening stone body mass, and the normal traffic of the pavement has been not affected. The practice has proved that it is effective to reinforce the overlying Subgrade in the shallow buried tunnel by the seepage grouting reinforcement method. Especially when the site which has a heavy traffic and complex terrain conditions is limited, the application of light construction machinery can make the construction convenient, fast and economical. At the same time, it is noted that the complexity of road traffic and buried pipelines should not be interrupted. During construction, close monitoring should be paid attention to, and the construction parameters should be adjusted at any time to keep safety and achieve the best results.

Acknowledgments

The work was supported by China Postdoctoral Science Foundation funded project (No.2016M600537). We would also like to express our sincere gratitude to the editor and the reviewers for their valuable contributions to this paper.

References

[1] Sturk R,Stille H.Design and excavation of rock caverns for fuel storage—a case study from Zimbabwe. Tunnelling and Underground Space Technology.1995,10(2):193-201

[2] Lee Y N,Yun S P,Kim D Y,et al.Design and construction aspects of unlined oil storage caverns in rock . Tunnelling and Underground Space Technology.1996,11(1):33-37

[3] Yang, Shang-yang, Li, Shu-cai, Xue, Yi-guo, Zhang, Qing-song. Finite element analysis on mechanical behavior of Yungui railway tunnel in shallow and unsymmetrical strata. International Journal of Hybrid Information Technology. 2016, 9(10): 21-30

[4] Yang, Shang-yang, Li, Shu-cai, Xue, Yi-guo, Zhang, Qing-song. Fluid solid coupling analysis of large underground oil storage caverns in containment of groudwater. International Journal of Hybrid Information Technology. 2016, 9(11): 415-424

IOP Conf. Series: Materials Science and Engineering **417** (2018) 012032 doi:10.1088/1757-899X/417/1/012032

Automatic Book Scanner

N Bano[1], M Aziz[2], U Ghani[3], M Taha[4] and H Ahmed[5]

[1,2,3,4,5]Department of Mechatronics, SZABIST 100, Block 5 Clifton, Karachi, Karachi City, Sindh 75600, Pakistan

nasreen.bano@szabist.edu.pk

Abstract. Preserving books and documents have become a hectic problem for today's world. Most of the documents and books were discarded due to insufficient physical storage and lots of maintenance. To overcome this problem, digitization of documents has become an important technology. The technical challenge is to realize an easy to use, high quality and high speed scanning system. The prototype is scanning large stacks of paper without using any human effort to flip pages i.e. it is using robotic arm to flip pages. The prototype is designed within small budget, to scan the standard size book i.e. (12 inches height and 9.5 inches wide), then converting the texts which are in human readable form to computer-readable digital image.

1. Introduction

BookScanning is the process of converting physical books into digital media Such as images, electronic text, or electronic books (e-books) by using an image scanner. Digital books can be easily distributed, reproduced, and read-on screen [1]. Over the years, many automatic book scanners have been designed, some examples of these scanners are [2] [3] [4] to mention a few.

The digitization of book requires a scanner. Scanning the bundles of pages would definitely increase the work load and consume lot of time of user. The problems like limited physical storage, scanning bundle of books by manual flipping and Stanford University Library Digitization [5] motivated us. Recently, many solutions and techniques have been developed to flip the page automatically but they all differs in specifications, design, speed and quality. Some are using Big Finger and Little Finger [6], mechanism of page thumbing [7] and Vacuum lift mechanism [8].

Finger flipping technique is used by M.T.Nikham [9], having roller and slider. The turning mechanism has to rotate about 45, 90, 135 and 180 degrees respectively. After the page turned the camera will capture the image and converted into several forms like PDF, CD, Word format, PC, laptops and Zerox of scanned copies.

Dany Qumsiyeh [10] who was an employee at Google developed a linear book scanner machine which uses air suction from an ordinary vacuum cleaner to isolate individual pages, scanning the front and back in one pass along the device's prism-shaped body. After a quick 40-second setup, it can digitize a 1000-page book in a little over 90 minutes.

Yoshihiro Watanabe [11] made an automatic page turner machine that turns the pages in a contactless manner by utilizing the elastic force of the page and an air blast. The experiment was carried out in different number of paper quality and it was found that average success rate of thin-paper book was about 100% while thick-paper book is about 98.4%.

The main function in the making of "automatic-book scanner prototype" is to make it reliable, faster and economic.

2. Methodology

The flipping technique for the prototype was chosen by keeping in mind reliability, speed and economic. The prototype is cantered on a 'robotic arm with a suction power' to lift the page and a 'slider' to drive the page on to the other side of the book. Prototype is comprised of four main components:

1. Pneumatic cradle actuator
2. Robotic arm with suction pipe (whose one end is attached to vacuum pump)
3. Pneumatic slider
4. Raspberry-pi Controller along with camera

The schematically working of automatic-book scanner is shown in figure 1.

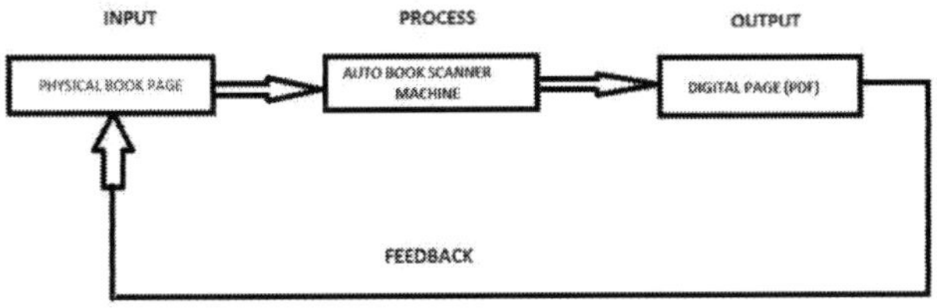

Figure 1. Overview of scanner's working.

3. Working principle

To execute the process of automatic-scanning, the first step is to place the book on cradle. Robotic arm with suction pipe attached to it will move in downward direction, in order to lift the current page of the book by source of vacuum suction. Then the slider will drive the lifted page on to the other side of book and after that the robotic arm will move in 90 degrees with respect to the initial position. So it doesn't come along on a track of camera, while capturing the pages of book .The actuator that is lay underneath the cradle will move upward, until the front page of book get stuck on platen. The camera that is mounted on top centre side of wall will take image of the entire page. Finally these images are fed into computer as PDF and thus the entire physical book is converted into e-book.
The flow of the approach is to run the book scanning process in a loop as shown in figure 2. For this purpose, a sensor is needed to sense the current process and give feedback back to controller. The sensor that is selected for this purpose is named "reed-switch". These sensors are attached along with cradle actuator, robotic arm and a slider. All the above mentioned processes are executed step by step. The sensor cannot be able to generate the signal, until the job that is assigned to each component won't be accomplished. These signals are fed back to controller which generate the signal to other components. After receiving the signal from feedback controller, the other component do its assigned job. As a result these sensors prevent the collision among the components.

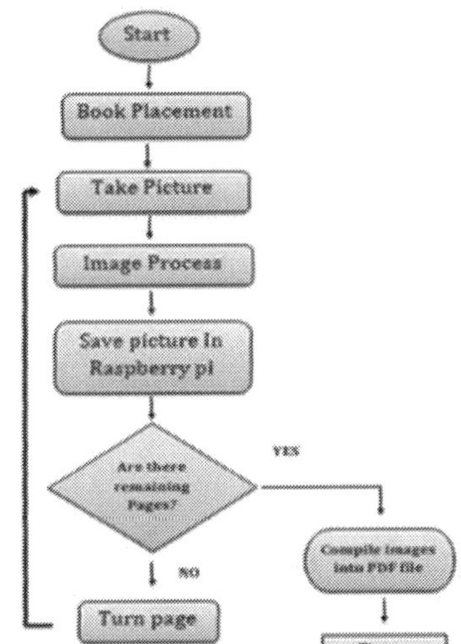

Figure 2. Working flow chart

IOP Conf. Series: Materials Science and Engineering **417** (2018) 012032 doi:10.1088/1757-899X/417/1/012032

4. Construction

The construction of system was first sketched in a solid works platform, in order to get an overview of the system. The top view and side view of the system that is sketched in solid works are shown in figure 3 and figure 4 respectively.

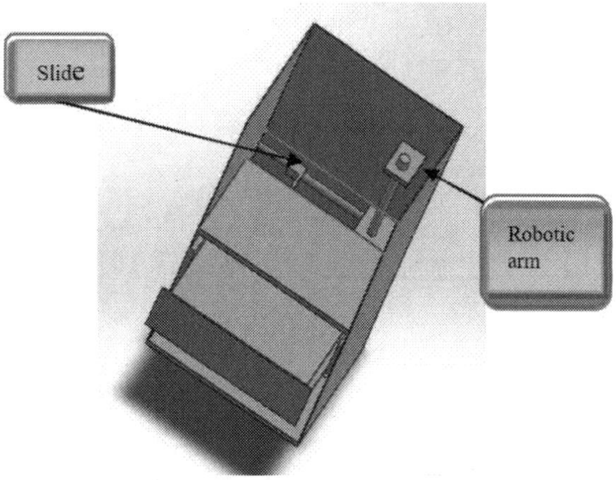

Figure 3. Top view of the structure.

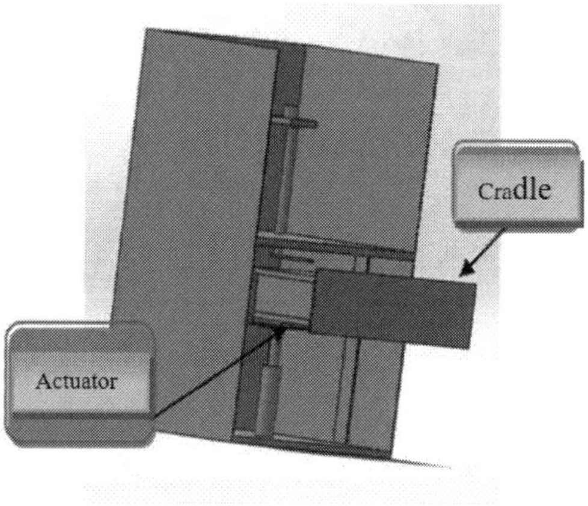

Figure 4. Side view of the structure

The finalize structure is about 30 inches in height and 18 inches in width as shown in figure 5.

Figure 5. Complete project.

5. Observations and calculations

The vacuum suction pressure is calculated as:

Diameter of Suction Pipe = **0.01m**

Radius of Pipe = **0.005m**

Cross-sectional area of pipe: $= \pi r^2 = \mathbf{0.00007855\ m^2}$

The pressure exerted on a page by vacuum strip is calculated by finding force (mass times gravity) and by using pressure (force per unit area) as:

$$\text{Pressure} = \frac{150\times10^{-3}\times9.81}{0.00007855} = \mathbf{18.75\ KPa}$$

IOP Conf. Series: Materials Science and Engineering **417** (2018) 012032 doi:10.1088/1757-899X/417/1/012032

Before starting the process, the timer was brought to zero seconds. Table 1 is representing relationship between number of pages and time.

Table 1. Relationship between Number of pages and time

S.No	No of pages	Time(s)
1	15	60
2	30	120
3	45	180
4	60	240
5	75	300

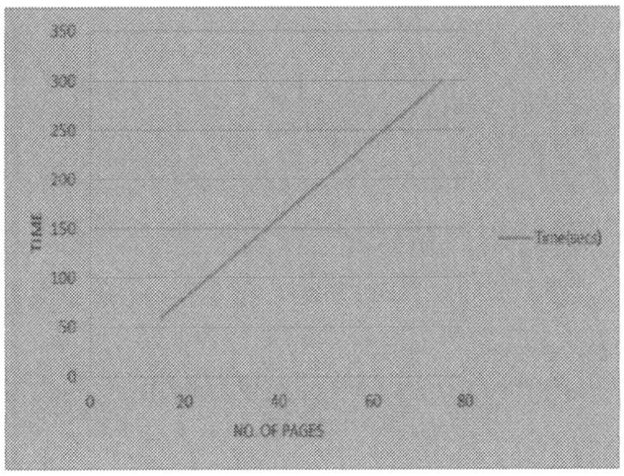

Figure 6. No of pages vs time.

The figure 6 shows the linearity behaviour between these two relations. On an average a book contains 900 pages in it, therefore the time required to completely scan 900 pages book will be approximately be about sixty minutes i.e. equal to an hour.

The scanned output are shown below in the figure 7-10:

Figure 7. The output of the scanned image without Cropping

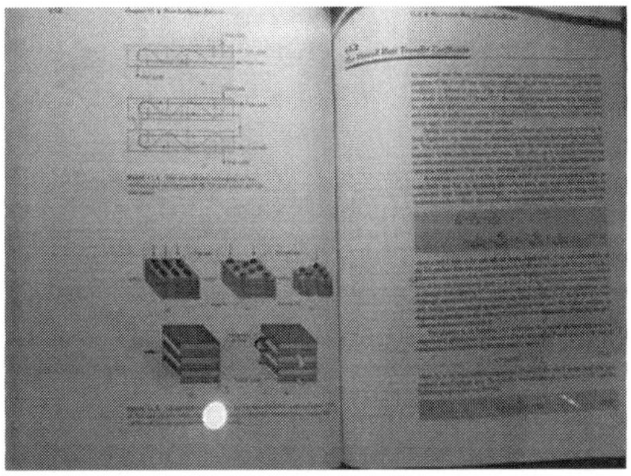

Figure 8. The output of the scanned image with Cropping

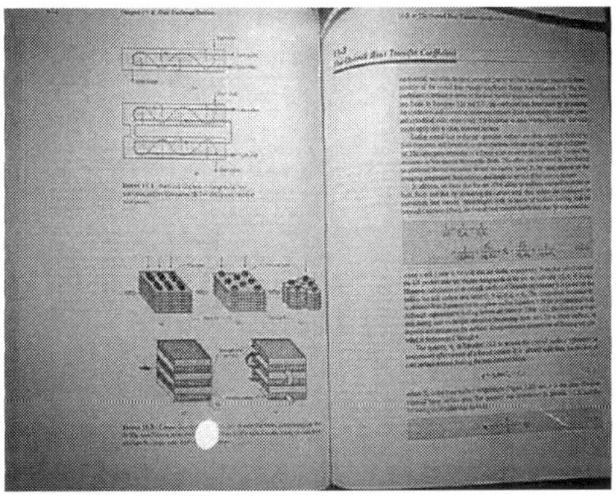

Figure 9. The output of the scanned image with high brightness and low contrast.

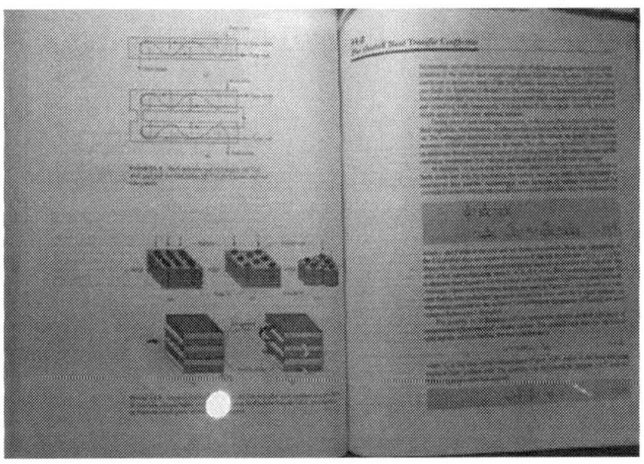

Figure 10. The output of the scanned image when expose perfectly.

6. Conclusion and recommendation

The prototype can turn 12 pages per minute. Due to the orientation of the book lying flat, pages turn best for larger books with sturdy bindings. The camera successfully captures consistently illuminated pages, due to the camera limitations, the resolution could not be improved to the desired DPI. By using two cameras (one for each page) or obtaining a better camera would remedy this issue. During the scanning the natural error of skipping and tiring is negligible and through this auto-flipping technique all the physical documents or book can easily be converted into digital form.

In future, the scanner would have additional features like i) the range of scanning, the number of pages that could be scan (as assigned by user), ii) may have a built-in touch screen, by which the user can give the instructions

Reference

[1] Pathway Innovations and Technologies, Inc., San Diego, CA (US), "METHOD, SYSTEM, AND APPARATUS FOR A DOCUMENT CAMERA BASED BOOK SCANNER AND READING MACHINE WITH AN AUTOMATIC PAGE TURNER", US 8,885,233 B2, 2013.

[2] "Kirtas KABIS III Automatic Book Scanner & robot book scanner range", Scanever.com, 2017. [Online]. Available: http://scanever.com/products/digitize/automatic-book-scanner/automatic-book-scanner.html. [Accessed: 1- Jun- 2017].

[3] "Book scanner ScanRobot® 2.0 MDS - The automatic page turner solution", Treventus.com, 2017. [Online]. Available: http://www.treventus.com/automatic-book-scanner-scanrobot.html. [Accessed: 13- Jun- 2017].

[4] "4DigitalBooks - the only real full automatic book scanners", 4digitalbooks.com, 2017. [Online]. Available: http://www.4digitalbooks.com/. [Accessed: 13- Jun- 2017].

[5] "Digitization", Stanford Libraries, 2017. [Online]. Available: http://library.stanford.edu/blogs/topic/digitization. [Accessed: 13- Jun- 2017].

[6] D. Reetz, "auto page turner prototype - DIY Book Scanner", Forum.diybookscanner.org, 2010. [Online]. Available: https://forum.diybookscanner.org/viewtopic.php?t=379. [Accessed: 13- Jun- 2017].

[7] "New Book Scanner Can Flip And Scan 250 Pages/Minute", Geekologie, 2017. [Online]. Available:http://geekologie.com/2013/01/new-book-scanner-can-flip-and-scan-250-p.php. [Accessed: 13- Jun- 2017].

[8] H. Yanagisawa, "Auto page turning Book Scanner part 1/3 Introduction", YouTube, 2004. [Online]. Available: https://www.youtube.com/watch?v=pkZ8OfgRig0. [Accessed: 13- Jun- 2017].

[9] M.T.Nikham, et al. ,"Automaticmatic Book scanning" International Journal of Advanced Research in Electronics and Communication engineering Volume 5, Issue 2, February 2016.

[10] Google Inc., Mountain View, CA (US), "LINEAR BOOK SCANNER", US 8,711,448 B1, 2011.

[11] Yoshihiro Watanabe,'' Automaticmatic Page Turner Machine for High-speed Book Digitization", IEEE/RSJ International Conference on Intelligent Robots and Systems (IROS) November 3-7, 2013. Tokyo, Japan

IOP Conf. Series: Materials Science and Engineering **417** (2018) 012033 doi:10.1088/1757-899X/417/1/012033

Bond Graphs as Mechatronic Approach for Supervision Design of Multisource Renewable Energy System

Belkacem OULD BOUAMAMA, Ibrahim ABDALAH, Anne-Lise GEHIN

Lille University, CRIStAL, UMR 9189, 59650 Villeneuve d'Ascq, France

Belkacem.ouldbouamama@polytech-lille.fr

Abstract. This paper deals with the bond graph methodology as a decision-making tool and multiphysic approach for supervision supervision system design. The developed algorithms have been applied to a mechatronic represented by Hybrid Renewable Energy System (HRES) which consists of Solar Photovoltaic Panels (PV) and wind turbines coupled to an electrolyser for hydrogen production used an energy vector for a fuel cell.

1. Introduction

Supervision is a set of tools and methods allowing the control of industrial processes in normal working as well as in presence of failures. At FDI (Fault Detection and Isolation) and FTC (Fault Tolerant Control) levels of the supervision and in the case of incidental or accidental situations, two kinds of information have to be displayed to operators. The first one concerns the detection and isolation of the fault, and the second deals with the way of continuing to control the process (reconfigurability). In order to help operators in those tasks, the integration of decision-making tools in the supervision system is needed. On one hand, supervision is associated with alarm processing to insure the safety based on dynamic model for fault indicators generation for online FDI. On the other hand the from human point of view, supervision system is seen by the operators in terms of functions and services (or missions) to be achieved. The set of services is organised into coherent subsets, called a User Operating Mode (USOM). At each time, and according to the technical specifications, a system carries on a coherent subset of missions. Each of these subsets is called operating mode. The supervision platform can be seen thus as a finite automaton $G(A,\tau)$ where A is a nodes representing a set of services and τ is the Boolean transition from one node to another controlled by the FDI system: a hardware failure (provided by FDI layer) implies the unavailability of some services and could put back the pursuit of some missions in question [1].

This paper deals with the BG methodology [2] as a decision-making tool for supervision plateform design. The BG as a unified, multi disciplinary and understandable approach is well adapted for those tasks. Indeed, in the normal situation the BG model shows clearly to the operator the physical phenomenon and the power flowing in the process [3]. Furthermore, in incidental or accidental situations, the causal properties of the bond graph allow the operator to resolve problems related with the safety of the supervised process. Compared with classical approaches [4] such as the hybrid automata, the Operating Mode Management (OMM) is much effective due to the separation between the continuous dynamics, represented by the BG continuous state, and the discrete state governed by a classical automaton.

The paper is organised as follows: The first part of the paper presentsBG as integrated design for not only dynamic modelling but also for supervision systems in terms of FDI. The second part concerns the real application of developed methodology to a complex HRES applied to a Hybrid Renewable

Energy Systems (HRES) which consists of Solar Photovoltaic Panels (PV) and wind turbines coupled to an electrolyser for hydrogen production used an energy vector for a fuel cell. and the third section concludes the paper.

2. Bond Graph as integrated design for supervision systems

2.1. Bond graph theory for modelling and diagnosis
Complex mechatronic systems are considered as a combination of different energy components which are exchanging several powers. A BG which is a graph, G(S;A) is a unified graphical language for multi-physic domains. The nodes S represent physical components, subsystems, and other basic elements called junctions. While the edges A, called power bonds represent the power exchanged between nodes. This power is labelled by two conjugated power variables, named effort (e) and flow (f). The set of components named BG element is: $S = \{R \cup C \cup I \cup TF \cup GY \cup Se \cup Sf \cup De \cup Df \cup J\}$ The R-element represents a passive energy dissipation phenomenon, while C, and I model the passive energy storage elements. (Se), and (Sf) are the sources of effort and flow, respectively. Sensors are represented by flow (Df), and effort (De) detectors. Finally, J, is used to connect the elements having the same effort (1-junction), or flow (0-junction). The conservative energy laws are obtained from the latter. TF, and GY are used to represent transformers and gyrators, respectively. From the BG model state space equations can be automatically generated, dedicated softwares are used for it. As shown in Figure 1(a), the power exchanged between two systems A and B indicated by a bond is the product of two variables - a potential variable (e.g. pressure, electrical potential, temperature, chemical potential, force, etc.) called effort (e) and a current variable (e.g. volume flow, current, entropy flow, velocity, molar flow, etc.) referred to as flow (f). One important structural property of the bond graph is its causality concept (algorithmic level of modelling). In the BG, it is denoted by the cross-stroke on the right indicating that the effort acts to the right, the side of the cross-stroke while the flow is in the reverse direction. As example in Figure 1(a), assigned causality means that system A imposes efforts on B. In the corresponding block diagram given by the Figure 1(b), the direction of action is indicated by an arrow on each connection as illustrated. Independently of the causality, the direction of the positive power is indicated by the half-arrow on the bond.

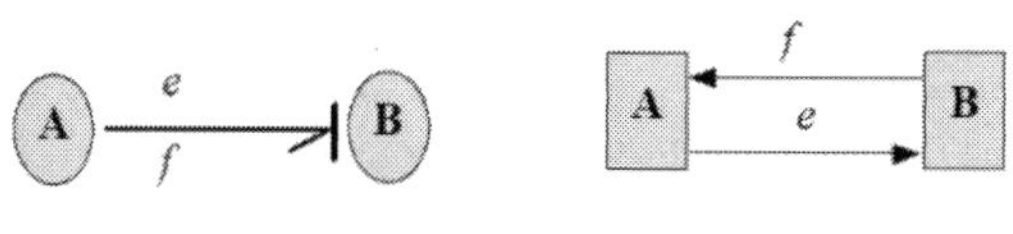

Figure 1. Causal Bond Graph representation (a) and corresponding bloc diagram (b).

Bond Graph modelling technique has been extensively used owing to the behavioural, structural and causal properties, that provide a systematic approach towards development of supervision and fault detection and Isolation (FDI) of highly non-linear and complex thermo-chemical systems. In BG framework, the model based FDI is mainly based upon ARRs as developed in [2].

3. *Application*

3.1. Description of the process
The proposed approach is applied to model a small size experimental HRES presented in Fig. 9b. The system main objective is to produce hydrogen. The system is composed of two sources Photovoltaic Panel (PV) and Wind Turbine (WT) connected through a common DC bus to a battery, an Electrolyzer (EL) and a Proton Exchange Membrane Fuel Cell (FC).

3.2. Bond Graph modelling

3.2.1. Word Bond Graph model (WBG). The technological level is represented by the word bond graph (Figure 3) which consists of decomposing the global system into subsystems. Comparing with classical bloc diagram, input and output of each subsystem are the exchanged power (not an information signal).

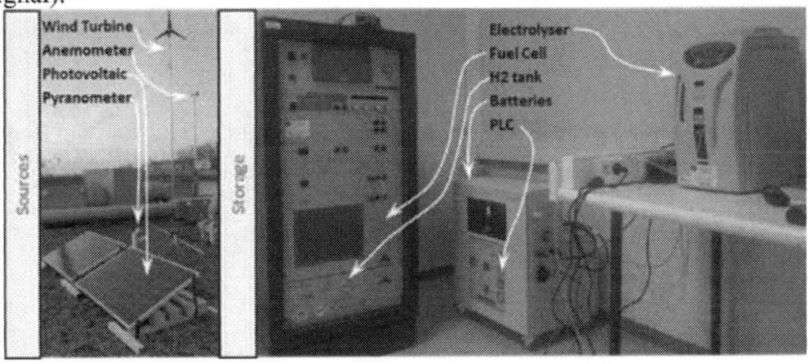

Figure 2. Overview of experimental multisource platform.

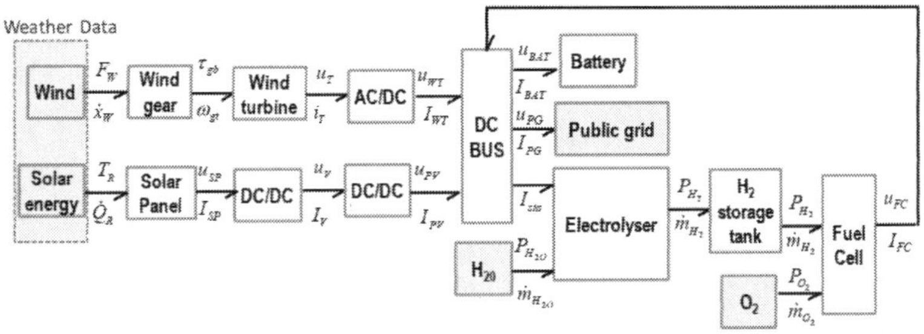

Figure 3. Word Bond Graph of the multisource platform.

3.2.2. Global bond graph model. The considered HRES is a real mechatronic system where occur not only electro mechanical energy but also chemical, thermofluidic, thermal and electrochemical phenomena. In the BG theory are used only a generic pair of power variables to represent all the phenomena. The used power variable (for pseudo and true bond graph) are the pair voltage-current (U,i) for electrical system, pressure-mass flow $(P,\dot{m})$ for hydraulic phenomena, temperature-enthalpy flow $(T,\dot{H})$ for thermal convection (Lagrangian point of view), temperature-thermal flow $(T,\dot{Q})$ for thermal conduction (Eulerian point of view), chemical potential-molar flow $(\mu,\dot{n})$ for chemical reaction and the pair chemical affinity-speed reaction (J,A) for chemical and electrochemical reaction. The global BG model is given Figure 4.

3.3. Supervision design of the system

As discussed before, the supervision platform can be seen as an automaton representing an operating modes management. Three operating modes are distinguished (Figure 5) *OM*1**: Low power** This mode is accessed when the power generated by the renewable sources does not cover the demand. In this case, the batteries are drained at first and then eventually the FC is triggered to use the stored hydrogen as a power back-up and prevent the power shortage. *OM*2: **High power** This mode is activated when the power generated by the renewable sources overcomes the required load. The power surplus is then stored as hydrogen using the Electrolyzer or/and as electricity using the batteries. *OM*3: **Safe power** This mode is activated when the system fails to provide the required power or when one or more faults occur which make some critical components to be unavailable. The availability of the hardware resources, used in each version, for each mission and then for each mode are evaluated by FDI system to control the transition signal τ_{ij} from the mode *i* to the mode *j*.

3.4. Simulation and Experimental results

The global system has been simulated using dedicated software ©20sim. Two scenario have been considered. Scenario 1 concerns the normal faultless behavior of the system represented by Figure 6.

The simulation uses 24 hours weather data of a sunny, average winds day1. The batteries are initially charged at 32% and the hydrogen pressure is about 93% of the maximum capacity of the tank. Between t=[00:00h] and t=[11:45h], the sum of both wind and the solar powers P_r were more than enough to satisfy the load $P_r > 200W$. The system starts then in OM_2: High power. The surplus is stored as hydrogen and electricity using both the electrolyzer and the battery. Between t=[11:45]h and t=[13:02h], the generated power of the sources drops but still more than the load demand 100W < Pr < 200W. The system still operating in OM2, the activation condition of the electrolyzer is then satisfied, as a result, the EL is deactivated and only the batteries are used to store power. After t=[13:02h], the generated power of the sources drops to less than load required power $Pr < 100W$. As a consequence, the system switches to low power mode $OM1$.

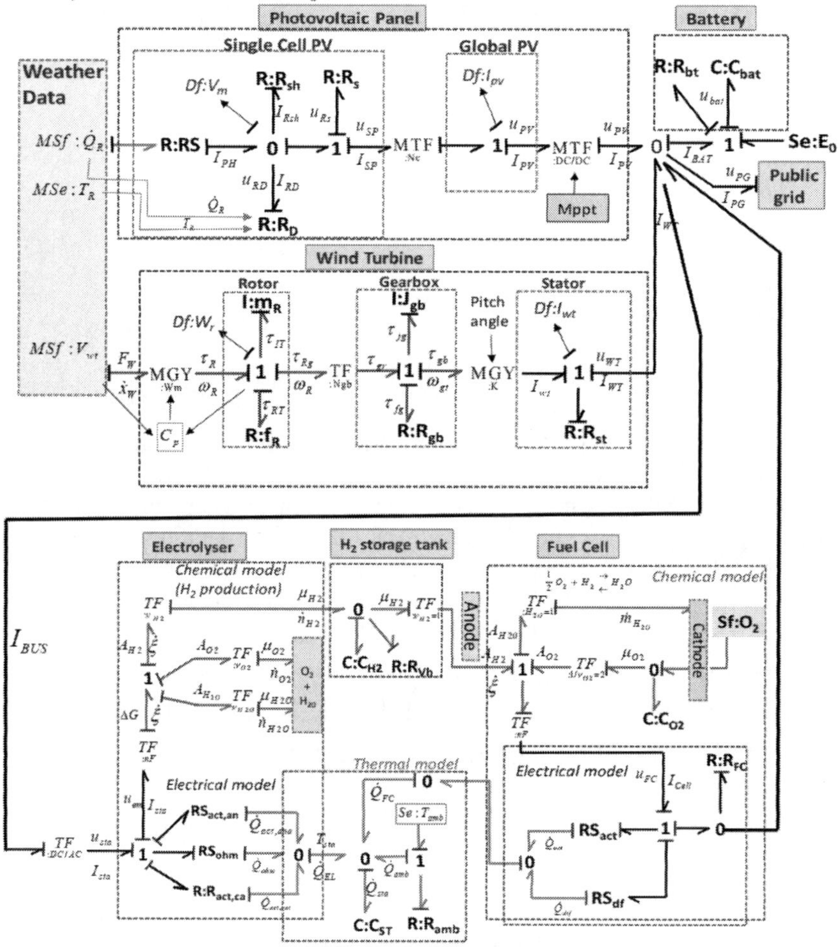

Figure 4. Bond Graph of the multisource platform

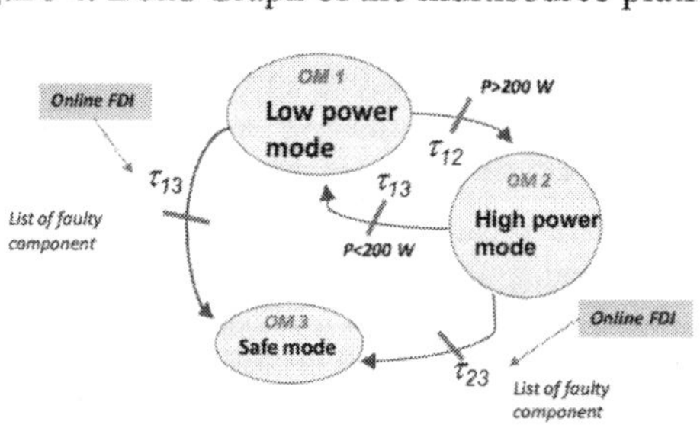

Figure 5. Operating Mode Management supervision.

Scenario 2: Leak in the Hydrogen tank. We consider a leak in the hydrogen tank between t=[10:30-12:30]h. The system, as before, starts in *OM2*, residual signal sensitive to this fault (Figure 7a) overpasses the thresholds indicating the leak detection. When the detection occurs at t=[10:30]h the hydrogen tank is marked as unavailable, consequently the unique version of the mission: secure the H_2 become unavailable. Consequently, *OM1* and *OM2* become inaccessible. According to the OMM, the system must switch to the safe mode *OM3* (Boolean variable τ_{23} or τ_{13} controlled by online FDI becomes 1) when the H_2 leak occurs, fault has been detected and isolated, both sources (wind and solar) are stopped, and their output powers are zero (Figure 6). The Figure 7b illustrates the generated and consumed powers of FC and EL. It shows that the EL is also shut-down after the detection, while the FC is activated in order to reduce the H_2 pressure.

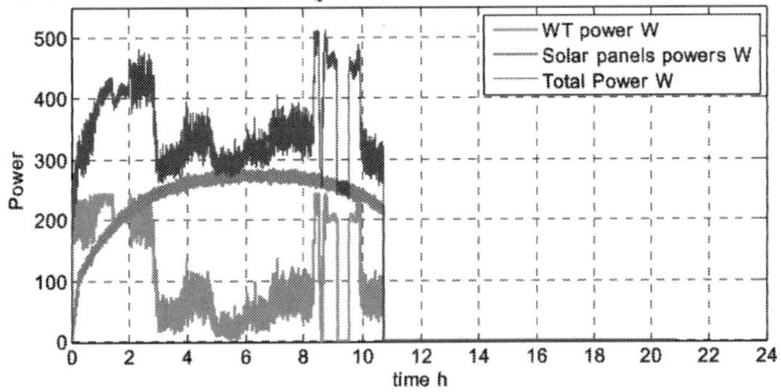

Figure 6. . Simulated generated power.

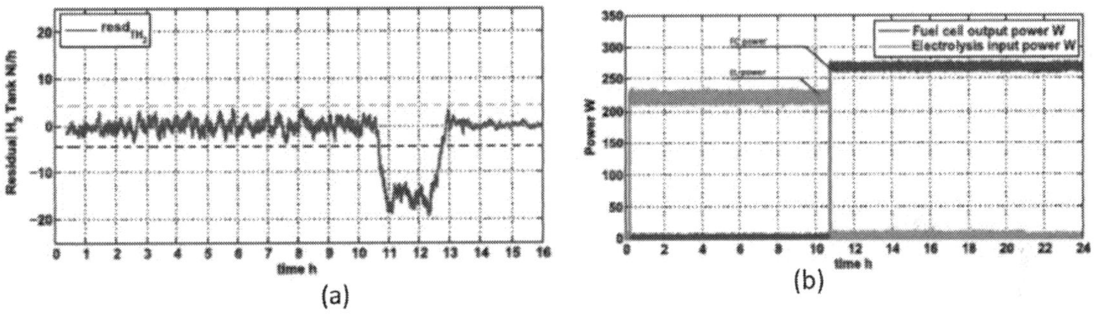

Figure 7. Residual signal reaction to the H_2 tank leak (a) and corresponding FC and EL powers (b) .

4. Conclusion

A Hybrid Renewable Energy System is a complex mechatronic system because of coupling of several kind of energies. Optimal operating mode management of such process need an unified language for modelling and supervision in terms of safety and control. For those tasks it is shown how the Bond Graph as an unified methodology and a real integrated tool is used not only for modelling but also for supervision design. The developed methodology is applied here online to a real multisource platform.

References

[1] Ibrahim, A.L. Gehin and B. Ould Bouamama , *Event Driven Hybrid Bond Graph for Hybrid Renewable Energy System,* Part I: Modelling and Operating Mode Management. International journal of hydrogen energy, DOI: 10.1016/j.ijhydene.2017.10.144 , December 2017

[2] Ould Bouamama, B. *et al. FDI of Smart actuators using Bond Graphs and external models.* Control Engineering Practice, vol. 13, n°2, pp. 159-175, 2005

[3] A.K. Samantaray and B. Ould Bouamama. *Model-based Process Supervision. A Bond Graph Approach.* Springer Verlag,, 490 p. ISBN: 978-1-84800-158-9, Berlin, 2008

[4] B. Ould Bouamama, *et al "Graphical methods for diagnosis of dynamic systems:* Review" Annual Review in Control doi:10.1016/j.arcontrol.2014.09.004, Vol. 38, Issue 2, 2014, pp. 199–219.

An Intelligent Exhaust Gas Processing System for Desktop 3D Printer

Ligang Cai[1,2], Shunlei Li[1,2], Qiang Cheng[1,2], Zhifeng Liu[1,2], Wei Cui[1,2] and Huirong Fu[1,2]

[1] Institute of Advanced Manufacturing and Intelligent Technology, Beijing University of Technology, Beijing 100124, China
[2] Beijing Key Laboratory of Advanced Manufacturing Technology, Beijing University of Technology, Beijing 100124, China.

lgcai321@yahoo.com.cn and lgcai321@aliyun.com

Abstract. With the rapid development of 3D printing technology, desktop 3D printers have gradually become popular. The mainstream 3D printing technologies are FDM（Fused Deposition Modeling）, SLA（Stereolithography）, and SLS（Selective Laser Sintering）. However, these 3D printing technologies will cause environmental problems during printing, and UFP (ultra-fine particles) and some harmful gases and dust will be emitted during work, which will affect human health. There is currently no treatment method for the pollution problem caused by 3D printing. In this paper, an intelligent exhaust gas processing system that can perform adsorption and catalysis is developed for desktop 3D printers such as FDM, SLA, and SLS . Using 51 single-chip microcomputers as the operation and data processing center, on the basis of manual control of the processor switches, a digital universal particle concentration sensor was adopted, and a new type of intelligent exhaust gas treatment system was designed based on the change of the internal particle concentration in the printer. In order to achieve the intelligence of the exhaust gas treatment.

1. Introduction

The 3D printer has brought revolutionary development to the manufacturing industry. It can generate objects of any shape directly from computer graphic data without the need of machining or molds. At present, it has become a trend that is rapidly developing and applied to all walks of life. Desktop 3D printers are gradually becoming popular. The mainstream 3D printing technologies are FDM, SLA, and SLS. However, these 3D printing technologies will cause environmental problems during printing, such as SLA(Stereo Lithography Apparatus) technology, the working principle of it is that a laser beam passes through a scanner controlled by a numerical control device and is irradiated onto a liquid photosensitive resin surface according to a designed scanning path. A layer within a specific area of the surface cures and produces a pungent smell during work. When 3D printing is performed using the FDM technique "melt deposition" technology, the heating head heats the hot melt material (ABS resin, nylon, wax, etc.) to a critical state, and UFP (ultrafine particles) and some harmful gases are emitted at this time. And dust, affect human health[1-3].

With the rapid development of control technology, intelligence has become more and more popular. Automatic control technology and sensor technology are of great significance in intelligent production. 3D printing is widely used in production. Research on how to achieve 3D printing exhaust gas

treatment control Automation and intelligence have very important practical significance and can also provide reference for the realization of other industrial intelligent control.
There is currently no treatment method for the pollution problem caused by 3D printing. In this paper, an intelligent exhaust gas processing system that can perform adsorption and catalysis is developed for desktop 3D printers. The use of 51 single-chip microcomputer for the operation and data processing center, so as to achieve the intelligent treatment of tail gas.

2. Design of control system

2.1. Selecting a Template
The system consists of a digital universal particle concentration sensor, a microcontroller, and a DC motor. This system chooses AT89C51 one-chip computer as the main control chip, AT89C51 one-chip computer has the characteristic such as low power consumption, high performance, low price, make the development cost of the system lower, the operation of the system is more reliable. The system uses a digital particle concentration sensor and uses laser scattering to achieve accurate measurements. The digital particle concentration sensor has the advantages of six-sided omni-directional shielding and stronger anti-interference performance, and its minimum resolution particle size is 0.3 μ m. The system uses a modular design to achieve control functions, control system overall design diagram shown in Figure 1, the main components of the microcontroller main control module, button control module, display module, particulate concentration acquisition module, motor control module, temperature module. Maintaining the Integrity of the Specifications[4].

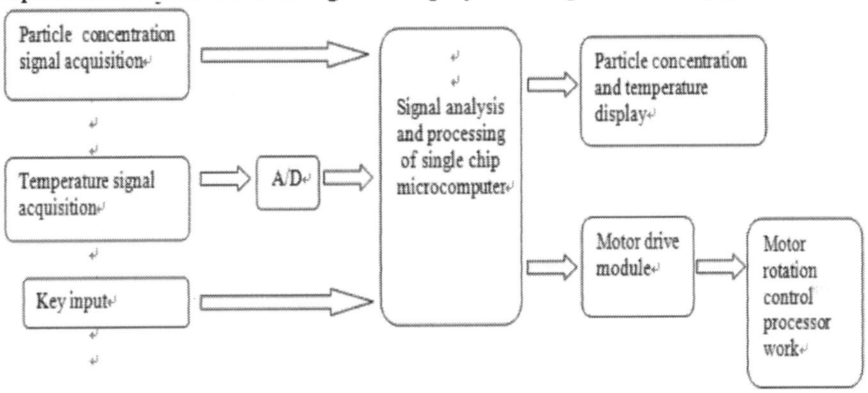

Figure 1. General design block diagram of control system

2.2. Hardware circuit design
The hardware circuit is mainly composed of a crystal oscillator circuit, a reset circuit, a data acquisition circuit, a display circuit, and a power transmission module. Control system hardware circuit shown in Figure 2. Among them, the power supply part through the external socket input AC 24V or DC 12V voltage supply, after the three-terminal integrated regulator voltage output 5V voltage to provide power for the display circuit, the use of 2 regular batteries to provide DC power.

2.2.1. Particle concentration acquisition module. Particle concentration sensor detects whether the concentration of particulates in the desktop 3D printer exceeds the lowest critical value harmful to the human body. The concentration of the detected particulates is compared with the preset concentration limits by the circuit. If the concentration exceeds the upper limit, the motor is turned on; if less than The lower limit, the motor is closed, if the user needs manual intervention and adjustment in special circumstances, can be manually controlled through an external switch to achieve air processor switch.

2.2.2. Temperature Acquisition Circuit Module. Through the circuit diagram shown in the figure below, real-time monitoring of the internal temperature of the 3D printer.

2.2.3. Display circuit module. In this paper, the display module uses LMO16L LCD display, which controls the lighting of that point by giving each level different levels to display the graphics or text in combination.

2.2.4. Motor Control Circuit Module. The motor control module mainly realizes the drive of the DC motor. The ULN2003A chip is used as the core of the drive circuit in this module. Due to the poor load capacity of the single chip microcomputer, this chip provides the power supply control of the motor through the single-chip microcomputer and allows the ULN2003A to assist the single-chip output high current. The motor voltage is satisfied to drive the motor.

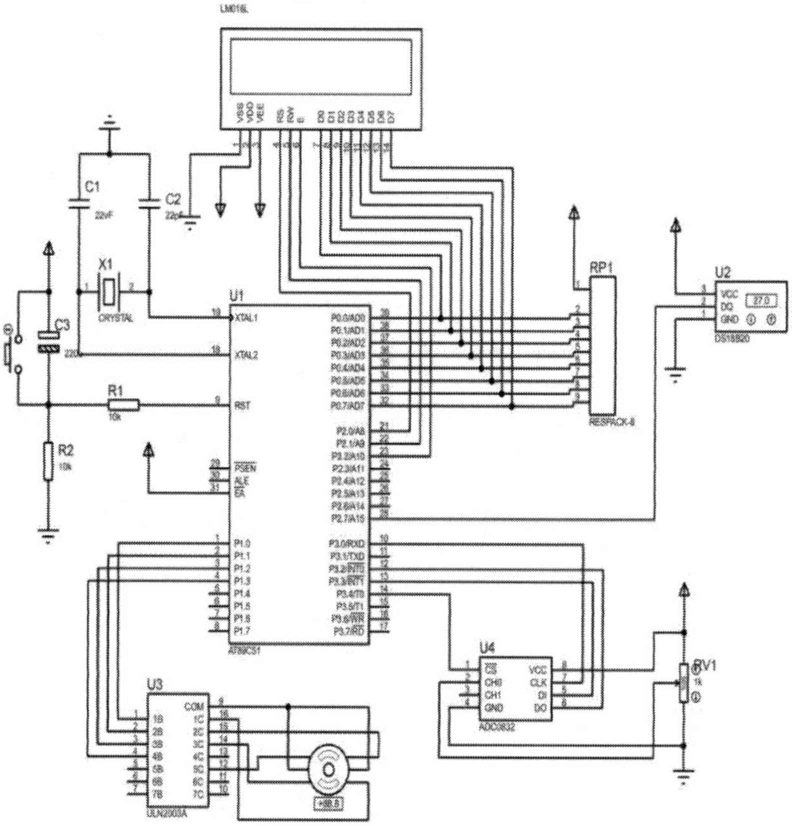

Figure 2. Control system hardware circuit diagram

3. System software design

The main task of the microcontroller microprocessor in this system is to complete the control and processing of the data collected by the particle concentration acquisition module and compare it with the preset standard value. When the value is greater than the standard value, the control motor is turned on.

The main program first completes the initialization, and then starts the particle concentration acquisition module and performs the corresponding output operation. The system flow chart is shown below. System software flow is mainly divided into system initialization, data acquisition, data processing three parts.

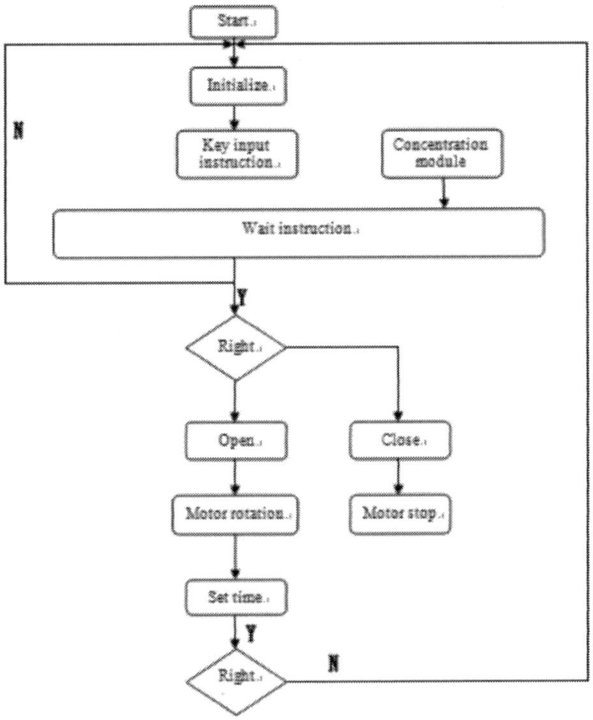

Figure 3. System overall software flow chart

4. Structure design of desktop 3D printer intelligent exhaust processor

4.1. Desktop 3D printer exhaust components

The mainstream application technology of 3D printing is FDM, and its market share is relatively high. When 3D printing is performed using FDM technology, "fusion deposition" technology, the heating head heats the hot melt material (ABS resin, nylon, wax, etc.) to a critical state, and UFP (ultrafine particles) and some harmful gases are emitted at this time. And dust, affect human health.

4.2. Tail gas treatment equipment

For the exhaust of the desktop 3D printer working process, due to its higher temperature, the adsorption equilibrium constant is small, that is, the adsorption volume per unit volume of adsorbent material is small, the required adsorption material is more, and the economy is poor. Therefore, It must be cooled before adsorption. After heating the desorption concentrated exhaust gas to a certain temperature, it catalyzes carbon dioxide and water, and discharges after cooling. The exhaust gas treatment equipment mainly includes an adsorption-desorption device and a catalytic burner. The main structure and function of the adsorption-desorption device are shown in Figure 4.

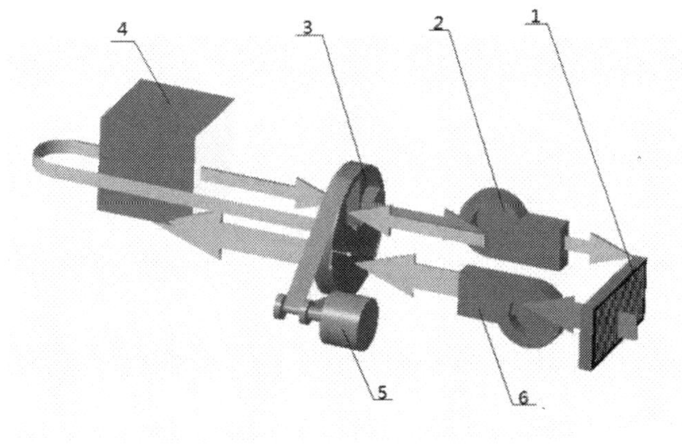

Figure 4. Tail gas treatment equipment structure diagram
1. Filter, 2, Fan (1), 3, Adsorption turntable, 4, Heater, 5, Motor, 6, Fan (2)

4.2.1. Cycling adsorption turntable. Adsorption pad 3 is filled with a shaped adsorbent, and the continuous operation can be realized by the pulley traction, adsorption zone, desorption zone and cooling zone.

4.2.2. Adsorption zone. Adsorption disc 3 VOCs-containing exhaust gas passes through the turntable to this zone for adsorption purification.

4.2.3. Desorption zone. The small amount of hot air blows through the desorption zone of the adsorption turntable 3, the VOCs adsorbed on the adsorption turntable are forcibly desorbed, and the adsorbent is regenerated.

4.2.4. Sorbent cooling zone. The desorption of the adsorbent is cooled with a small amount of low-temperature airflow to ensure the effect of adsorption of the adsorption zone in the turntable 3 .
The main function of the catalytic combustion device is to catalytically combust the desulfurized high-concentration tail gas to produce harmless carbon dioxide and water vapor. The choice of catalyst plays an important role in catalytic combustion. In this project, honeycomb ceramics loaded with noble metals Pd and Pt are used as catalysts, cordierite is the first carrier, and alumina is the second carrier. The catalyst has the advantages of high activity, wide application, low ignition temperature, low poisoning, long service life, and the like. In order to reach the catalyst activation temperature, the honeycomb catalyst was installed in a silicon carbide electric heater to form an adsorption-catalytic combustion composite reactor. In order to avoid the "heat island effect" caused by the high-temperature gas after catalytic combustion on the external environment, an efficient plate-fin heat exchanger is installed behind the composite reactor to reduce the temperature of the burned gas to the atmosphere.

5. Conclusion

The exhaust gas produced by the desktop 3D working process contains solid particles and harmful gases, which will cause some damage to the environment and people. Although it is suitable for FDM desktop 3D printers, this paper can also be used on other desktop 3D printers such as SLA and SLS, and develop an adsorption and catalytic combustion composite reactor based on 51 single-chip microcomputer based intelligent coupling adsorption desorption and catalytic combustion, which can realize 3D printing. The harmless treatment of exhaust gas fills the blank of 3D printing waste gas treatment, and greatly expands the application of desktop 3D printer.

References

[1] Ceretti, E., P. Ginestra, P. I. Neto, A. Fiorentino and J. V. L. Da Silva (2017). Multi-layered scaffolds production via Fused Deposition Modeling (FDM) using an open source 3D printer: process parameters optimization for dimensional accuracy and design reproducibility. *3rd Cirp Conference on Biomanufacturing*. A. Shih and J. Cao. **65**: 13-18.

[2] Corsini, E., S. Ozgen, A. Papale, V. Galbiati, G. Lonati, P. Fermo, L. Corbella, G. Valli, V. Bernardoni, M. Dell'Acqua, S. Becagli, D. Caruso, R. Vecchi, C. L. Galli and M. Marinovich (2017). "Insights on wood combustion generated proinflammatory ultrafine particles (UFP)." *Toxicology Letters* **266**: 74-84.

[3] Gumperlein, I., E. Fischer, G. Dietrich-Gumperlein, S. Karrasch, D. Nowak, R. A. Jorres and R. Schierl (2018). "Acute health effects of desktop 3D printing (fused deposition modeling) using acrylonitrile butadiene styrene and polylactic acid materials: An experimental exposure study in human volunteers." Indoor air.

[4] Tan, H., Q. Yan and Iop (2017). Design of Water Temperature Control System Based on Single Chip Microcomputer. *1st International Conference on Frontiers of Materials Synthesis and Processing*. **274**.

IOP Conf. Series: Materials Science and Engineering **417** (2018) 012035 doi:10.1088/1757-899X/417/1/012035

Investigation on Influence of Geometry on Performance of a Cavity-less Pressure Sensor

Kevin Amith Mathias[1] and **S M Kulkarni**[2]

[1]Department of Mechanical and Manufacturing Engineering, Manipal Institute of Technology Manipal, Manipal Academy of Higher Education, Manipal -576104, Karnataka State, India
[2]Department of Mechanical Engineering, National Institute of Technology Karnataka, Surathkal, PO Srinivasnagar-575025, Karnataka State, India

kevinamith@gmail.com

Abstract. In recent years the automobile comfort and safety greatly improved by incorporation of specialized sensors. These sensors monitor the satisfactory working of different parts of an automobile by providing timely information to the driver. Tyre inflation pressure sensor is one of the important devices whose satisfactory working will improve tyre life as well fuel economy. The present work is focussed on modeling of a Cavity-Less pressure sensor for sensing inflation tyre pressure. Dielectric soft material for the sensor is taken as Silicone rubber. The sensor is modeled using finite element technique. The influence of width to thickness ratio of dielectric material for square geometry is studied. It is observed that change in capacitance increases with the applied pressure. The highest sensitivity of 4.79 fF/kPa is obtained for a sensor with width to thickness ratio of 100.

1. Introduction

Tyres of an automobile play a vital role in providing better comfort and control while driving. Under inflation of tyre pressure may lead to increased rolling resistance, reduced fuel economy and reduce tyre life [1, 2]. Tyre Pressure Monitoring System (TPMS) is a system that monitors tyre air inflation pressure at any instant of time [3]. Direct TPMS makes use of pressure sensor to measure air inflation pressure whereas Indirect TPMS is based on wheel speed measurements [4]. It provides the information about individual tyre pressure more accurately compared to Indirect TPMS [5]. The Direct TPMS consists of Micro-Electro Mechanical System (MEMS) pressure sensor that made up of a thin elastic diaphragm placed above a cavity that is filled with vacuum [6]. The deflection of diaphragm is transducted to measure of tyre inflation pressure.

The repeated deflections of the thin elastic diaphragm, leads breaking the cavity and therefore, shortens the sensor life [7]. One of the alternative to this problem could be a Cavity-Less Pressure Sensor (CLPS), which is not fully explored is discussed in the subsequent section.

The performance of MEMS pressure sensors is generally studied by modeling using Computer Aided Design tools based on Finite Element Analysis (FEA) such as ANSYS, Intellisuite, COMSOL, etc. [8-11]. These software packages are widely used due to the availability of multiphysics modules for analysis of sensor materials. Limited literature is available on performance analysis CLPS using FEA. Thus the following study is an attempt on such modeling and analysis of a CLPS pressure sensor.

IOP Conf. Series: Materials Science and Engineering **417** (2018) 012035 doi:10.1088/1757-899X/417/1/012035

In the present work, a CLPS is modeled using finite element modeling technique using COMSOL Multiphysics software. Silicone rubber material is used as a thin soft dielectric material for the sensor. The influence of Width to Thickness ratio (W/T) of sensor material on sensitivity CLPS is analyzed.

2. Methodology

Working of CLPS is similar to a parallel plate capacitor, where the bottom electrode is fixed and the top electrode is subjected to the pressure with a soft dielectric material placed in between. The intermediate soft material undergoes compression under the application of pressure. This intern increases the capacitance due to reduced gap between the electrodes. This change in capacitance is a measure of the applied pressure [12].

An attempt is made to determine the sensitivity of the CLPS in terms of geometry parameter. The sensitivity is defined as the smallest change in the output capacitance for a given change of the applied pressure. Sensitivity of CLPS depends upon dielectric constant of material, area of the electrode and thickness of the material. In present study influence of geometry on the sensitivity of the sensor is studied.

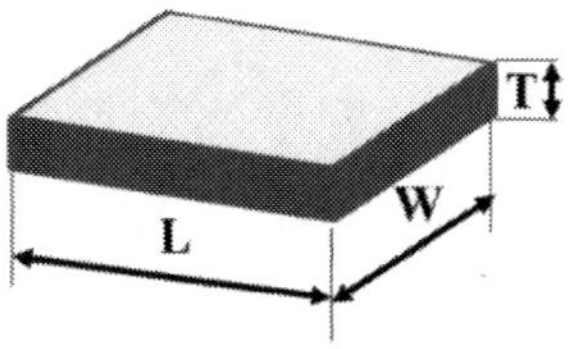

Figure 1. Dimensions of the Sensor.

2.1. Modeling

Modeling of CLPS is done on COMSOL Multiphysics 5.2. It's a commercial finite element modeling and analysis software for analysing domains involving multidiscipline. Three dimensional model of CLPS with square geometry having L x W dimensions as 10 mm x 10 mm (figure 1) is developed as shown in Figure 2 with material properties as shown in table 1. The top and bottom volumes are specified as electrodes. The geometry is converted to finite element model with hexahedral elements as shown in figure 3. The bottom volume is defined with fixed boundary condition whereas top volume is applied with a pressure varying from 0 to 300 kPa in steps of 25 kPa. The bottom volume is grounded and top volume is applied with a voltage of 1 V.

Table 1. Material properties of sensor

Parameters	Values
Young's Modulus	1 MPa
Poisson's ratio	0.47
Dielectric constant	2.9
Density	1100 Kg/m^3

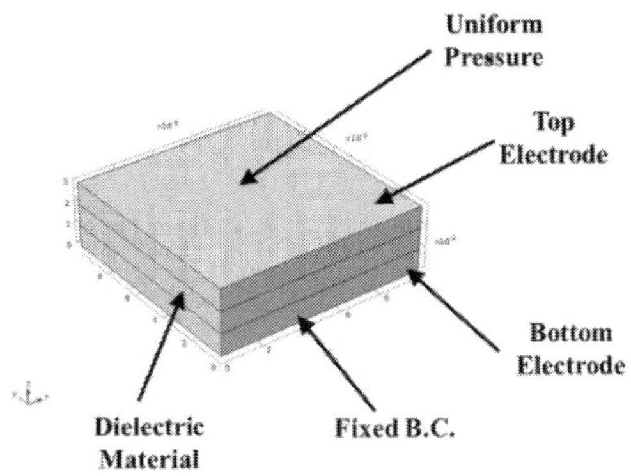

Figure 2. Geometry of CLPS for W/T=5

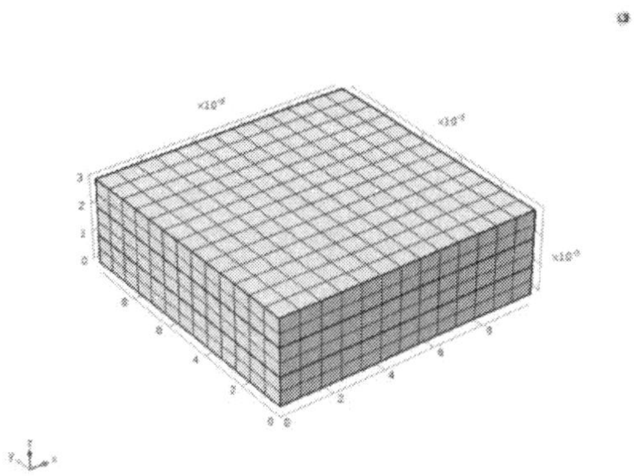

Figure 3. Finite element model of CLPS for W/T=5.

The sensor performance is analysed by coupling three physics in COMSOL Multiphysics software namely, solid mechanics, moving mesh and electrostatics. The moving mesh and electrostatics modules defined in a frame to allow the mesh to move. Initially, solid mechanics module solves for the deflection of dielectric material for the applied pressure. The recalculation of the mesh is done using moving mesh module using the dependent variables of solid mechanics module. Later for solving the electric field, the electrostatics module is used. The electrostatics module uses deformed mesh to calculate the capacitance of the sensor. Change in capacitance is evaluated by taking the difference between the capacitance before and after application of pressure.

3. Results and discussion

The performance of CLPS is analysed by varying W/T ratio as 1, 5, 10, 50 and 100. The plot of deflection and capacitance obtained from COMSOL software for the sensor material is as shown in figure 4 and 5.

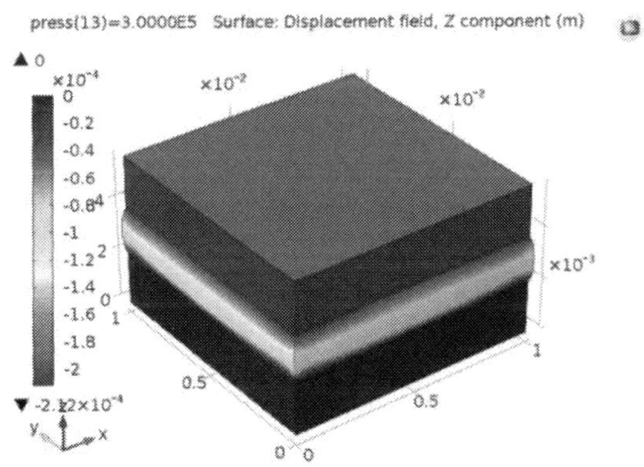

Figure 4. Deflection plot for the pressure of 300 kPa for W/T =5

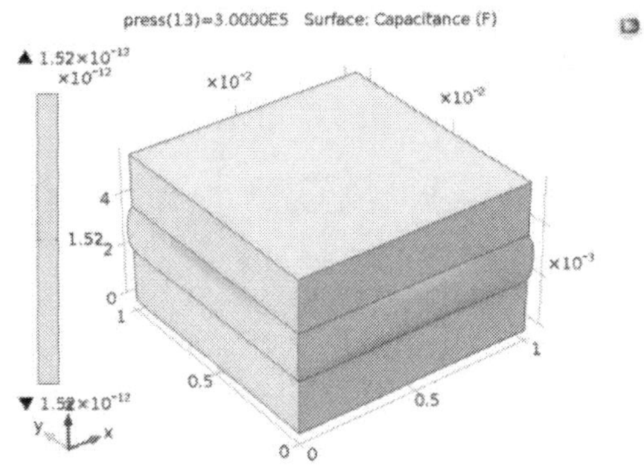

Figure 5. Capacitance plot for W/T=5

The variation of change in deflection of dielectric material with respect to pressure is shown in figure 6. When pressure is applied to the sensor, the dielectric material undergoes deflection. It is observed that deflection gradually increases with the increase in the applied pressure. It is noted that deflection is lesser for higher W/T ratio of the dielectric material. This could be due to the fact that, as dielectric material becomes thinner, it becomes more resistant to deformation.

Figure 7 shows the plot of the change in capacitance with the applied pressure for different W/T ratios. It is observed that the change in capacitance gradually increases with increase in the applied pressure. It is also noted that the W/T ratio of 100 gives the highest change in capacitance. The slope of the graph is less for lower W/T ratio. Therefore, it is observed that the CLPS with higher W/T ratio of the dielectric material is more sensitive. This could be mainly due to the fact that the capacitance is inversely proportional to the thickness of the dielectric material. Highest sensitivity 4.79 fF/kPa is obtained for a sensor with W/T ratio of 100.

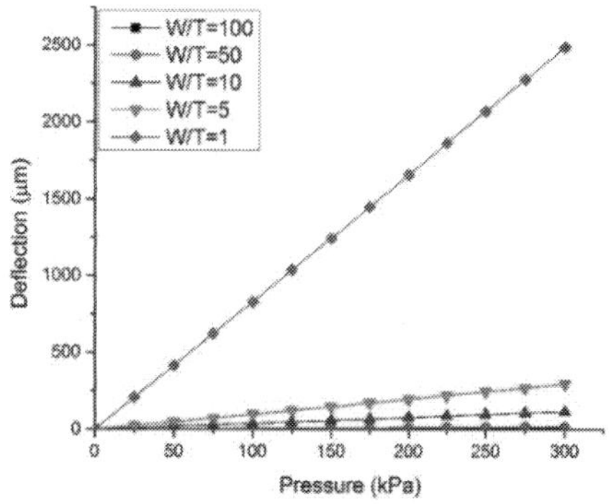

Figure 6. Variation of deflection with pressure

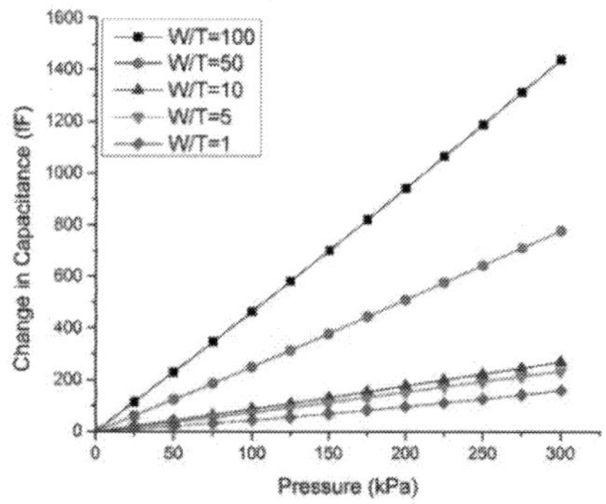

Figure 7. Variation of change in capacitance with pressure

From figures 6 and 7, it is observed that as sensor material becomes thin, the change in capacitance increases whereas deflection decreases. The variation of deflection and change of capacitance are plotted for different W/T ratios for a pressure of 300 kPa as shown in figure 8. It is observed that, both the graphs cross each other for a value W/T ratio ranging from 5 to 10. This implies that, there exists a value of W/T ratio for which change in deflection and change in capacitance are maximum. In the present case for a sensor with W/T ratio of nearly 6.5 exhibits optimum performance.

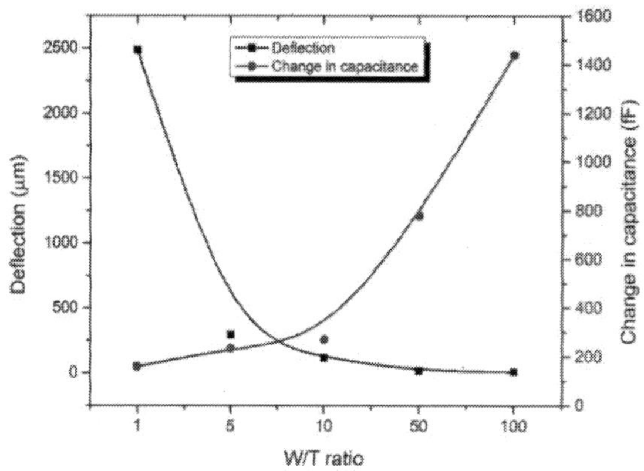

Figure 8. Variation of Deflection and Change in capacitance for different W/T ratio for the pressure of 300 kPa

4. Conclusion

The CLPS successfully modeled and analyzed using COMSOL Multiphysics software. From the above study, it can be concluded that for lower values of W/T ratio the deflection of the sensor material is higher. For higher values of W/T ratio change in capacitance of the sensor is higher that intern increases the sensitivity of the sensor. Highest sensitivity of 4.79 fF/kPa is obtained for a sensor with dielectric material W/T ratio of 100 which is higher than the sensitivity of TPMS sensor cited in [13]. Hence proposed sensor could be a possible alternative for the diaphragm type of pressure sensor used in TPMS. Also, it is observed that as sensor material becomes thinner, the deflection and change in capacitance exhibit trade-off the relationship. The CLPS performance is found to be optimum for W/T ratio of nearly 6.5.

References

[1] Grugett BC, Reineman ME and Thompson GD 1981 *SAE Tech. Pap.*
[2] Suender R, Prokop G and Roscher T 2015 *SAE Int. J. Passeng. Cars. - Mech Syst.* **8** 110–8.
[3] Tian B, Zhao Y, Jiang Z, Zhang L, Liao N, Liu Y and Meng C 2009 *Sensors* **9** 1382–93.
[4] Velupillai, S and Guvenc L 2007 *Control Syst. IEEE* 22–5.
[5] Li T-L, Wu Z, Hu H and Zheng L 2010 *In Acoustic Waves InTech.* Chapter 15 pp 341-58.
[6] Sharma A and Singh J 2013 *In. Control, Automation, Robotics and Embedded Systems IEEE* pp. 1-5.
[7] Takahata K and Gianchandani YB 2008 *Sensors.* **8** 2317–30.
[8] Jeong T 2014 *Int. J. Distrib. Sens Networks.* **10** 902976.
[9] Sathyanarayanan S and Vimala Juliet A 2011 *J. Nanotechnol. Eng. Med.* **2** 034502-3.
[10] Hrairi M and Ahmed M 2008 *In. Dielectric Materials for Photovoltaic Systems* North African Workshop IEEE 168–72. pp 1-4.
[11] Suja KJ and Komaragiri R 2015 *Photonic Sensors* **5** 202–10.
[12] Mathias KA and Kulkarni SM 2016 *Manipal J. Sci. Technol.* **1** 12–20.
[13] Kolle C, Scherr W, Hammerschmidt D, Pichler G, Motz M, Schaffer B, Forster B and Ausserlechner U, 2004, *In. Sensors, 2004 Proceedings of IEEE*, pp 244-247.

Effect of Hole-uniformity of Shrouds on Vortex Shedding Suppression behind a Circular Cylinder

A M Azmi[1], H Jamil[1] and T. Zhou[2]

[1]Faculty of Mechanical Engineering, Universiti Teknologi MARA Selangor, 40450 Shah Alam Selangor, Malaysia .
[2]School of Civil, Environmental and Mining Engineering, The University of Western Australia, 35 Stirling Highway, Crawley, WA 6009, Australia.

Corresponding author: azlinma@gmail.com

Abstract. A passive flow control method over a circular cylinder using perforated shrouds was computationally investigated to manipulate the unsteady wake behind the circular cylinder for vortex shedding suppression. The perforated shrouds were of 33% porosity with uniform and non-uniform holes. The computation was performed at low Reynolds number (based on the circular cylinder diameter), *Re*=300. The objective of the study is to examine the effect of hole-uniformity of the perforated cylinders on the vortex shedding behind the circular cylinder. It was found that both perforated shrouds (uniform and non-uniform holes) significantly suppressed the vortex shedding behind the cylinder, although with an increased drag due to normalization according to cylinder diameter. The perforated shrouded cylinder with non-uniform holes has longer vortex formation length, suppressing vortex better than that of the uniform holes distribution. The intensity of vortices behind the shrouded cylinder wake with non-uniform holes was also considerably smaller than that of the uniform holes.

1. Introduction

Flow over circular cylinder is one of the most studied in fluid dynamics and relevant to many engineering applications. At critical Reynolds number, this flow is characterized by flow separations and vortex shedding behind the cylinder wake. The shedding of vortices results from the rolling up of the shear layers originated at the separation point. These vortices may lead to unwanted vibration when the shedding frequency matches the resonance frequency of the cylinder. This phenomenon is called Vortex induced Vibration (VIV). VIV can accelerate fatigue damage to the cylinder.

A number of passive control techniques have been explored and suggested to reduce drag, suppress fluctuating lift and mitigate vortex shedding from circular cylinders. These techniques include helical strakes, shrouds, splitter plates, guiding vanes and base bleed [1]. Passive control via perforated shrouds was first studied by Price [2]. Price demonstrated that VIV suppression could be achieved by the shrouds due to the delay in the formation of vortex street. Every et al. [3] reported that shrouds with the optimum dimension gave a 50% reduction in oscillations of a plain cylinder in suppressing flow induced vibrations. In the recent years, shrouds using screens have also been studied by researchers for VIV suppression [4, 5]. Interestingly, Huera-Huarte [6] found that about 95% VIV attenuation in addition to 20% drag reduction was achieved using wire meshes.

Mitigation in vortex shedding via shrouds also led to noise suppression. Ikeda and Takaishi [7] showed that suppression of Aeolian tone was achieved due to the jets emitted from the holes of perforated cylinder at regular interval. Boorsma et al. [8] discovered that the associated broadband

noise level is reduced remarkably by using perforated fairings due to the breakdown of the vortex shedding process.
In the turbulent wake regime, vortex shedding occurs in cells along the spanwise direction for Re>200 [9]. These cells are also out of phase, thus reducing the maximum resultant force acting on the cylinder. The average length of the cells may be termed the correlation length. Azmi et al. [10] found that a low correlation length was associated with small scale weak vortices indicating an enhanced three dimensionality in the flow. It is conjectured that by making the holes size of the perforated shroud non-uniform, a greater three dimensionality of the wake can be achieved, hence further suppressing the vortex shedding and reducing lift force. Therefore, this paper aims to investigate the effect of hole- uniformity of the shrouds on the vortex shedding behind a circular cylinder at low Reynold number. Two different porosity representing porosity lower than 50% and below 50% was studied, but only results for the 33% porosity are reported preliminary in this paper.

2. Methodology

The creation of model was done in the Ansys 16 Workbench. The model of the circular cylinder with the perforated cylinder (hereafter is referred to as perforated shrouded cylinder) is shown in Figure 1 and 2. The diameter ratio between the perforated cylinder and the circular cylinder, D/d was set at 2. The Reynolds number, Re was 300 based on the circular cylinder diameter, d. The computational domain was large to eliminate the effects of the blockage and outlet pressure. The streamwise spacing of 40D was chosen behind the cylinder and a 20D spacing was created on each side of the cylinder. Using the porosity definition, β = void area/ total area, the perforation area of the perforated cylinder was fix at an angle of 10° for the uniform holes while three different perforation angles of 5°, 10°, and 15° were used to model the non-uniform holes as shown in Figure 2. The simulation was performed for two different porosity values: 33% and 55%. However, only results using 33% porosity are presented in this paper.

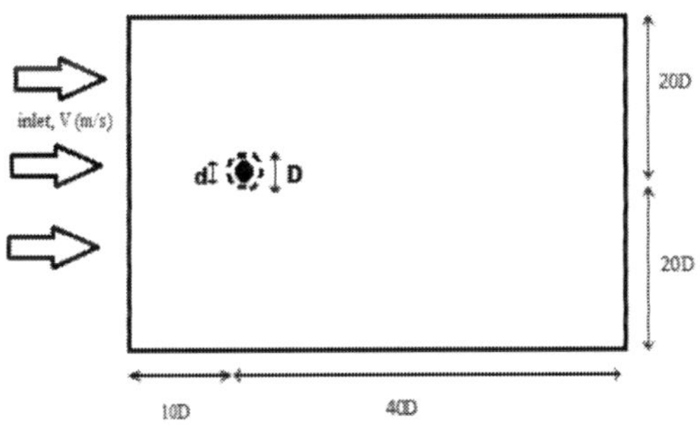

Figure 1. Two dimensional flow over perforated shrouded cylinder.

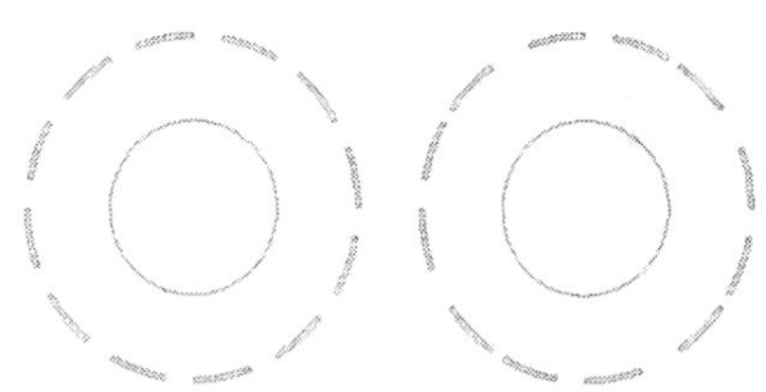

Figure 2. 33% porosity perforated cylinder model (uniform and non-uniform holes).

Figure 3 shows the mesh used in the two dimensional 2-D simulation. The employed mesh was structural mesh. A grid independency study is conducted to ensure a grid–independent solution with high quality mesh. The study was done by increasing the numbers of cells and nodes of the mesh around the cylinder. The mesh elements and the results of Strouhal number and drag coefficients are summarized in Table 1. The grid convergence results reveal that M2 and M3 mesh provide similar outcomes in terms of the Drag Coefficient, however the Strouhal number is slightly changing. Therefore, mesh M3 is considered to give good grid resolution for the models. The grid independence study was only conducted for the shrouded cylinder of 55% porosity. A similar fine mesh scheme was used for the 33% porosity.

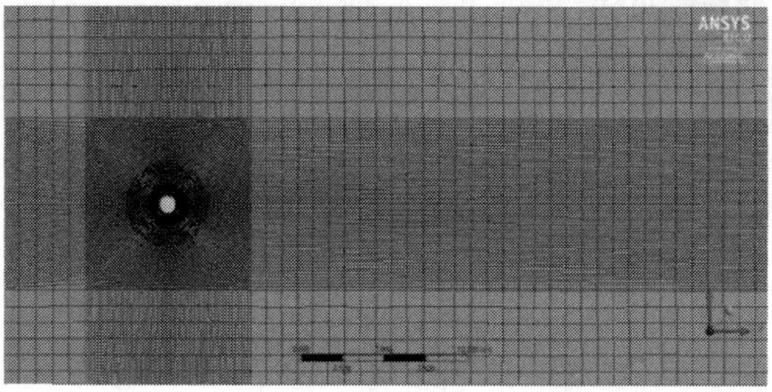

Figure 3. Mesh structure around perforated shrouded cylinder.

The simulation was performed using viscous laminar model with transient time for unsteady flow at Re=300. The wall and cylinder setup was no-slip condition and stationary wall boundary conditions. The second implicit solution was utilized. The simulation used hybrid initialization and the time step size was 0.001s for 1500 number time with 50 maximum iterations per time step to capture the flow structure at the wake region of the cylinders.

Table 1. Grid independence study

55% uniform perforated shrouded cylinder				
Mesh	Nodes	Cells	Drag Coefficient, C_d	Strouhal Number, St
M1	38209	37808	1.35	0.080
M2	39454	39019	1.35	0.075
M3	42602	42114	1.36	0.070
55% non-uniform perforated shrouded cylinder				
Mesh	Nodes	Cells	Drag Coefficient, C_d	Strouhal Number, St
M1	35591	35187	1.33	0.100
M2	36658	36214	1.33	0.090
M3	39383	38901	1.33	0.085

3. Results

Results are presented in terms of force coefficients, Strouhal number, velocity contours, velocity vector plot and vorticity contours. Comparison are made between the wakes of uniform and non-uniform holes shrouded cylinders on vortex shedding behaviour and suppression.

3.1 Model validation

The simulation was first validated by comparing result of the flow over a circular cylinder with experiment and other simulation. The validation was done to ensure the right simulation setup. Both simulation and experimental results were in good agreement with each other, a difference of between 1% and 10% as shown in Table 2.

Table 2. Strouhal and drag coefficient at Re=300

Circular cylinder		
	Drag Coefficient, C_d	Strouhal Number, St
2D- computation Mital and Balachandar, 1997 [11]	1.26	0.213
2D-Numerical simulation [12]	1.37	0.215
Experiment [13]	1.22	-
Experiment [14]	-	0.203

Present study	1.28	0.21
33% porosity perforated shrouded cylinder		
Uniform	1.43	0.10
Non-uniform	1.52	0.08

3.2 Drag and Lift Coefficient

Figure 4 shows the time history of the drag and lift coefficient, C_l and C_d, respectively for the circular cylinder and perforated shrouded cylinders (uniform and non-uniform holes). The periodic sinusoidal nature of the flow in the circular cylinder is obvious due to the presence of correlated vortex shedding. This is in contrast to the flow in the perforated shrouded cylinders where the oscillation are significantly damped signifying very low correlation of vortex, especially observed in that of the non-uniform holes. The drag coefficients for both shrouded cylinders are quite similar, with the non-uniform holes cylinder gives slightly larger value than that of the uniform-holes, but both values are about 20% higher than that of a circular cylinder as tabulated in Table 2. This is due to the normalizing the parameter based on the circular cylinder diameter.

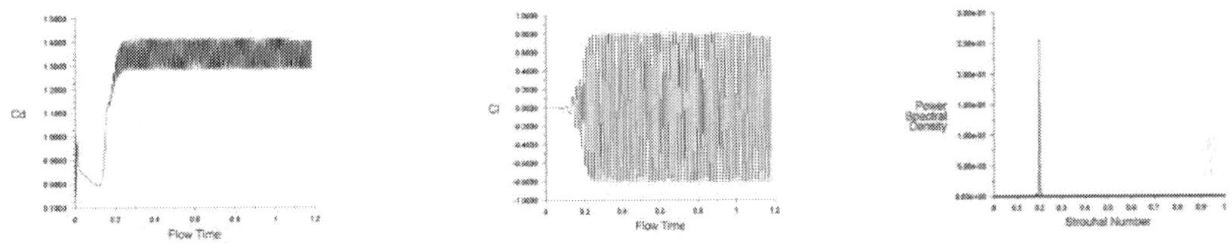

a) Circular cylinder

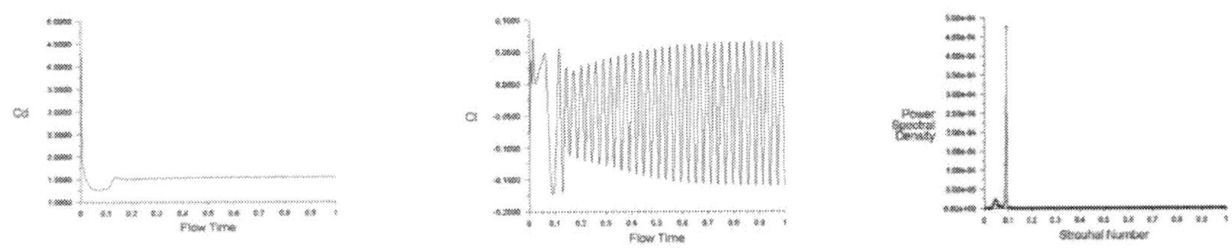

b) Shrouded cylinder with uniform holes (β=33%)

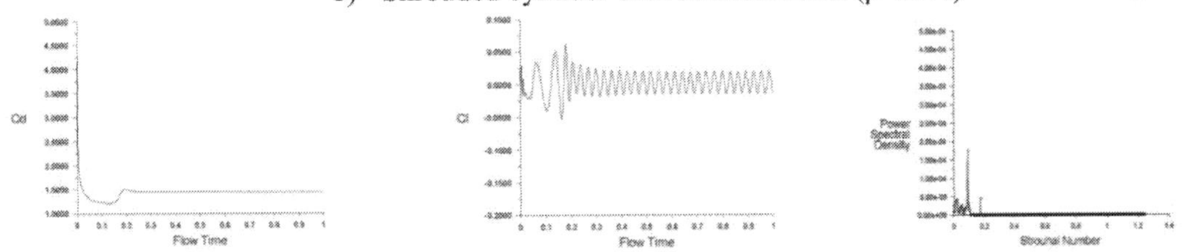

c) Shrouded cylinder with non-uniform holes (β =33%)

Figure 4. Time history of Drag and Lift coefficient and Strouhal number for circular cylinder and perforated shrouded cylinder at Re=300.

3.3 Strouhal number

Figure 4 also shows the Spectral Analysis of Lift Convergence for the circular cylinder and perforated shrouded cylinder, obtained using Fast Fourier Transform. The amplitude spectrum of the Lift is plotted as a function of the dimensionless frequency. A sharp peak of large amplitude is observed for the circular cylinder wake at St=0.2, whereas several peaks of substantially low amplitude are evident for the shrouded cylinder wake with the highest peaks occur at St of around 0.1, indicating vortex has been suppressed. A considerably lower amplitude is observed by the non-uniform holes in comparison to that of uniform holes indicating the former suppress vortex better than the latter.

3.4 Velocity contour and velocity vector

Figure 5 shows the normalize velocity contour and velocity vector for the circular cylinder and perforated shrouded cylinders. The circular cylinder shows a reverse flow immediately downstream, circulating in the wake, whereas both the shrouded cylinders reveal that circulation is mainly observed in the annular region between the cylinder and the perforated cylinder. At the central region, region of significantly low velocity is observed clearly in both shrouded cylinder wakes. This region is elongated farther down the wake of the non-uniform holes in comparison to that of the uniform holes. It seems that this 'void' region prevents early interaction of the shear layers to form the large coherent vortices in the wake.

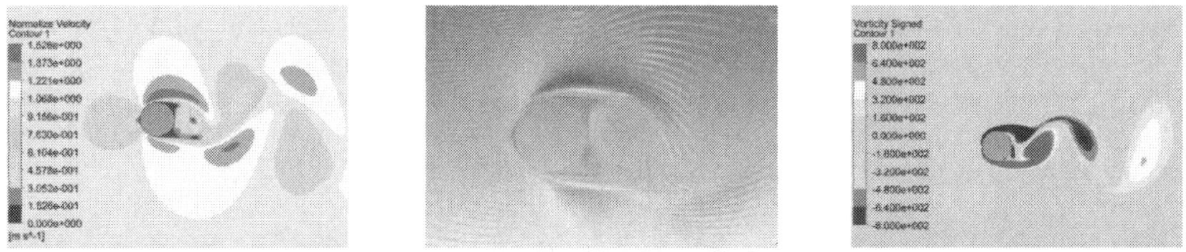

a) Circular cylinder

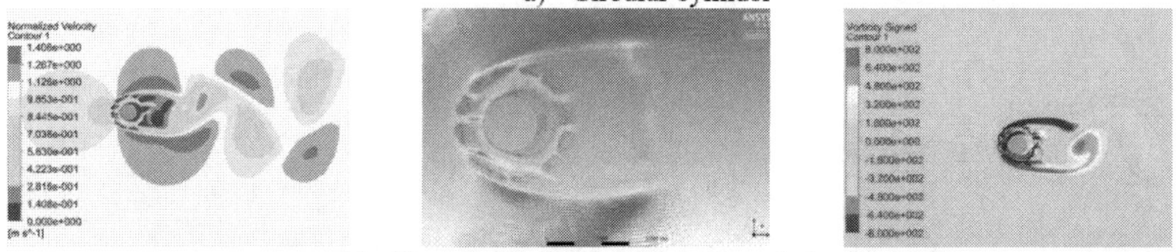

b) Shrouded cylinder with uniform holes (β=33%)

c) Shrouded cylinder with non-uniform holes (β =33%)

Figure 5. Normalized velocity contours , velocity vectors and vorticity contours.

3.5 Vorticity contour

Figure 5 also shows the vorticity contour in the wake of all cylinders. The red color indicates clockwise vortices while the blue indicates counterclockwise vortices. Formation of vortices of high intensity occurs immediately behind the circular cylinder while low intensity vortices occurs quite a distance downstream in the wake of the shrouded cylinders. The vortex formation length for the shrouded cylinders is longer for the non-uniform holes in comparison to that of the uniform holes, consistent with the Lift, Strouhal and velocity contour results discussed previously.

4. Conclusion

Numerical simulations for two dimensional flow over perforated shrouded cylinders at Reynolds number of 300 were performed to understand the effect of hole-uniformity in controlling the wake behind a circular cylinder. Results showed that the shedding of vortex was controlled significantly in the wake of the circular cylinder using both perforated cylinders with uniform and non-uniform holes of 33% porosity, but superior suppression was achieved with that of non-uniform holes, indicated by the low intensity vortices and long vortex formation length. This suppression however, does not take into account the effect in the spanwise direction which would further enhance the vortex suppression effect.

Acknowledgments

AMA wishes to acknowledge financial support from Universiti Teknologi MARA and the Ministry of Higher Education Malaysia (600-RMI/FRGS 5/3 (0100/2016)).

References

[1] Zdravkovich MM 1981 Review and classification of various aerodynamic and hydrodynamic means for suppressing vortex shedding *J. Wind Eng. Ind. Aerodyn.* **7(2)** 145-189

[2] Price P 1956 Suppression of the fluid-induced vibration of circular cylinders *J. Eng. Mech.* **82(3)** 1-22

[3] Every MJ, King R, Weaver DS 1982 Vortex-excited vibrations of cylinders and cables and their suppression *Ocean Eng.* **9** 135–157.

[4] Azmi AM, Zhou T, Cheng L, Wang H, Chua LP 2012 On the Effectiveness and Mechanism of Vortex-induced Vibration Suppression using a Screen Cylinder *The Twenty-second International Offshore and Polar Engineering Conference* International Society of Offshore and Polar Engineers

[5] Azmi AM, Zhou T, Zhou Y, Chen J, Cheng L 2015 The effect of a screen shroud on vortex-induced vibration of a circular cylinder and its wake characteristics *The Twenty-fifth International Ocean and Polar Engineering Conference* International Society of Offshore and Polar Engineers

[6] Huera-Huarte FJ 2017 Suppression of vortex-induced vibration in low mass-damping circular cylinders using wire meshes *Mar Struct.* **55** 200-213

[7] Ikeda M, Takaishi T 2004 Perforated pantograph horn aeolian tone suppression mechanism Q. Report of RTRI, **45(3)** 169-174

[8] Boorsma K, Zhang X, Molin N, Chow LC 2009 Bluff body noise control using perforated fairings *AIAA J.* **47** 33-43

[9] Sumer BM 2006 *Hydrodynamics Around Cylindrical Structures* (World scientific)

[10] Azmi AM, Zhou T, Zhou Y, and Cheng L 2016 Statistical analyses of a screen cylinder wake *Fluid Dyn. Res.* **49(1)** 015506

[11] Mittal R, Balachandar S 1997 On the inclusion of three-dimensional effects in simulations of two-dimensional bluff-body wake flows *Proc. of the ASME Fluids Eng. Div. Summer Meetg. (Vancouver)*

[12] Rajani BN, Lanka HS and Majumdar S 2005 *Laminar flow past a circular cylinder at reynolds number varying from 50 to 5000* NAL PD CF 0501

[13] Wieselsberger C New data on the law of hyfro and aerodynamic resistance 1992 NACA TN 84

[14] Williamson CHK 1996 Vortex dynamics in the cylinder wakes *Ann. Rev. Fluid Mech.* **28** 477-539

IOP Conf. Series: Materials Science and Engineering **417** (2018) 012037 doi:10.1088/1757-899X/417/1/012037

Heating lines Arrangement Method of Hull Complex Surface Based on Interpolation Principle[1]

Shuai Wang [1], Kaifa Zhou [2], Jiacan Sun [2], Yifan Yu [3], Zhixiang Wang [4]

[1]Department of Mechanical Engineering, Chongqing Vocational College of Transportation, Chongqing, 402247,China;
[2]College of Civil Engineering, Chongqing Jiaotong University, Chongqing, 400074, China;
[3]International College, Chongqing Jiaotong University, Chongqing 400074,China;
[4]Research Centre of Special Ship Digital Design and Manufacturing Engineering Technology, Chongqing Jiaotong University, Chongqing, 400074,China.

Corresponding author e-mail address:kfzhou@cqjtu.edu.cn.

Abstract. The technology of hull complex surface is an important part in the ship-building industry. It has been widely concerned by the researchers about how to make the complex surface forming better and faster. In this study, an interpolation principle is proposed and utilized to substitute for the conventional manual process. In the proposed process, the new heating line arrangement methods are adopted to conFigure an equivalent forming arc plate, S plate, sail plate and saddle plate according to the interpolation principle. And they are transferred to a numerical simulation model using COMSOL Multiphysics, and finite element analysis is conducted to check the formability of the process. Further, experimental investigations are carried out to verify the feasibility of the process using the new heating lines. The results obtained from the finite element analysis are in good agreement with results of experiments. Consequently, it is confirmed that four kinds curved thick plate could be obtained by interpolation principle heating lines arrangement methods.

1. Introduction

Large complex curved plate is widely used to construct hull structures in the ship-building industry. Complex curved surface forming technology is a hard issue in ship-building for a long period. On the one hand, the ship manufacturing technology is becoming more and more advanced, such as the demand of luxury cruise ships are increasing, the hull surface shape is more complex than before, the research on the hull surface rapid precision forming method and manufacturing equipment also has attracted much attention. On the other hand, this procedure, has low productivity, is ineffective, and is economically taxing because all the processes involved in the procedure are done by experts based on their experience, and the traditional forming method cannot meet the precision requirement of ship manufacturing.

Therefore, accurate hull surface forming technology can greatly improve the efficiency and precision of the manufacturing process, and it has a positive impact on improving the competitiveness of ship manufacturing industry. As for hull complex curved surface forming technology, researchers

[1] This work was supported by the Science & Technology research project of the Chongqing Municipal Education Committee (CQKJ) grant funded by the Chongqing government (No. KJ1605403).

have made some positive exploration in hull surface forming method, optimization on machining route arrangement and the automation of hull surface forming system.

Chen J predicted the hull plate bending deformation by the temperature field of FE simulation, and used FE model to calculate the shape of steel plate deformation, the simulation results was consistent with the experimental results [1]. Ishiyma M etc. in Japan IHI heavy industry co., LTD used finite element method to simulate the plate elastic-plastic deformation under static induction heating, and compared the simulation results with the experimental results, finally they confirmed that the finite element modeling was reliable [2]. Imatani S has studied and discussed the thermodynamic mathematical model of plate bending deformation under high frequency induction heating, and proved that the mathematical model was feasible combined FE simulation and the experiments [3]. A study on a continuous thermal processing program applied contour processing methods was developed by Ishiyma M [4]. Reutzel E W used finite element method to simulate the deformation of metal plate at different geometric sizes. The deformation of metal plate was obtained by the method of progressive approximation and the computation time was optimized greatly [5]. Jong G S and his partner set up three-dimensional finite element simulation analyze modeling to study the high frequency induction heating plate temperature field and stress field of the thermal stress bending forming, and put forward the processing algorithm of saddle plate high frequency induction heating thermal stress forming, then the artificial neural network method was applied to parameter prediction [6].

A simplified thermodynamic model is used to study the forming process of metal plate under high frequency induction heating with finite element simulation method, and the angular deformation of sheet metal has been predicted [7]. A set of system IHI-Alpha which can automatically design and process the program was developed by Yu G [8]. Professor Jang C D used electromagnetic - thermal - plastic large deformation structure coupling analyze method to simulate of the plate bending process under high frequency induction heating, and established a high frequency induction heating plate forming finite element model and computer integrated manufacturing system [9]. Nguyen T T et. put forward a kind of artificial neural network to establish a method which can simplify the relation between deformation and heating parameters, the results showed that this method can easily to calculate the heating parameters in the deformation of the hull plate in the heating process [10]. Jan C D proposed an iterative method of the hot-line position under machining control mode through intelligent robot [11]. The system of flame heating lines heating plate computer auxiliary design developed by Ueda Y etc., and the relationship between induction heating conditions, inner heat source distribution by the high frequency heating planking forming experiments is also studied [12]

Jan C D proposed an iterative method of the hot-line position under machining control mode through intelligent robot [10]. Heo S C In this study proposed a flexible forming process, and it is confirmed that a double-curved thick plate could be obtained by the flexible forming process [13]. Seong W J etc. developed an automation system to increase not only the plate forming productivity, but also product quality and establish the relationship between the bending angle and the radius of the curvature of the deformed curve shape[14]. Pankaj B found initial plate templates to be applied in the desired shape of the ship plate and decide the line heating parameters[15]. Lee I K proposed a hot multi-point forming process. The hot multi-point forming die can form the various curved surface by controlling the height of each punch. In addition, the cooling system was installed inside each punch and can be selectively used as a punch or a cooling nozzle [16].

Since then, the hull plate curved surface forming has been extensively explored and it is still under investigation as well in methodological aspects as in concrete applications. In this study, a relatively simple heating lines of interpolation algorithm is developed, and the heating lines of arc plate, S curved plate, sail plate and saddle plate is obtained in this method. And the nonlinear finite element simulation prediction of plate forming is carried out by COMSOL Multiphysics software. Furthermore, experimental investigations are taken to verify the feasibility of the hull complex curved surfaces forming process. From the results, displacements of finite element simulation and experimental results are compared.

IOP Conf. Series: Materials Science and Engineering **417** (2018) 012037 doi:10.1088/1757-899X/417/1/012037

2. Complex hull surface craft design

2.1. Interpolation principle

Interpolation is a method of constructing new data points within the range of a discrete set of known data points. One of the simplest methods is linear interpolation. That is the method of calculating the intermediate point between the known points according to some algorithm, also known as "the density of data points".

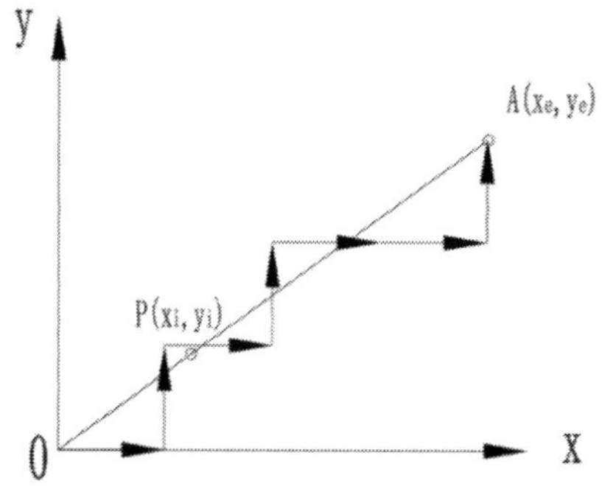

Figure 1. Linear interpolation principle.

Figure 1 is the line in first quadrant. The end point coordinate is *A (Xe, Ye)*, starting coordinate is origin point (*0, 0*), random point coordinate is *P (Xi, Yi)*, which plots around the line. There are three possibilities: *P* is above the line, *P* is on the line, *P* is below the line. In order to determine the relationship between *P* point and the line, give the following judgment:

$$\frac{Y_i}{X_i} = \frac{Y_e}{X_e}, \tag{1}$$

then, $$Y_i \cdot X_e - Y_e \cdot X_i = 0 \tag{2}$$

Here we definite deviation discriminant function is *Fi*, and its equation is as follow:

$$F_i = Y_i \cdot X_e - Y_e \cdot X_i, \tag{3}$$

the relationship between *P* and the line is determined by F_i:

- When $F_i = Y_i \cdot X_e - Y_e \cdot X_i > 0$, *P* is above the line;
- When $F_i = Y_i \cdot X_e - Y_e \cdot X_i = 0$, *P* is on the line;
- When $F_i = Y_i \cdot X_e - Y_e \cdot X_i < 0$, *P* is below the line;

If $F_i \geqslant 0$, move one step along $+X$;

If $F_i < 0$, move one step along $+Y$, in a recursive form, until the end.

In the interpolation principle, the heating lines of sail plate and the saddle plate is redesigned. However, the arc and S curved plate are designed according to the method of line heating. In the design process, the heating lines are obtained based on conventional heating lines of sail plate and saddle plate in the literature [17]. The conventional heating lines (arcs in the Figures) of sail plate and saddle plate are shown in Figure 2 and Figure 3. The number "1" in the Figures expresses heating lines on the top surface, and the number "0" is the under surface.

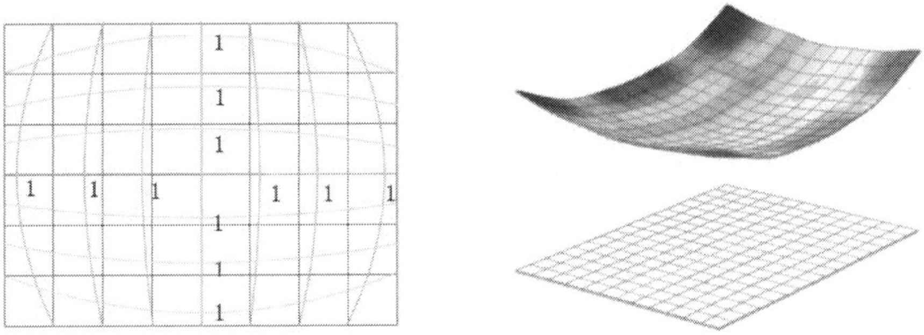

Figure 2. Sail plate heating lines and approximate implementation.

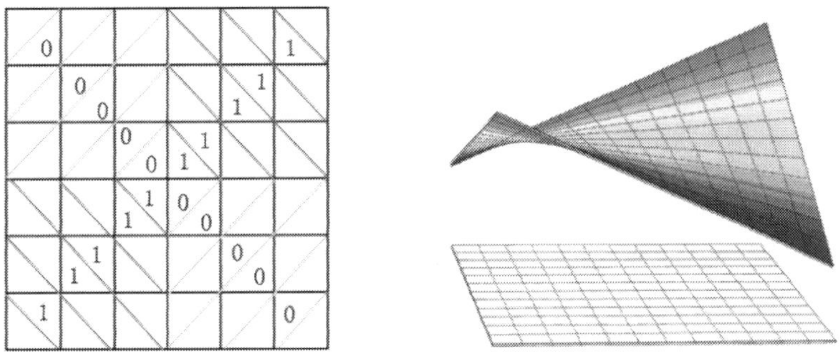

Figure 3. Saddle plate heating lines and approximate implementation.

2.2. The heating lines plan of hull surface

The heating lines of the sail plate and saddle plate, which marked a number in Figure 2 and Figure 3, are rearranged by using the interpolation algorithm. The size of the hull plate is 1000mm * 600mm* 12mm in designing process. The steps are as follow: firstly, the hull plate is divided into squares, and the size of each squares is 80mm*80mm, then the last one at the end is 100mm*80mm. Secondly, the heating lines in each square are replaced to the horizontal and vertical straight lines. All of the Numbers "1" in Figure 4 represents heating on the upper surface of plate, and the number "0" represents heating on the lower surface of plate in Figure 5.

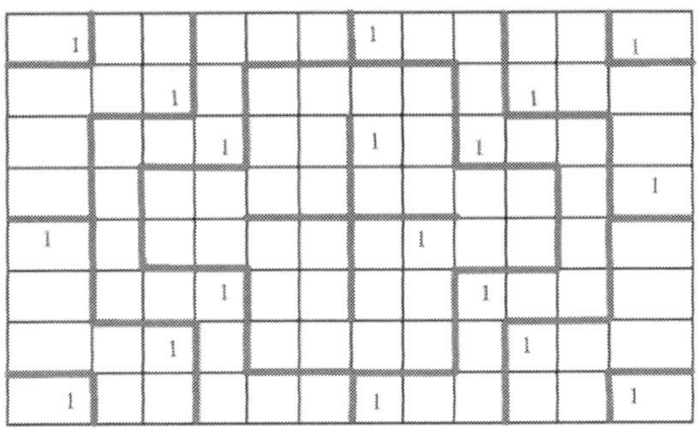

Figure 4. Sail plate heating lines arrangement

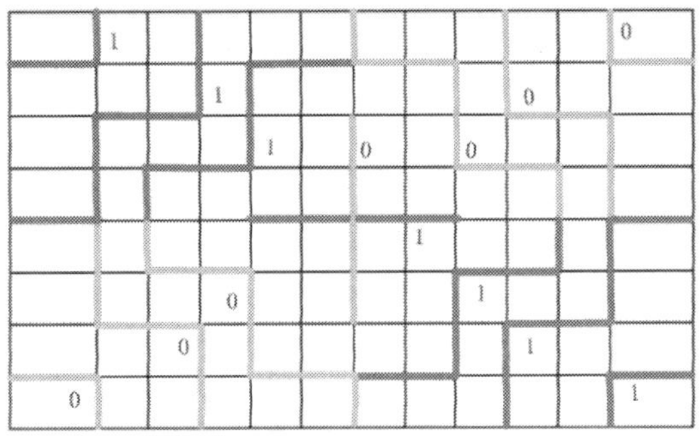

Figure 5. Saddle plate heating lines arrangement

Heating lines of arc plates and S curved plates is designed according to the thermal stress forming rule of hull plate, which is rearranged in Figure 6 and Figure 7. The heating sequence of the arc plate is appeared in Figure 6 and the numbers represents the heating sequence, and the number "1 "represents the plate is heated in the first time, the number "2" represents the second time, and the number "3" represents the last heating. And the number "1" is middle line of the plate, and the number "2" located at 1/4 of half plate. The number "3" located at 1/2 position after first and second times. The S cured plate heating lines are shown in Figure7. The numbers "1" and "2" represent heating sequence. The solid line represents the upper surface heating lines, and dash line represents the lower surface heating lines. The number "1" and "2" are located at 1/4 and 3/4 of the hull plate.

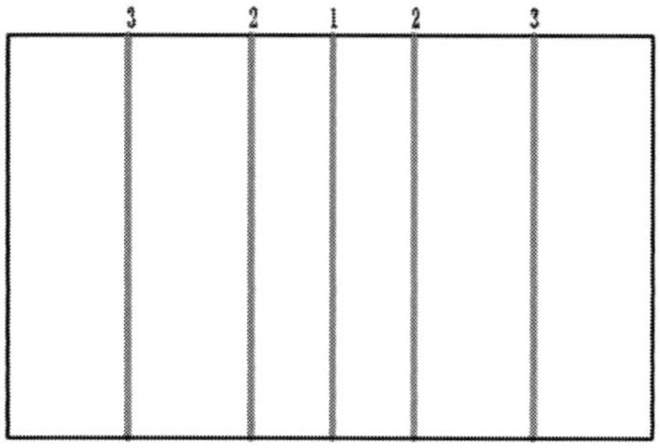

Figure 6. Arc plate heating lines arrangement

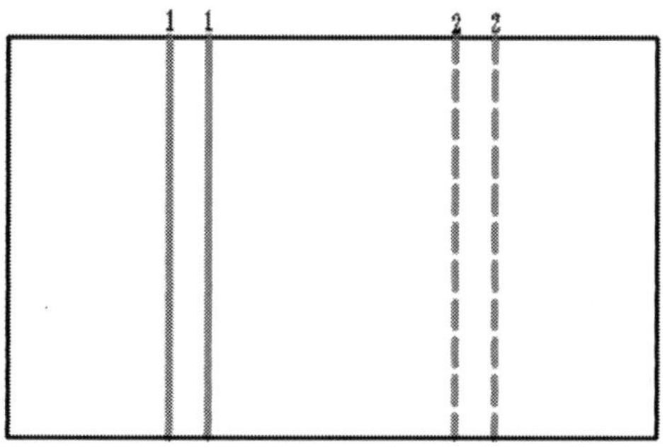

Figure 7. S urved plate heating ling arrangement

For the interpolation principle is the data intensive process, the length of sail plate and saddle plate heating lines is larger than the length of conventional heating lines. However, the conventional heating lines of hull plate are intensive, and the interpolation form heating lines are optimized. Heating lines actual distance along the lengthwise and widthwise directions were investigated to confirm the forming accuracy. The interpolation heating lines actual distance in Figure 4 was measured as 5,740mm, which has about 58.48% error with regard to the conventional heating lines. In Figure 5, the interpolation heating lines actual distance was also measured as 5,740mm, with about 75.65% error. These errors could be considered as an ineviTable characteristic of the "interpolation form" forming process. The results are shown in the Table 1.

Table 1. The comparison of heating lines and actual distance.

	Heating lines actual distance of conventional form (mm)	Heating lines actual distance of interpolation form (mm)	Percentage took by the interpolation form (mm)
Sail plate	9815	5740	58.48%
Saddle plate	7588	5740	75.65%

In Table1, sail plate and saddle plate interpolation forms heating lines distance are shorter than conventional forms obviously. It means that interpolation heating lines are able to shorten the processing time greatly, and improves machining efficiency.

3. Finite element prediction of hull surface forming

3.1.Finite element simulation scheme

To predict the deformation of large hull plate, the heating lines based on interpolation method is calculated by finite element (FE) simulation. The finite element prediction model of arc plate, S curved plate, sail plate and saddle plate is established by using the COMSOL Multiphysics software. The current is 3200A in heating process, the power is 35kw, and the heating time is 10s. Air cooling method is used in the cooling process and the cooling time is 9000s. The finite element simulation scheme is shown in Table 2.

Table 2. Finite element simulation scheme

Arc plate	S curved plate	Sail plate	Saddle plate
Heat 10s， cool 9000s	Heat 10s， cool 9000s	Heat 10s， cool 9000s	Heat 10s， cool 9000s

3.2. Finite element simulation of hull surface

3.2.1. Finite element simulation process

The arc and s curved plates are simulated according to the heating lines in Figure 4 and Figure 5. The 2 dimensional plane modes are used for the simulation of the sail plate and the saddle plate. Firstly, the plate is divided into 2D planes along the width direction and takes 1/4 of plate as the object. Secondly, the arc lines are replaced by the straight line, and the electromagnetic induction heating and cooling calculation is carried out. Thirdly, each displacement data of hull plate simulation is recorded, and the whole plate forming data can be got. Finally, the 3 dimensional curved surface of whole plate can be drawn by the FE simulation results. The finite element simulation process is shown in Figure 8.

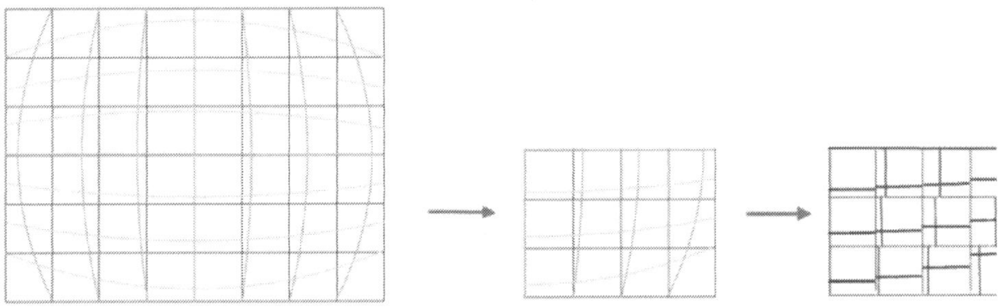

Figure 8. Finite element simulation method.

In the simulation process of high frequency induction heating, the FE geometry model size of hull plate is 1000mm * 600mm * 12mm, and the material is steel plate. The heating coil material is copper, The clearance between the hull plate and the induction coil is 2mm. The hull plates is around by air , which is used for cooling. Because of the symmetry of the model, 1/2 plate is used as the object, and the 2D model is shown in Figure 9.

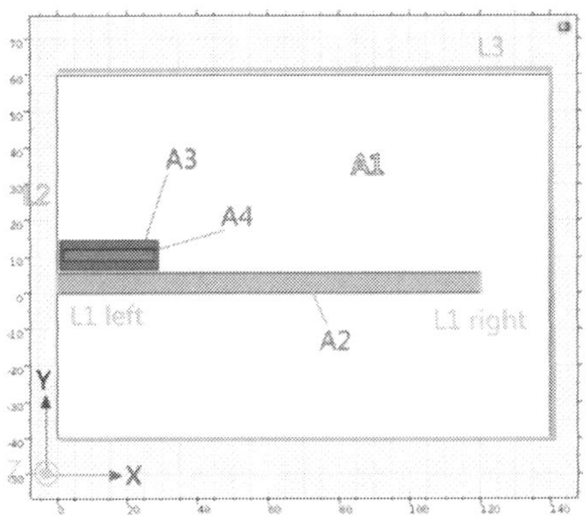

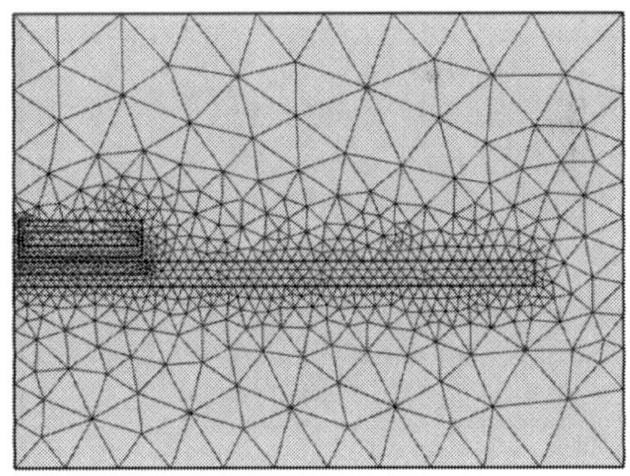

Figure 9. Geometrical model of finite elements. **Figure 10.** The model of finite element mesh

In Figure 9, where: *A1* is air, *A2* is steel plate, *A3* is copper, and *A4* is cooling water.

The finite element boundary conditions are given in the Figure 9. Where: the constraint of *L1* left is *Ux = 0*. The constraint of *L1* right is *Uy = 0*, and *L2* is the symmetry plane .*L3* is initial boundary temperature whose value is 20 ℃. All material initial temperature is 20 ℃.

3.2.2. Material parameters

The physical parameters of steel plate are nonlinear with temperature variation. In the COMSOL Multiphysics software, the interpolation function is used in the parameters definition. The high

IOP Conf. Series: Materials Science and Engineering **417** (2018) 012037 doi:10.1088/1757-899X/417/1/012037

frequency induction coil is pure copper material, and the cooling water in it. Therefore, the coil temperature is 20 ℃ during the heating process.

3.2.3. Mesh
The model is divided into uneven meshes, which is benefit to reduce the operation time. Due to skin effect, the heat is mainly focus on upper surface of hull plate. So the grids of the upper surface are dense, and the grids of the lower surface are loose. The mesh results are shown in Figure 10.

3.2.4. Multi-physical field coupling calculation
The physical field of magnetic field, solid heat transfer and solid mechanics were used in the analysis. The electromagnetic heat source, boundary magnetic - thermal coupling, electromagnetic heat source, thermal expansion and temperature coupling is used in multi-physical field coupling calculation.

3.2.5. Finite element prediction results of hull surface
The Z direction displacement of hull plate is obtained through the simulation of induction heating and cooling. For the arc plate and the S curved plate are 2D flat, whose y direction displacement is recorded in Table 3 and Table 4. Then the Z direction displacement results of sail plate and saddle plate is got. The simulation results are shown in Table 5 and Table 6.

Table 3. Simulation results of arc plate (unit/cm)

	X1	X2	X3	X4	X5	X6	X7	X8	X9	X10
Y	4.86	3.30	2.25	1.64	0.00	0.00	1.54	2.49	3.51	4.83

Table 4. Simulation results of S curved plate (unit/cm)

	X1	X2	X3	X4	X5	X6	X7	X8	X9	X10
Y	0.00	1.41	3.23	1.76	1.07	0.68	0.00	0.73	1.85	3.11

Table 5. Simulation coordinate results of sail plate (unit/cm)

	X1	X2	X3	X4	X5	X6	X7	X8	X9	X10	X11	X12	X13
Y1	0.000	0.915	1.390	1.840	2.135	2.220	2.260	2.220	2.135	1.840	1.390	0.915	0.000
Y2	0.260	1.260	1.740	2.125	2.325	2.510	2.620	2.510	2.325	2.125	1.740	1.260	0.260
Y3	0.480	1.510	2.055	2.555	2.815	2.925	2.985	2.925	2.815	2.555	2.055	1.510	0.480
Y4	0.510	1.845	2.390	2.735	2.970	3.190	3.290	3.190	2.970	2.735	2.390	1.845	0.510
Y5	0.520	1.995	2.540	2.915	3.115	3.320	3.550	3.320	3.115	2.915	2.540	1.995	0.520
Y6	0.510	1.845	2.390	2.735	2.970	3.190	3.290	3.190	2.970	2.735	2.390	1.845	0.510
Y7	0.480	1.510	2.055	2.555	2.815	2.925	2.985	2.925	2.815	2.555	2.055	1.510	0.480
Y8	0.260	1.260	1.740	2.125	2.325	2.510	2.620	2.510	2.325	2.125	1.740	1.260	0.260
Y9	0.000	0.915	1.390	1.840	2.135	2.220	2.260	2.220	2.135	1.840	1.390	0.915	0.000

Table 6. Simulation coordinate results of saddle plate (unit/cm)

	X1	X2	X3	X4	X5	X6	X7	X8	X9	X10	X11	X12	X13
Y1	-2.1070	-1.5470	-1.2250	-0.9030	-0.5957	-0.2947	-0.2889	0.2947	0.5957	0.9030	1.2250	1.5470	2.1070
Y2	-1.5680	-1.1620	-0.9170	-0.6755	-0.4410	-0.2170	-0.1708	0.2170	0.4410	0.6755	0.9170	1.1620	1.5680
Y3	-1.0430	-0.7700	-0.6076	-0.4452	-0.2863	-0.1358	-0.0098	0.1358	0.2863	0.4452	0.6076	0.7700	1.0430
Y4	0.5229	-0.3857	-0.3031	-0.2198	-0.1365	-0.0574	-0.0014	0.0574	0.1365	0.2198	0.3031	0.3857	0.5229
Y5	0.0392	0.0308	0.0245	0.0175	0.0105	0.0028	0.0000	0.0028	0.0105	0.0175	0.0245	0.0308	0.0392
Y6	0.5229	0.3857	0.3031	0.2198	0.1365	0.0574	0.0014	-0.0574	-0.1365	-0.2198	-0.3031	-0.3857	-0.5229
Y7	1.0430	0.7700	0.6076	0.4452	0.2863	0.1358	0.0098	-0.1358	-0.2863	-0.4452	-0.6076	-0.7700	-1.0430
Y8	1.5680	1.1620	0.9170	0.6755	0.4410	0.2170	0.1708	-0.2170	-0.441	-0.6755	-0.9170	-1.1620	-1.5680
Y9	2.1070	1.5470	1.2250	0.9030	0.5957	0.2947	0.2889	-0.2947	-0.5957	-0.9030	-1.2250	-1.5470	-2.1070

In Table 5, the maximum Z direction displacement of the sail plate is 3.55 cm, which is located at the center of the hull plate. The maximum displacement of the saddle plate is 4.214cm in Table 6, which appears in the diagonal position. To check the deformation tendency of sail plate and saddle plate, the finite element simulation data is drawn by MATLAB surface fitting algorithm. The fitted space surfaces of the sail plates and saddle plates are shown in Figure 11 and 12. In the two Figures, the surface shape is consistent with the space shape in Figure 2 and 3. The heating lines of sail plate and saddle plate which is arranged by interpolation method is feasible and reliable.

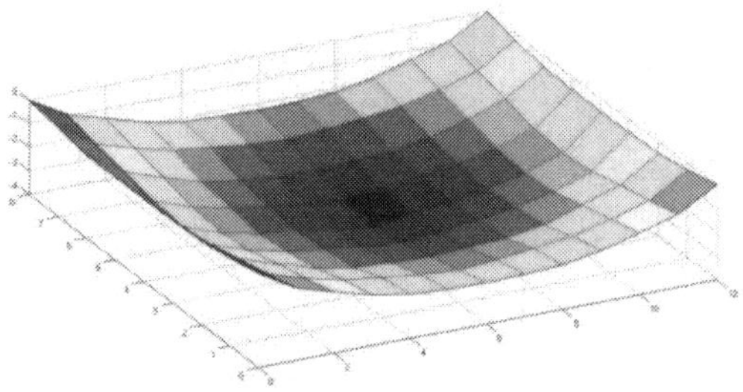

Figure 11. Data fitting shape of sail plate.

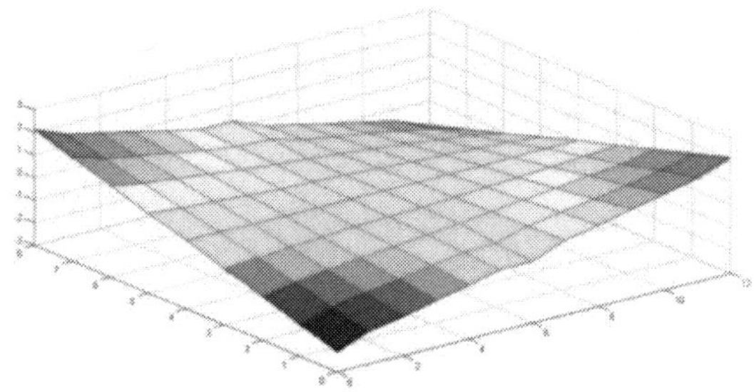

Figure 12. Data fitting shape of saddle plate

4. Experiment of hull surface forming

4.1. Test materials and parameters

A forming experiment using high frequency induction heating was carried out to manufacture a large steel plate. The test material is Q345. The processing parameters are shown in Table 7.

Table7. Heating parameters

Power (kW)	Frequency (kHz)	Heating time (s)	Heating mode	Cooling way
35	20	10s	Static heating	Water cool

4.2. Experiment process

In the experiment, the surface temperature of the plate is heated to 500 ℃. When it is heating, the coil moves along the prefabricated heating lines. In the heating process, the steel plate is heated the upper surface firstly, then it is cooled by water. After the temperature drops to room temperature, the other surface starts to heating and repeat the process， until all the heating lines are completed. The experiment process is shown in Figures 13 and 14.

Figure 13. Heating lines of arc plate.

Figure 14. Heating lines of sail plate

The arc plate, s curved plate, sail plate and saddle plate are completed according to the interpolation heating lines, and the deformation of plates is shown in Figure 15 、Figure 16 、Figure 17 and Figure 18. Thus, it was verified that the simulation results matched those that obtained from the experiment

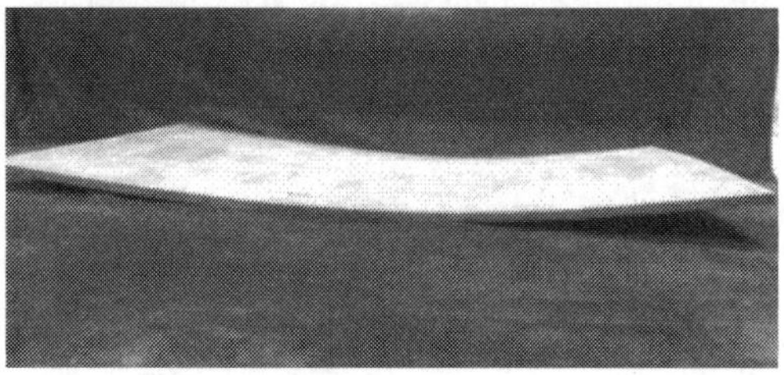

Figure 15. Forming test for arc plate.

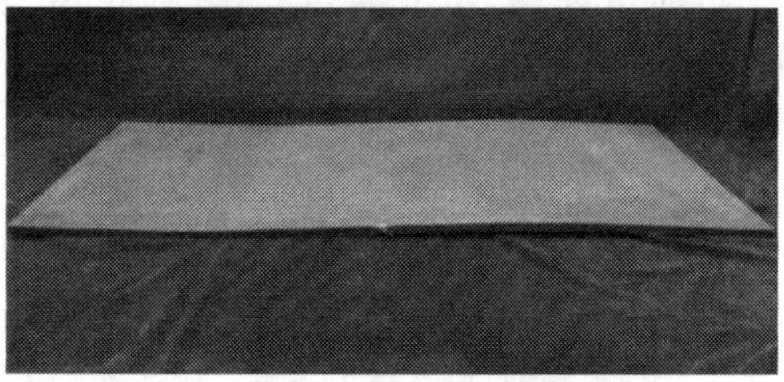

Figure 16. Forming test for S curved plate.

Figure 17. Forming test for sail plate.

Figure 18. Forming test for paddle plate.

To compare forming test results of the sail plate and saddle plate with the simulation results, the Z direction displacement of the forming test were also measured, and the error of experiment process is calculated, which is shown in Table 8. The results from Finite element simulation are compared with those from the experiments, the average relative percentages of deformation errors are most within 5%.

The result reveals that the interpolation heating lines arrangement method can well be applied for the manufacture of the hull curved plate in ship-building.

Table 8. Comparison of prediction displacement and experimental displacement of sail plate

Number	Finite element displacement (cm)	Experimental displacement (cm)	Error	Number	Finite element displacement (cm)	Experimental displacement (cm)	Error
1	0.915	1.000	8.50%	10	1.390	1.400	0.70%
2	1.390	1.400	0.70%	11	0.915	0.900	1.70%
3	1.840	1.900	3.20%	12	0.260	0.300	13.30%
4	2.315	2.400	3.55%	13	0.480	0.500	4.00%
5	2.220	2.200	1.00%	14	0.510	0.500	2.00%
6	2.260	2.300	1.70%	15	0.520	0.600	13.30%
7	2.220	2.300	3.40%	16	0.510	0.500	2.00%
8	2.135	2.200	3.00%	17	0.480	0.600	2.00%
9	1.840	1.800	2.20%	18	0.260	0.3	13.30%

Table 9. Comparison of predicted displacement and experimental displacement of saddle plate

Number	Finite element displacement (cm)	Experimental displacement (cm)	Error	Number	Finite element displacement (cm)	Experimental displacement (cm)	Error
1	-2.1070	-2.000	9.00%	12	1.547	1.600	3.30%
2	-1.5470	-1.700	12.50%	13	2.1070	2.200	4.20%
3	-1.2250	-1.400	9.70%	14	-2.1070	-2.000	5.30%
4	-0.9030	-1.000	0.70%	15	-1.5680	-1.600	2.00%
5	-0.5957	-0.600	1.70%	16	-1.0430	-1.000	4.30%
6	-0.2947	-0.300	3.70%	17	-0.5229	-0.600	12.50%
7	-0.2889	-0.300	1.70%	18	0.0392	0.040	2.00%
8	0.2947	0.300	1.70%	19	0.5229	0.600	12.80%
9	0.5957	0.700	14.90%	20	1.0430	1.000	4.300%
10	0.9030	0.900	0.30%	21	1.5680	1.600	2.00%
11	1.2250	1.300	5.80%	22	2.1070	2.200	4.20%

5. Conclusions

In this study, a method of interpolation principle is applied for the manufacture of curved surfaces thick plates used in hull structures in ship-building. With this method, the heating lines of the arc plate, the S curved plate, the sail plate and the saddle plate are rearranged. The induction heating is simulated by the multiphysics coupling software. Finally, experiment tests are carried out using a 12mm thick steel plate. The conclusions are as follows:

(1) Based on the interpolation principle, the heating lines of sail plate and the saddle plate are designed by this method, then the distance of conventional heating lines and interpolation heating lines distance are compared, the results show that the interpolation heating lines reduce distance greatly, which save time and improve efficiency.

(2) Multiphysical fields' software COMSOL Multiphysics is used to simulate the steel plate deformation under the interpolation heating lines. Displacements of induction heating after cooling are obtained by finite element analysis, then whole hull surface is obtained using the simulation point displacement by MATLAB surfaces fitting algorithm. And simulation hull surface is consistent with the theoretical sail plate and saddle plate curved surface, so interpolation heating lines for the hull plate curved surfaces forming are proved to be reasonable.

(3) The experiment tests of the hull plate are carried out with the interpolation heating lines. The experimental results are in accord with the finite element simulation results, and the error is within 5%. It shows that the finite element simulation is reliable, and proves that the interpolation heating lines method is feasible. This method can not only contribute to the hull surface forming, but also simplify the numerical control program and facilitate learning.

References

[1] Chen J, Qi Y, Shi Y and Bi Z 2010 *Proc. Ins. Mech. Eng. C. J. Mech. Eng. Sci.* **224** 655–60
[2] Ishiyama M and Tango Y 2000 *J. Ship Production* **16** 121–32
[3] Neki I, Ogawa J I and Kamichika R 1993 *J. Soc. Naval Architects of Japan* **174** 535–42
[4] Ishiyama S, Tango Y and Nakamura Y 1998 *J. Society Naval Architects of Japan* **6** 335–42
[5] Reutzel E W, Zhang L and Michaleris P 2006 *Int. J. Mech. Sci.* **48** 1046–62
[6] Jong G S, Jang H L and Sung K P 1999 J. Ship Production 15 10–20

[7] Yoshihiko T, Morinobu I and Nagahara S 2004 *J. Soc. Naval Architects of Japan* **193** 85–95
[8] Yu G, Anderson R J, Maekawa T and Patrikalakis N M 2001 *Int. Mech. Sci.* **43** 2349–70
[9] Jang C D, Kimh K and Ha Y S 2002 *J. Ship Production* **18** 226–36
[10] Nguyen T T, Yang Y S, Bae K Y and Choi S N 2009 *J. Mech. Sci. Tech.* **23** 1211–21
[11] Jang C D and Moon S C 1998 *J. Ship Production* **14** 238–45
[12] Ueda Y, Murakawa H and Mohamed R A 1993 *J. Soc. Naval Architects of Japan* **173** 409–31
[13] Heo S C, Seo Y H, Park J W, Ku T W, Kim J and Kang B S 2010 *J. Mech. Sci. Tech.* **24** 137–40
[14] Seong W J, Ahn J, Na S J, Han M S and Jeon Y C 2010 *J. Mater. Proc. Tech.* **210** 1811–20
[15] Pankaj B, Mandal N R and Sha O P 2009 *Comput. Mater. Sci.* **45** 167–75
[16] Lee I K, Lee S Y, Jeong M S, Hwang S K, Cho Y J, Kim B M, Ko D C, Lee S B and Lee S K 2017 *Int. Conf. on the Technology of Plasticity* (17-22 September, Cambridge, United Kingdom) vol 207 (American Elsevier) pp 1165–69
[17] Zhou H, Luo Y and Jiang Z Y 2012 *Curved Plate Forming Technology of Ship Hull Surface Based on High Frequency Induction* (Beijing: China Communications Press)

IOP Conf. Series: Materials Science and Engineering **417** (2018) 012038 doi:10.1088/1757-899X/417/1/012038

Finite Element Analysis on Loosen Teeth using Fibrous Periodontal Splint Restoration

Chengshui Ding[1], Xue Shi[2], Yang Song[1], Lulu Qiu[1], Xinyue Li[2] and Chunqiu Zhang[1]

[1]Tianjin Key Laboratory of the Design, Intelligent Control of the Advanced Mechatronical System, National Demonstration Center for Experimental Mechanical and Electrical Engineering Education, School of Mechanical Engineering, Tianjin University of Technology, Tianjin 300384, China
[2]Periodontitis Department, Tianjin Stomatological Hospital, Tianjin 300384, China

272625001@qq.com

Abstract: Objective To investigate the effect of fibre periodontal splint restoration on mandibular anterior teeth displacement and periodontal membrane stress. **Methods** Micro-CT scanning technology, combined with Mimics, Geomagic studio and SolidWorks were used to establish 3D dental models of mandibles with teeth and the models fibre periodontal splint restoration; the models of mandibular anterior loosen teeth was constructed under 30 degree oblique 25N loading on the labial and lingual sides respectively. Then, under different angles force with or without the splint, the analysis on dental mechanical state are given. **Results** It was determined that the sum of the maximal displacement of the loosen tooth model at the lingual and labial sides was 1 mm, which represented the degree I loosen model. After splint repair on the model, the maximal displacement of loosen tooth decreased 47.7%, 85.1% and 85.8% at 0 degree, 15 degree and 30 degree forces respectively, and the maximum stress of periodontal membrane decreased by 22.3%, 61.2% and 74.9%. **Conclusion** Fixation of loosen teeth through fibre periodontal splints can reduce the maximum stress of periodontal membrane and the maximal displacement of the teeth, thus ensuring the stability of loosen tooth.

1. Introduction

Periodontal disease is a tissue disease caused by many factors. Dental plaque, periodontal inflammation, traumatic occlusion and other factors usually cause periodontal support tissue damage and alveolar bone absorption, which greatly affect the bearing of the teeth ability [1]. In addition, the alveolar bone absorption makes periodontal tissue around the loosen teeth reduce. Even if normal bite occurs, it will also lead to tooth movement and occlusion dislocation. With the development of the disease, the degree of the teeth mobility gradually increases, eventually leading to shedding or removal [2]. The tooth mobility refers to the degree of activity of natural teeth under external force, which is to evaluate the status of dental support, the development of treatment options and prognostic indicators [3].

In clinical, to achieve the purpose of dental stability, loosen teeth are treated by combining periodontal treatment and occlusal correction [4-5]. Using the fibre periodontal splint (FPS) to fix the loosen teeth can achieve the target that redistribute the force when the patient chews food [6], mobilized the potential reserve of periodontal tissue and improved the bite force, so it is benefit to restore the loosen teeth [7].

In recent years, many oral scholars have made a lot of pathological researches on alveolar bone resorption. However, there is little research on the displacement and the stress of periodontal membrane in the repaired single loosen teeth and healthy tooth connected with FPS. In this paper, we used the three-dimensional finite element method to model the periodontal tissue and the FPS in the human mandible. From the biomechanical point of view, we investigated the effect of fibre periodontal splint repaired loosen tooth on mandibular anterior teeth displacement and periodontal membrane stress under different angles force.

2. Materials and Methods

In this study, selecting patients with I degree loosen tooth (You1) while other teeth were healthy and normal. The establishment of three-dimensional finite element models include: a total of six teeth from the left canine to the right canine, mandible model, and the periodontal membrane. The structure of these models were mainly derived from tomography (Micro-CT) acquisition of computer software. Then using Mimics software to establish models of the mandibles with teeth, whose surface were smoothed and optimized by Geomagic studio software. In addition, the model of the periodontal membrane was founded by applying a positive offset of 0.25 mm based on the middle-lower part surface model the tooth. Finally, all the models processed above were imported into SolidWorks and transformed into solid models (Fig. 1), and the FPS model was designed based on the shape of tooth surface. In this paper, the splint size is 2.5 * 80mm, while the actual length is based on the physician's use.

All solid models were imported into the finite element analysis software and divided into quadratic tetrahedral units (Fig. 2). Then setting the rotation of the X-axis, Y-axis and Z-axis of the bottom surface of the alveolar bone were completely restrained; and the bonding constraint was applied between the tooth and the periodontal membrane between the periodontal membrane and the alveolar bone, as well as between the tooth and the FPS. In addition, the angle between the oblique lingual side and tooth axis was set as 0 degrees, 15 degrees and 30 degrees, and the load force was 25N. In the simulation process, the teeth, periodontal membrane, alveolar bone and FPS were all isotropic elastic materials. Referencing to the previous scholars' research results, material parameters are shown in Table 1 [8-10].

Firstly, on the reference of some teeth mobility tests, we reduced the elastic modulus of the alveolar bone model that peripheral loosen teeth about 0.5mm, then apply the loading of 25N in inclined 30 degree to determined that the sum of the maximal displacement of the loosen tooth model at the lingual and labial sides was 1 mm. Then, under 0 degree, 15 degree and 30 degree oblique 25N loading, the effects of loads in different directions on loosen tooth were evaluated. Finally, adding FPS in the anterior teeth the effects on model at the same loading conditions were determined.

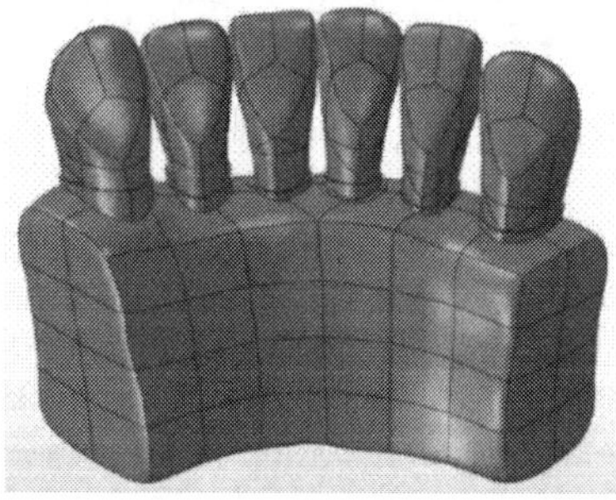

Figure 1. 3D solid model of mandibles with teeth with curved surface.

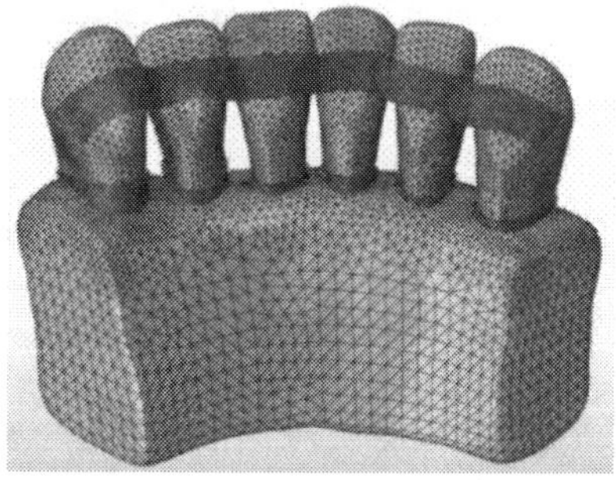

Figure 2. Finite element model mandibles with teeth after FPS repair.

IOP Conf. Series: Materials Science and Engineering **417** (2018) 012038 doi:10.1088/1757-899X/417/1/012038

Table 1. Material parameters in finite element model of teeth.

Material	Elastic Modulus（MPa）	Poisson's ratio
Enamel	41 410	0.30
Dentin	18 600	0.32
Periodontal Membrane	50	0.45
Alveolar Bone	13 800	0.3
Fibre Periodontal Splint	28 000	0.3

3. Results

The sum of the tooth displacements on the lingual and labial sides meets I degree loosen conditions. So we can make sure the model is I degree mobile model. The loosen teeth displacement nephogram are shown in Figure 2.

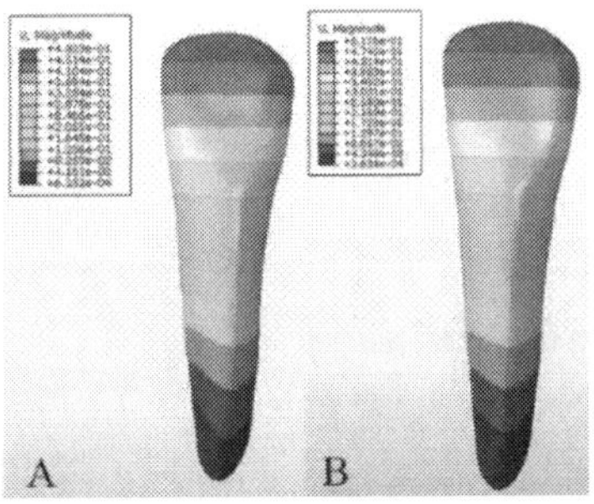

Figure 3: The loosen teeth displacement nephogram under the labial and lingual loading.
Figure 3A: Under labial side loading.
Figure 3B: Under lingual side loading.

The maximal displacement of teeth and the maximum von Mises stress of periodontal membrane are shown in Figure 3 and Figure 4 under different angles applied force with or without the splint. The range of maximal displacement and maximum von Mises stress were reduced under same angle load before and after FPS repair. For example, under the vertical load before FPS repair, the maximum von Mises stress ranged from 0.36 to 2.7 MPa and the maximal displacement of the teeth ranged from 0.0032 to 0.017 mm. After FPS repair, maximum von Mises stress ranged from 0.40 to 2.1 MPa, and maximal displacement ranged 0.0030 to 0.0091mm.

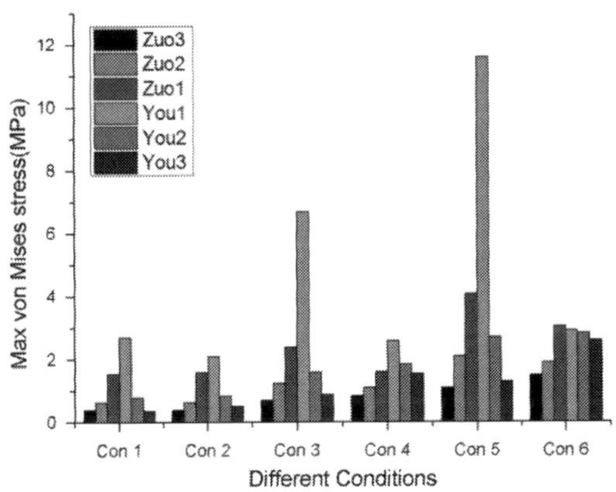

Figure 4. Max von Mises stress of periodontal membrane under different conditions.

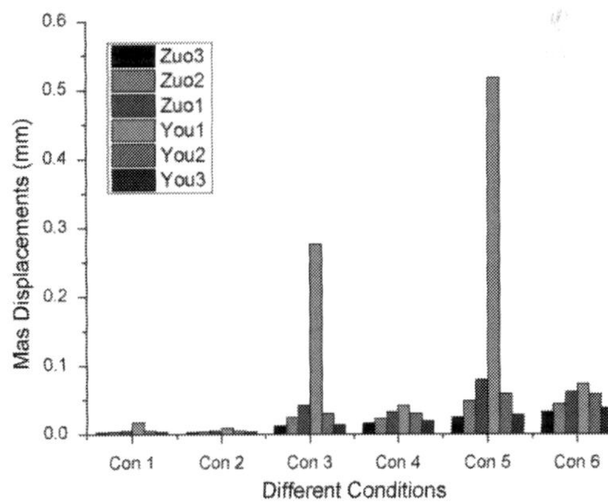

Figure 5. Max displacement of teeth under different conditions.

Zou3, Zou2, Zou1, You1, You2, You3 represent: Left canine, Left lateral incisors, Left central incisors, Right central incisors, Right lateral incisors, Right canine.

Con 1 through 6 represent: 0 degree without FPS (Con 1), 0 degree with FPS(Con 2), 15 degree without FPS (Con 3), 15 degree with FPS (Con 4), 30 degree without FPS (Con 5), 30 degree with FPS (Con 1).

Under 15 degree oblique load of force, with or without FPS, the nephogram of teeth displacement are shown in Figure 6 and Figure 7. The distribution area of the maximal displacement of teeth shifted from the uniform state to the oblique direction of the near middle.

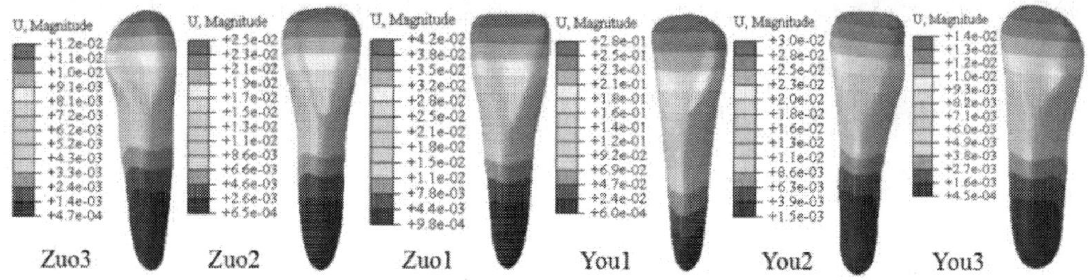

Figure 6. Teeth displacement nephogram under 15 degree oblique load before FPS repair.

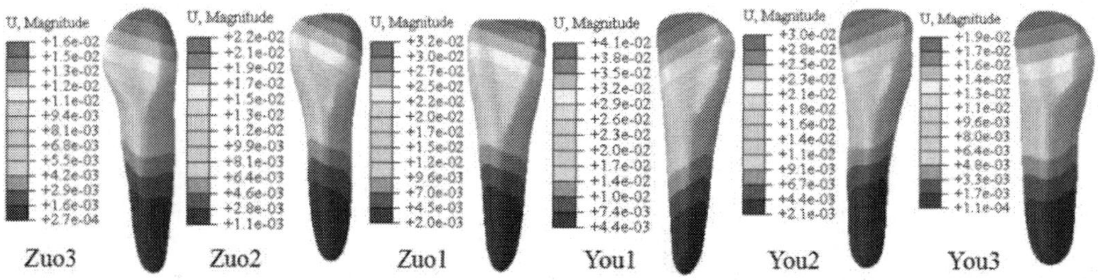

Figure 7. Teeth displacement nephogram under 15 degree oblique load after FPS repair.

4. Conclusion

In this paper, through the finite element simulation of mandibles with teeth and the models FPS restoration, the law of tooth displacement and periodontal membrane stress under different angles force was suggested. No matter there is a FPS or not, the maximum stress of periodontal membrane and maximal displacement of the teeth are increased with loading angle increase. Under same angle load with FPS repair, maximal displacement and maximum stress of loosen tooth decreased significantly. The displacement nephogram of each tooth was obviously changed under 15 degree oblique load after repairing with splint. The analysis showed the FPS restoration played the role of bridge, and healthy teeth shared the displacement of loosen tooth. Therefore, the use of FPS will transfer occlusal force of the loosen tooth to other healthy teeth. It can reduce the maximal displacement of loosen teeth and the maximum stress of periodontal membrane. These can ease the injury of the teeth chewing force on the loosen tooth, which improves the efficiency of periodontal tissue repair and provides the possibility of recovering the chewing ability of the teeth as soon as possible.

Acknowledgments

The project was partly supported by the National Natural Science Foundation of China (Nos. 11702190, 11672208, 11432016).

References

[1] Soares PB, Fernandes Neto AJ, Magalhães D, Versluis A and Soares CJ 2011 *Archives of Oral Biology.* **56** 1373-81
[2] Shuhuan Shang, Chengzhang Li, Qin Qian and Li Liang 2005 *Journal of Biomedical Engineering.* **22** 725-29
[3] Forabosco A, Grandi T and Cotti B 2006 *Minerva Stomatologica.* **55** 87-97
[4] Van SD, Rosenberg D, Naert IE, Van den Bossche L and Nys M 1995 *Journal of Periodontology.* **66** 165-70
[5] Bernal G, Carvajal JC and Muñoz-Viveros CA 2002 *Journal of Contemporary Dental Practice.* **3** 10-22

[6] Ai bi bu la·ai mai er 2017 *Cardiovascular Disease Journal of integrated traditional Chinese and Western Medicine.* **5** 45-45
[7] Geramy A, Adibrad M and Sahabi M 2010 *Journal of Dental Sciences.* **5** 1-7
[8] Stasinopoulou I, Manda M, Galanis C and Koidis P 2013 *Journal of Prosthetic Dentistry.* **110** 303-12
[9] Yongjia Xu, Liying Yu, Qiong Chen, Shumin Wang and Huanguo Xiong 2010 *International Journal of Stomatology.* **37** 525-29
[10] Lijuan zheng, Weidong Ma and Bingzhi Chen 2012 *Journal of Practical Stomatology.* **28** 691-94

Energy Dissipated in Tapping Mode Atomic Force Microscopes due to Humidity

Xiaoting Zheng, Yan Sun, Zheng Wei ,

College of Mechanical and Electrical Engineering, Beijing University of Chemical Technology, Beijing 100029, China

Weizheng@mail.buct.edu.cn

Abstract. Phase imaging is an important imaging model of atomic force microscopes (AFMs). It provides many sample properties that the height image does not. The phase information reflects the energy dissipation of the probe and sample interactions. Therefore, the energy dissipation between the tip and sample in the tapping mode is of great significance for understanding the mechanism of phase imaging experimentally and theoretically. In this paper, we propose a new method for measuring energy dissipation in tapping mode AFMs. The formation and rupture processes of liquid bridges are studied by this method. Finally, the experimental results are compared with the theoretical model. The comparison shows that the method of experimental measurement and the formation mechanism of liquid bridges in the tapping mode are reliable.

1. Introduction

Atomic force microscopes (AFMs) originated from the scanning tunneling microscope (STM), which was invented by Nobel laureates Binnig and Rohrer. The AFM and STM are members of a family called scanning probe microscopes (SPMs). In contrast to an STM, which monitors the electron tunneling currency between the tip and the conductive samples, an AFM essentially is essentially a force sensor that feels the interaction between the tip and the samples, which does not to be conductive [1,2]. Since its invention in the 1980s, the powerful capacities of AFMs give them a broad range of use in various areas such as materials science, life science and the semiconductor industry.

Based on vibration theory, the TM (Tapping Mode) AFM phase image mostly depends on energy dissipation between the tip and the sample [3].Currently, measuring the energy dissipated in TM AFMs is based on the method proposed by Cleveland and coauthors [3,4]. We can measure the rupture energy of the liquid bridge by measuring the force curve, but this method is not suitable for the rupture of the liquid bridge in the tapping mode because thefrequency of the force curve measurement is much less than the frequency of the tapping mode. Because of the limitation of the experimental method, the dynamics of the formation and rupture of the liquid bridge between the tip and sample remains aninteresting field. There is no conclusive description of this dynamic process[5,6,7].

In this paper, we present an indirect method to measure the dissipation energy between the tip and sample in a TM AFM. Meanwhile, the relationship between the rupture energy of the liquid bridge and the relative humidity is theoreticallyanalyzed for the tapping mode. By comparing the experimental data with the theoretical results, we demonstrate the feasibility of this method. We also verify the mechanism of the formation of the liquid bridge in this AFM operation mode.

2. Experimental results and discussion

2.1. Tuning the cantilever in air

When the AFM tip is far away from the sample, we perform the auto tuning operation on the cantilever. The amplitude-frequency response curve (AFRC) is shown in Figure 1. The natural frequency and damping of the structure can be read from the AFRC. If we turn the front thumb wheel of the AFM scanning head counterclockwise to lower the cantilever towards the sample surface, the amplitude of the AFRC starts to decrease when the tip reaches the area of the tip–sample interaction. The amplitude of the AFRC continuously decreases until the tuning operation fails. In this experiment, we only change the distance between the tip and sample, leaving the excitation voltage applied to the piezoelectric tube unchanged.

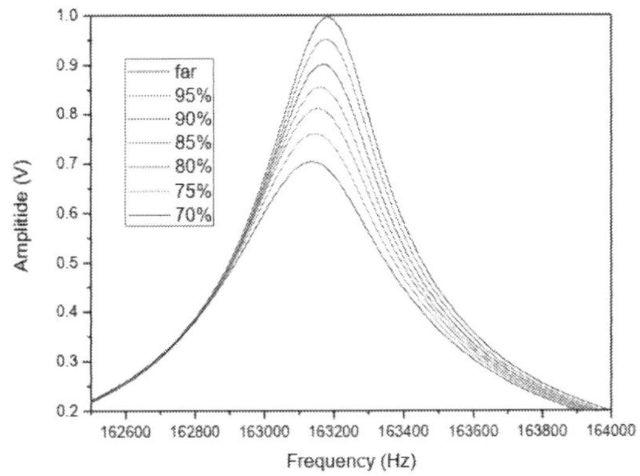

Figure 1. Tuning test for the AFRC for different distances between the AFM tip and sample at 45% relative humidity. The percentage in the inset key representsthe relative positions of tip and sample, whilethe ratio of the maximum response amplitude to the free vibration amplitude at the resonance point is the percentage of the curve.

2.2. Quality factor of the cantilever at different positions and relative humidities

Wet nitrogen gas and dry nitrogen gas are mixed in a long tube, and the mixed gas is introduced into the humidity chamber. In order to achieve the desired humidity, two mass controllers are used to control the flow of the wet and dry gasses. At different locations and relative humidities, we obtain a series of AFRCs.Then, we obtain the corresponding quality factor, Q. The quality factor at different relative humidities and positions is shown in Figure 2. No matter what the relative humidity is, the quality factor decreases gradually when the tip approaches the sample. However, when the relative position is fixed, and the relative humidity is changed, the change in the quality factor is complicated. The quality factor changes little with humidity at positionsof 100%, 90% and 85%, but after the position reaches 80%, the quality factor decreases with increases in relative humidity.

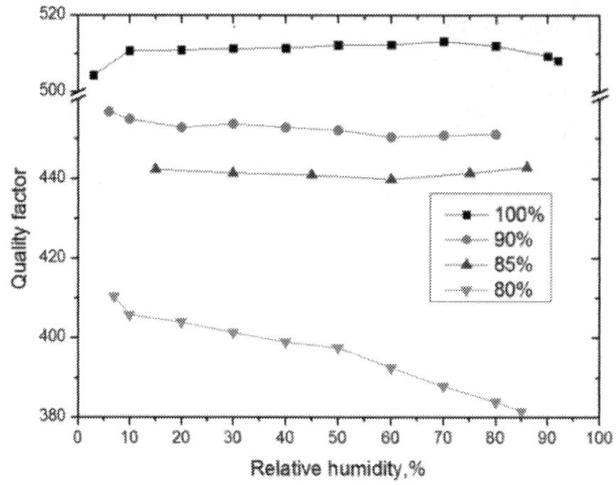

Figure 2. Quality factor variation of the cantilever at different positions and different relative humidities. The black curve represents the quality factor, Q , of the cantilever when the tip is far away from the sample.

3. Theory

3.1. The oscillation system of TM AFM

As shown in Figure 2, when the tip approaches the sample so closely that the amplitude of the cantilever decreases to 80% of the free amplitude A, the quality factor decreaseswith the increase in relative humidity, unlike the upper three lines representing greater distances. Thus, the effect of the liquid bridge should be taken into consideration in this SDOFsystem. In the operation of the TM AFM, when the tip contacts the sample, a liquid bridge is formed. As the tip is drawn from the sample, the liquid bridge ruptures. Ignoring the mass of the liquid bridge, the elastic and damping effect of the liquid must be introduced into the SDOF system.

Here, we briefly analyze the vibration ensemble of the cantilever–tip–liquid bridge (CTLB) in the tapping mode. The vibrationensemble model is shown in Figure 3. Now, the dissipation energy of draw of the liquid bridge is taken into account as damping, so the effective quality factor of the SDOF system is

$$Q_{eff} = 2\pi \frac{E_0}{\Delta E_1 + \Delta E_2} \tag{1}$$

Where ΔE_0 is the energy of the SDOF system, ΔE_1 the internal dissipation energy in the cantilever, ΔE_2 is the dissipation energy due to the liquid bridge. Thus,

$$\frac{\Delta E_2}{E_0} = 2\pi \frac{Q_1 - Q_{eff}}{Q_1 Q_{eff}} \tag{2}$$

where Q_{eff} represents the experimental results shown in the 80% line of Figure 2. Q_1 represents the experimental results shown in the upper three lines of Figure 2.

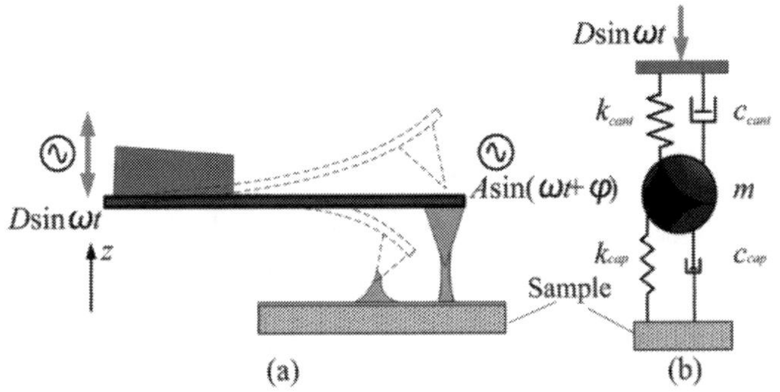

Figure 3. The cantilever oscillation ensemble and its simplified SDOF model.

3.2. The volume and shape of the liquid bridge in TM AFMs

Numerous experimental results prove that humidity has a strong effect on the vibration behavior of cantilevers. Thus, the liquid bridge must play its role in an oscillatory cycle. The squeezed model of liquid bridges is illustrated by Figure 4.

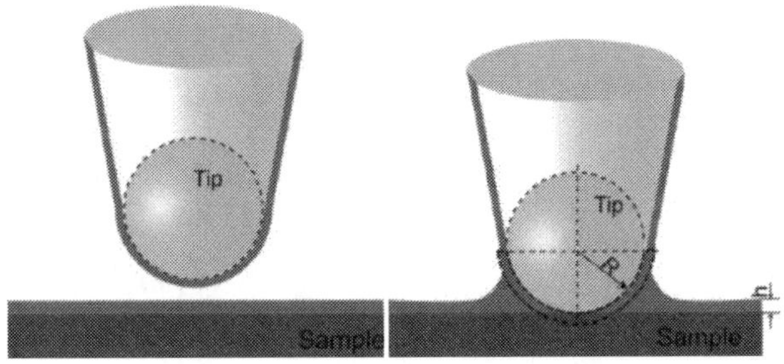

Figure 4. Squeeze model. Owing to the existing thin water film on the surfaces of tip and sample, when the AFM tip approaches to thesample, some water is squeezed out from the contact zone and subsequentlyforms a liquid bridge. The volume of the squeezed water is the overlappingregion shown in the figure. The thickness of the thin water film is h .Generally, the tip radiusis $R >> h$.

In equilibrium, the Young-Laplace pressure in the liquidbridge is equal to the disjoining pressure in the liquid film. In addition, the meniscuscurvature ($1/r_a + 1/r_m$) is associated with the relative vapour pressure (relative humidity for water) according to the well-knownKelvin equation. Based on the above physical facts, we obtain the thickness of a thin waterfilm over the surface as [5,8]

$$h = (\frac{A_H v_m}{6\pi k_B T \ln H_r})^{1/3} \tag{3}$$

3.3. Energy dissipation due to the liquid bridge in TM AFMs

In the extension of the liquid bridge, the capillary force, as a non-conservative force, will do the dissipation work. As shown inFigure 5, the capillary force is composed of an attractive force acting on the cross-section of the liquid bridge and a tension force acting on the contour of the cross-section[9].

$$F = F_{Y-L} + F_T \tag{4}$$

The attraction force caused by the Young-Laplace pressure on the cross- section is

$$F_{Y-L} = \pi r_a^2 \gamma \left(\frac{1}{r_a} + \frac{1}{r_m} \right). \tag{5}$$

So, the capillary force due to the liquid bridge is

$$F = \pi r_a^2 \gamma \left(\frac{1}{r_a} + \frac{1}{r_m} \right) + 2\pi r_a \gamma . \tag{6}$$

Based on the force–distance curve of AFMs, it appears that the rupture distance is roughly proportional to the cube root of the volume of the liquid bridge [10,11],

$$D_r = \frac{1}{2}\left(1 + \frac{\theta_1 + \theta_2}{4} \right) V^{1/3}. \tag{7}$$

Then, the dissipation energy for breaking the liquid bridge is

$$\Delta E = \int_0^{D_r} F dD . \tag{8}$$

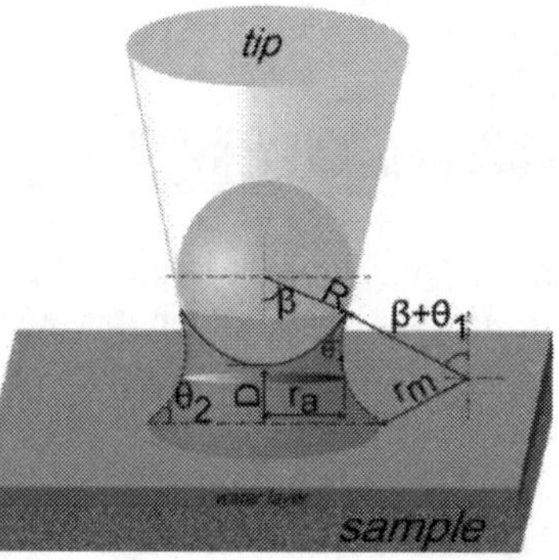

Figure 5. The geometry of the liquid bridge between the AFM tip and samples. θ_1 and θ_2 are the contact angles of the tip and sample, respectively.

Eqs.6-8 must be solved simultaneously. Because of the complicated relationship between the geometric parameters, we apply numeric methods. The parameters are assigned as follows: $\theta_1 = 30^\circ$, $\theta_2 = 8^\circ$, $R = 50\text{nm}$ and $A_H = 8.7\times10^{-21}\text{J}$ [12]. Here, two growth models of liquid bridge are considered, one is the squeeze model, another is the equilibrium model[13].

After normalizing the dissipation energy due to the liquid bridge, the relationship between the energy dissipation and relative humidity is shown in Figure 6. Because the predicted thickness of Eq. (3) has a large deviation from experiments carried out in lower relative humidity, we chose the relative humidity at 60% as the standard for energy normalization. Figure 6 shows that the experimental results agree well with the theoretical results. It proves that the method of tuning the cantilever is feasible for investigating energy dissipation between the tip and sample and is also feasible for investigating the dynamic growth and rupture processes of the liquid bridge.

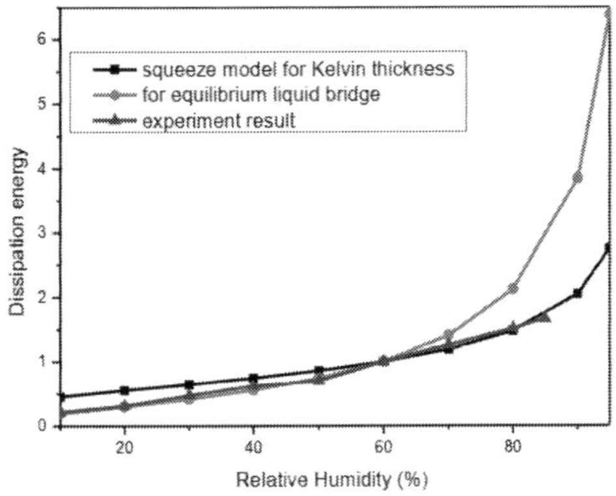

Figure 6. Energy dissipated due to the rupture of the liquid bridge vs. the relative humidity.

4. Conclusion

In conclusion, we have shown a method for measuring the relative energy dissipated in TM AFMs. The quality factor, Q, will be given in various test conditions by the tuning operation. The energy dissipation in the TM AFM results from analysis of the oscillation system of the TM AFM. Of course, if the stiffness and amplitude of the cantilever are known, we can also measure the absolute dissipated energy. By this method, it has been proven that in the tapping mode, the liquid bridge is mainly formed by the extrusion effect.

Acknowledgments
This work was supported by the National Natural Science Foundation of China (NSFC, No. 11572031)

References

[1] BinnigG, Rohrer H. Scanning tunneling microscopy: from birth to adolescence. Rev Mod Phys.1987;59:615-625.
[2] Binnig G, Rohrer H. In touch with atoms. Rev Mod Phys. 1999;72:s324-s330.
[3] Cleveland JP, Anczykowski B, Schmid AE, et al. Energy dissipation in tapping-mode atomic force microscopy. ApplPhys Lett. 1998;72:2613-2615.
[4] Anczykowski B, Gotsmann B, Fuchs H, et al. How to measure energy dissipation in dynamic mode atomic force microscopy. Appl Surf Sci. 1999;140:376-382.
[5] Wei Z, Sun Y, Ding WX, et al. The formation of liquid bridge in different operating modes of AFM. Sci China-PhysMech Astron. 2016;59:694611.
[6] Szoszkiewicz R, Riedo E. Nanoscopic friction as a probe of local phase transitions. ApplPhys Lett. 2005;87:084502.
[7] SahagúnE,García-Mochales P, Sacha GM, et al.Energy Dissipation due to Capillary Interactions: Hydrophobicity Maps in Force Microscopy. Phys Rev Lett. 2007;98:176106.
[8] Asay DB, Kim SH. Evolution of the adsorbed water layer structure on silicon oxide at room temperature. J PhysChem B.2005;109:16760-16763.
[9] Butt HJ, Kappl M. Normal capillary forces.Adv Colloid Interfac.2009;146:48.
[10] Lian G, Thornton C, Adams MJ, et al. A Theoretical Study of the Liquid Bridge Forces between Two Rigid Spherical Bodies. J Colloid Interface Sci. 1993;161:138-147.
[11] Lucel S, Robert SA, Riedo E, et al. Volume of a Nanoscale Water Bridge. Langmuir. 2006;22:1093-1098.
[12] Israelachvili JN. Intermolecular and Surface Forces 3rd. Pittsburgh: Academic Press; 1985.
[13] Wei Z, Sun Y, et al. The formation of liquid bridge in different operating modes of AFM, Sci.China-Phys.Mech.Astron.2016 ; 59, 694611.

IOP Conf. Series: Materials Science and Engineering **417** (2018) 012040 doi:10.1088/1757-899X/417/1/012040

MEMS Capacitive Sensor for Wound Monitoring Applications

Trystan Duell, Max Muehlbauer, Timothy Seitzinger, James Westfall, and Reza Rashidi*

Mechanical Engineering Technology, State University of New York, Alfred State College, Alfred, NY, USA, 14802

*rashidr@alfredstate.edu

Abstract. This paper presents development of a capacitive microsensor to be used in monitoring of wound healing process. Wounds will heal more quickly if they are kept under a bandage and presently bandages must be removed to track the healing process. This could potentially delay the healing process. A sensor can be used to detect the presence of blood in the wound by measuring the capacitance. Blood has a higher permittivity than air or any other substance that may be in the wound and will significantly increase the total measured capacitance in the device. The presence of blood in the wound under the bandage and so in the device would signify that the wound is not fully healed. The sensor was fabricated using standard MEMS (Microelectromechanical Systems) techniques in cleanroom. The sensor was tested in present of no solution, DI water and saline solution, and capacitance was measured to be 1 pF, 35 pF and 53.3 pF, respectively.

1. Introduction

There have been very few breakthroughs in wound care over the past century. During the 1960's, it was discovered that a wound would heal at a faster rate when it was kept humid and moist [1]. Therefore, when a wound is bandaged, it will heal quicker than when it is un-bandaged. The concept behind wound monitoring sensors is that the bandaging will not need to be removed, but the wound can still be monitored for healing progress.

Currently, a wound can be monitored in a variety of ways. The most common method is to measure pH in the wound area [2-3]. Healthy skin has a pH value in the range of 5.5 to 6.5, while infected wounds can exhibit pH levels of 7.4 or more [4]. Since early 2000, wearable bio-sensing devices have been used for health management and allow the caretaker to avoid unnecessary removal of wound dressing [5-6]. They are made of a material which integrates sensors as a wearable part of the bandage. The sensors are used to detect infections by measuring the pH in the wound. The pH can be measured using optical signals which measure the refractive index of the surface of the skin using a built in LED and spectrometer. A potential issue with measuring pH as a method of wound monitoring is that pH can change drastically with temperature, thus making the temperature of the skin and ambient air temperature uncontrolled variables.

Another method for measuring wound healing progress is by use of wound mapping technology, which measures the electrical properties of the wound tissue throughout the healing process [7]. Specifically, the impedance of the skin will be measured, allowing for the wound progress to be determined. This method may not be able to determine whether the wound has healed completely, as many factors can affect impedance, including blood flow in the skin and potentially skin temperature [8].

As another alternative, capacitance can be used to simply detect the presence of blood, which will allow for the track of a healed wound. This would simply act as a notification that bandages may be removed without disrupting the healing process. The sensor will consist of a small enclosed channel, which will attract blood due to the capillary effect when it is present. The blood will flow through the channel, which has electrodes on both sides. The blood will change the capacitance between the electrodes due to different permittivity, thus signalling that the wound has not healed fully.

This paper covers the design, fabrication and experimentation of the capacitive sensor for wound heal monitoring. The microsensor consists of glass and silicon wafers bonded to each other. It works based on the detection of saline as a substitute for blood around the wound by measuring the capacitance around an open wound. The capacitance will change when the particles travel between aluminum electrodes through the microfluidic channel. The microsensor is able to read whether the fluid is adequate for the wound and if it is bleeding.

2. Design

Schematic perspective and top views of the layers in the sensor are shown in Figure 1. The working principle of the capacitive sensor for wound monitoring relies on a gap between electrodes on an etched area in a silicon chip and glass substrate. This will allow for the capillary effect to take place, drawing the blood into the device. To allow for a reading of capacitance to be taken, aluminum is deposited on the bottom of the etched groove in the silicon chip and in the corresponding area on the glass substrate. Aluminum pads are connected to the aluminum electrodes in order to create the capacitor. The electrodes are then passivated with insulting material in order to avoid electrical current passing between them and enable capacitance readings.

The aluminum pad on the silicon extends to one side, allowing for wire to be bonded to the pad without interfering in the glass layer. The aluminum pad on the glass is offset to the other side, allowing wire bonding without interfering with the bonding of the glass and silicon as shown in Figure 1. In addition the aluminum electrode on the glass is slightly smaller than the electrode on the silicon to allow for slight misalignment without affecting the capacitance value. The dimensions of key features on the designed capacitance microsensor are presented in Table 1.

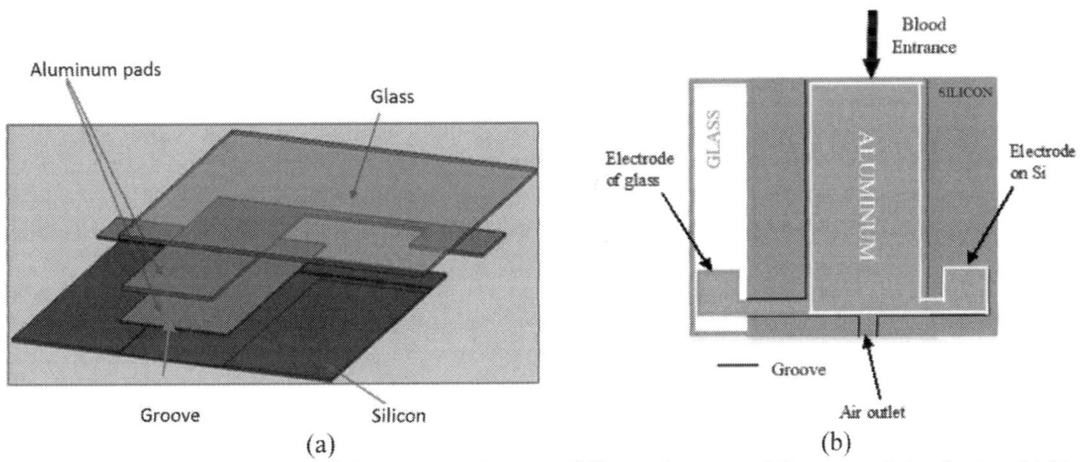

Figure1. a) Perspective view of the sensor, showing different layers and features of the device. b) Top view of the entire design.

Table 1. Dimensions of key features on the designed capacitance microsensor

Overall Area of Microsensor	1 cm^2
Depth of Groove	60 μm
Length of Main Channel	8700 μm
Width of Main Channel	4000 μm
Width of Exit Channel	1000 μm
Effective Length of Al Electrodes	8500 μm
Effective Width of Al Electrodes	3000 μm

A capacitance will increase significantly when a liquid, in this case blood, is present in the gap between the aluminum electrodes on the glass and the silicon. There is a small groove at the end of the etched area on the silicon chip to allow air to escape. This will allow blood to be evenly dispersed through the gap, maximizing the capacitance outputs.

The microsensor was designed to measure the capacitance between the aluminum electrodes. The equation used to calculate capacitance of the sensor is:

$$C = \varepsilon_0 \varepsilon_r \frac{A}{d}$$

where C is capacitance, ε_o is the permittivity of free space (constant), ε_r is the relative permittivity of the medium between the electrodes, A is the projected area of the electrodes and d is the distance between the electrodes. Both the area of the electrodes and the distance between them are fixed in this design. The only parameter that changes the capacitance is the relative permittivity of the medium which depends on the type of fluid present in the groove. The relative permittivity of air that is supposed to be present in the gap when the wound is completely healed is 1 while the relative permittivity of blood which enters the gap when the wound is not healed is over 500, resulting in a significant drop in capacitance after healing. In order to make the experiments easier, a 0.01 M saline can be used as a substitute for the blood. The saline has a relative permittivity of about 90. It has a slightly lower electrical conductivity than blood at this concentration [9-10].

3. Fabrication

Fabrication of the device was performed based on common MEMS techniques in the cleanroom. Figure 2 shows the fabrication steps of the sensor, including silicon wafer fabrication, glass wafer fabrication and bonding them. The silicon wafer was a double side polished n-type wafer with initial resistivity of 1-10 Ω.cm. The wafer was first oxidized in an oxidation furnace to form a 0.5 μm thick oxide layer on both sides of the wafer.

Defining the groove was done by depositing photoresist on each side of the wafer and soft baking. After developing the photoresist on the front side of the wafer, the photoresist was hard baked on the hotplate. In order to pattern the oxide layer as an etch mask, the wafer was dipped in hydrofluoric acid (HF) which opened up the places where bulk etching will take place. To finalize the groove, a wet etch of TMAH heated to roughly 90°C was used to create a 60 μm deep grooves. The remnants of oxide layer was then stripped by dipping the wafer in HF.

After cleaning the wafer with piranha and RCA clean techniques, another 0.5 μm thick oxide layer as an electrical passivation film was grown on all surfaces in the oxidation furnace. The wafer was then loaded into the physical vapor deposition (PVD) thermal evaporator to deposit an aluminum layer over the entire wafer. The lithography process was used to pattern the photoresist on the aluminum layer. The wafer was then submerged into a commercial aluminum etchant to form aluminum electrodes, conductors and pads. After rinsing and drying the wafer, the wafer is covered with acetone to remove the remaining photoresist.

A double sided polished borosilicate glass wafer was used as the mating part of the device. The processes on the glass wafer were similar to those on the silicon wafer, however only an aluminum layer was deposited onto the glass. The glass was simply cleaned first with soap. It was then placed into the PVD thermal evaporator to deposit an aluminum layer on it. The aluminum layer was then patterned using lithography process and subsequent aluminum etch in order to form electrodes, conductors and pads on the glass wafer. Both silicon and glass wafers were cleaved to small individual pieces for further processing

The aluminum electrodes on both silicon and glass wafers required to be insulated from the medium. A dual purpose process was designed to passivate the electrodes and bond the chips in a single step. This step was performed by spinning and patterning a 0.7 µm thick photoresist on the chips, covering electrodes and conductors. The same wet photoresist layers were used as a bonding interface between the two chips. In order to bond silicon and glass chips, they were aligned and pushed against each other on a 135°C hotplate for five minutes. The photoresist was baked in this process and made a bond between two chips while the aluminum electrodes were still passivated by the thin photoresist layer.

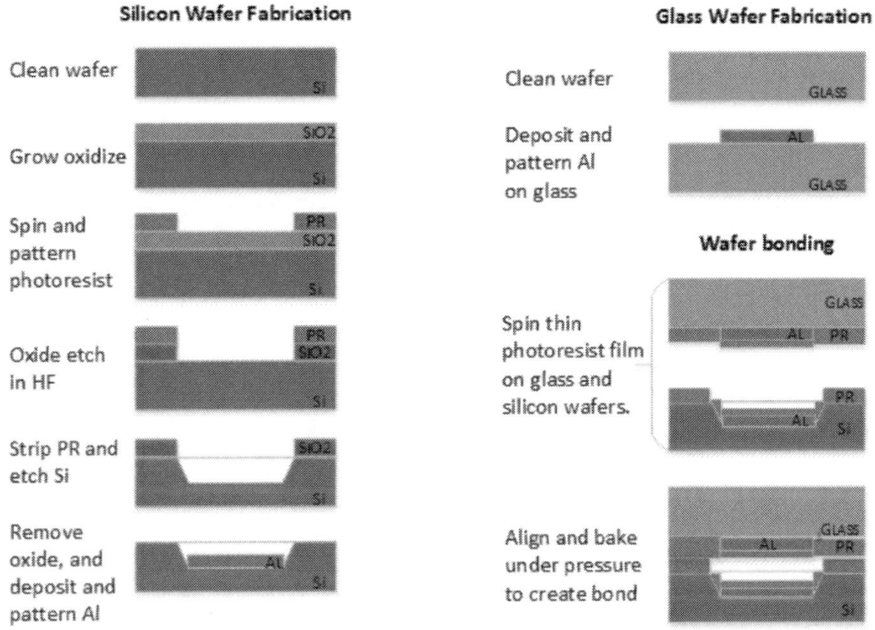

Figure 2. Fabrication steps of the capacitive microsensor

4. Results and Discussion

The capacitive sensor was characterized in present of air, deionized (DI) water and 0.01 M saline solution. The saline solution was used to represent blood, as blood is a biohazard and unsafe to be used in this experiment. The sensor electrodes were connected to a capacitance meter for measurement. A syringe was used to deposit each solution into the channel on the sensor.

The completed experiment yielded results as expected. Air yielded a capacitance reading across the electrodes of 1 pF as a baseline. This is because the permittivity of air is as low as ~1. The capacitance reading in present of the DI water was measured significantly higher at 35 pF. The saline solution yielded the highest reading at 53.3 pF. These results are summarized in Figure 3. The capacitance reading when the saline solution was present in the channel was much higher than that of the air, meaning that the presence of a small amount of blood can be easily detected. Although the readings detected during experimentation did not completely match the design values, the design was still theoretically successful. The inconsistent values are likely due to misalignment of the electrodes which result in a drastic area difference of the electrodes. Since the permittivity value of blood is significantly higher (5-10 times) than that of saline solution, it is reasonable to think that the presence of blood in a wound would result in a significantly higher reading that could easily contrast that of a bloodless environment.

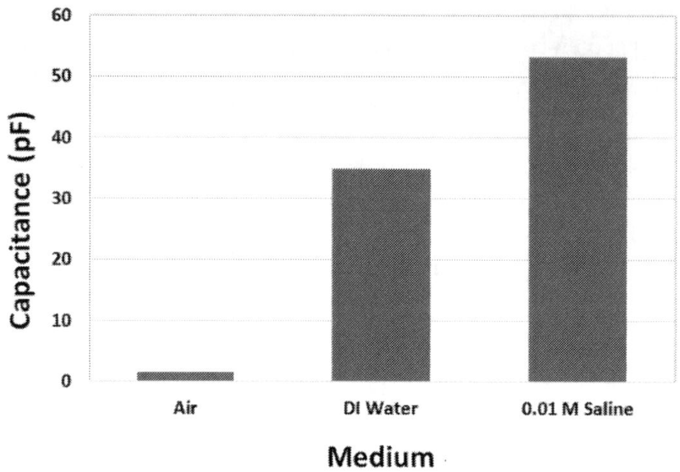

Figure 3: Comparison of capacitance readings of each fluid tested; Air, DI Water and 0.01 M Saline solution.

5. Conclusions

This paper described the development steps of a highly sensitive microsensor for wound monitoring applications. The microsensor design is highly plausible as a way to monitor a wound healing progress. The sensor could be used to detect the presence of blood under a bandage, which would therefore eliminate the need to remove a bandage to check on the healing process. Since the bandage could stay in place during the monitoring process, wounds would be able to heal quicker. This sensor can be integrated into bandages if needed or used as a disposable sensor. The results showed that the device can accurately detect the present of blood in the wound, monitoring the healing process.

References

[1] Winter G D 1962 Formation of the scab and the rate of epithelization of superficial wounds in the skin of the young domestic pig *Nature* 193(4812)293

[2] McColl D, Cartlidge B and Connolly P 2007 Real-time monitoring of moisture levels in wound dressings in vitro: an experimental study *International Journal of Surgery* 5(5)316-22

[3] Guinovart T, Valdés - Ramírez G, Windmiller J R, Andrade F J and Wang J 2014 Bandage - based wearable potentiometric sensor for monitoring wound pH *Electroanalysis* 26(6)1345-53

[4] Rahimi R, Ochoa M, Parupudi T, Zhao X, Yazdi I K, Dokmeci M R, Tamayol A, Khademhosseini A and Ziaie B 2016 A low-cost flexible pH sensor array for wound assessment *Sensors and Actuators B: Chemical* 28(229)609-17

[5] Pasche S, Angeloni S, Ischer R, Liley M, Luprano J and Voirin G 2008 Wearable biosensors for monitoring wound healing *Advances in Science and Technology* (57)80-87

[6] Stoppa M and Chiolerio A 2014 Wearable electronics and smart textiles: a critical review *Sensors* 14(7)11957-92

[7] Weber S A, Watermann N, Jossinet J, Byrne J A, Chantrey J, Alam S, So K, Bush J, O'Kane S and McAdams E T 2010 Remote wound monitoring of chronic ulcers *IEEE Transactions on Information Technology in Biomedicine* 14(2)371-7

[8] Buono M J, Burke S, Endemann S, Graham H, Gressard C, Griswold L and Michalewicz B 2004 The effect of ambient air temperature on whole-body bioelectrical impedance *Physiological Measurement* 25(1)119

[9] Stiles D K and Oakley B 2003 Simulated characterization of atherosclerotic lesions in the coronary arteries by measurement of bioimpedance *IEEE Transaction of Biomedical Engineering* (50)916-21

[10] Peyman A, Gabriel E H 2007 Complex permittivity of sodium chloride solutions at microwave frequencies *Bioelectromagnetic* (28)264-74

3D Printed Micro Check Valve for Biomedical Applications

Abdou Mbaye, Colton Kreamer, Lukas Zink, Mitchell Fredenburg and Reza Rashidi*

Mechanical Engineering Technology, State University of New York, Alfred State College, Alfred, NY, USA, 14802

*rashidr@alfredstate.edu

Abstract. This paper presents development of a novel three dimensional imprinted micro check valve for biomedical applications. Due to its rapid and cost effective fabrication techniques, micro check valve devices are widely used in several areas of the biomedical industry such as intravenous (I.V.) fluid transfusion. The fabrication method consists of SolidWorks modelling and 3D printing steps for prototyping. The micro check valve is a self-controlled valve that relies on pressure change for operation. In this particular design, a sphere was used to prevent any backward flow allowing the pressure change to only initiate a one-way flow. The tests showed that the check valve design allowed for zero backward flow while also allowing flow through the device in the proper direction at a rate of 98.6 μl/sec.

1. Introduction

Check valves have been used for the past 50 years and keep improving as technology advances. A check valve allows flow in one direction and prevents flow in the reverse direction. They can be designed with a spring mechanism that opens when flow goes through and closes when the flow stops. Check valves have been implemented in different areas of the biomedical engineering such as surgical procedures [1], in vitro tests [2-3], drug delivery [4-5] and microfluidic applications [6-13]. They have also been used to facilitate procedures to treat glaucoma [14-15]. Different materials such as silicon-on-insulator wafers [16], shape memory alloys [17], piezoelectric materials [18-19], graphene [20] and hydrogels [21] have been used in fabrication of micro check valves. One type of the check valves works based on spring loaded design and uses a plunger style stopper with a rubber grommet that stops backward flow by pressing the stopper against the housing walls. This type of check valve provides quick interruption of flow and positive sealing at a pressure below the cracking pressure of the valve. In another study, a polymeric micro check valve used springs to channel fluid on a wafer. The results showed that as flow increased through these check valves, the pressure increased [22].

This paper presents the design, fabrication and testing of a micro check valve used in I.V. fluid transfusion. The goal was to design a check valve that could be classified as a microsystem and be fully functional when being applied in an I.V. transfusion. The check valve has a free floating ball inside on tracks. When fluid is pushed through the valve, the ball will move to open the flow chamber, allowing fluid to pass. When the flow stops, the ball will slide down on the tracks to stop any backward flow. The check valve was modelled using SolidWorks and printed on a 3D printer.

2. Design

The check valve was required to be designed such that it could be applied for I.V. transfusions. The initial design of the valve included ball and socket parts. The ball would block the smaller channel from any possible backward flow and sit into a perforated socket to allow the fluid to pass. A redesign was created and involved bars that would contain the ball in the chamber and a cap-type bottom to cover the smaller inlet channel. The perforated socket containing the ring remained the same, and small tabs were added and connected to the bars to ensure that the ball would not rotate. In order to improve the valve performance, a three part valve was designed and included a perforated ring, a containing ring with bars and a ball with grooves cut into it, as shown in Figure 1. The ball still needed to contain the capped bottom to allow for proper backward flow prevention, while the bars needed to remain to contain the ball. When fluid comes from the inlet of the valve, the ball will be pushed up into the perforated ring, allowing for fluid to freely flow around the ball through the small holes in the ring. Once the fluid stops flowing, the back pressure on the outlet of the valve will push the ball flush against the inlet of the valve, ensuring that no fluid will be able to flow backwards (Figure 2). The cap on the bottom of the ball ensures that a tight seal is made on the inlet of the valve to completely prevent backward flow. While most designs have a spring loaded ball, the design in this paper is based entirely off of the laws of pressure flow.

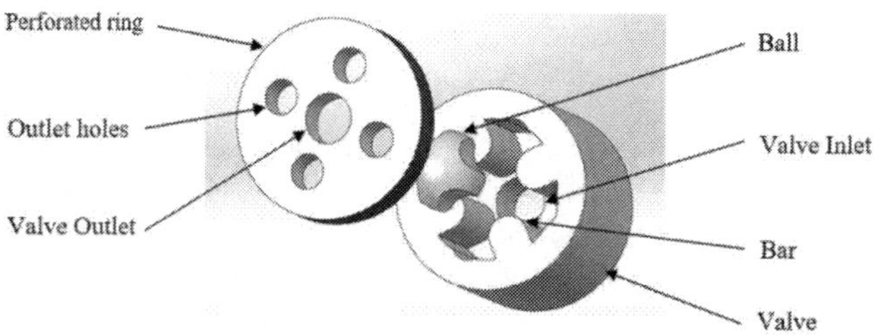

Figure 1. Exploded view of the check valve assembly.

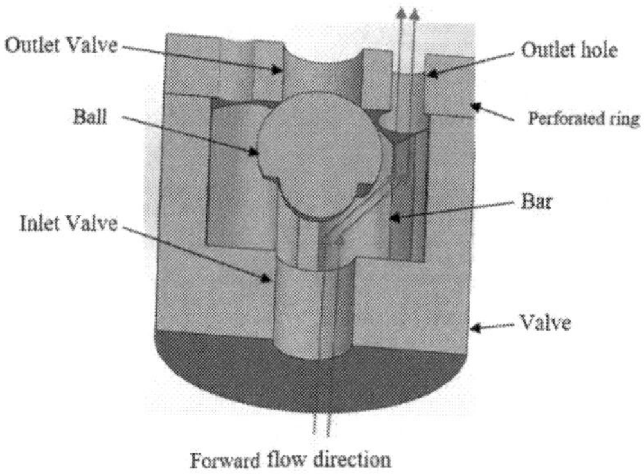

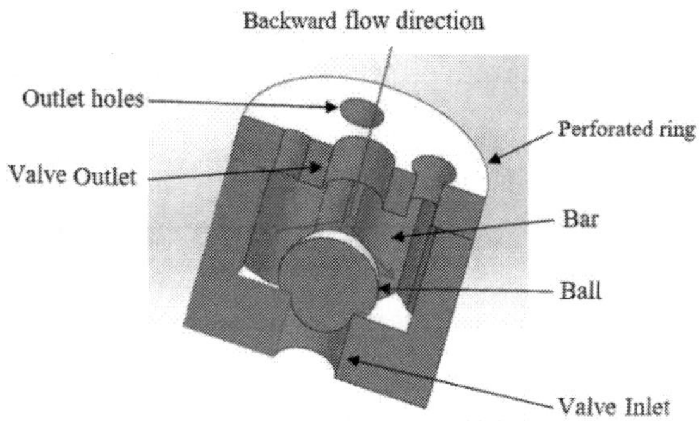

Figure 2. Cross sectional view of the assembly in the forward flow positon (a) and backward flow positon (b), with the flow shown as red arrows.

The ball itself has a diameter of 2mm where the valve inlet and outlet have diameters of 1.25 mm. The perforated ring has the same outer diameter of the valve outlet, but the inner diameter is 1.25 mm. This inner diameter needed to be smaller than the ball because the ball required to be contained within the flow chamber. The diameter of the cap on the bottom of the ball is 1.25 mm. The reason for this being larger than the valve inlet is to ensure that the ball can adequately cap the inlet. Further dimensions of the parts are presented in Table 1.

Table 1. Dimensions of the developed valve parts

Part	Dimension (mm)
Ball diameter	2
Valve inlet diameter	1.25
Valve outlet diameter	1.25
Valve length	4
Perforated ring outside diameter	3
Perforated ring inside diameter	1.25
Perforated ring thickness	1
Outlet holes	0.875
Ball smaller diameter	0.6
Ball radii cut	1.8
Bar radii	0.5
Bar width	1
Overall length	5

3. Fabrication

To fabricate the check valve, a Formlabs Form 2 3D printer was used to make parts with an ABS plastic resin material. This fabrication method was quick and simple, which facilitated the process. To use the 3D printer, the design was first modelled in SolidWorks. The check valve was printed in three parts: the ball, the valve inlet and the perforated ring outlet, which were all printed at the same time. This printer uses a laser that heats resin in a tank and solidifies the resin. The resin is attached to a base plate hanging above the resin tank. As it solidifies, the base plate moves upward pulling the already solidified layer out of the tank so the next layer can be solidified by the laser. It continues this process until the part is completed and entirely pulled out of the resin base. The 3D printer allowed a print resolution of 100 μm per layer which made for very clean and precise features in the parts.

After the printing process, the parts were placed in isopropyl alcohol for 10 min, rinsed in water for 10 min, and dried in ultraviolet (UV) light. The mounts that the parts were printed on were then removed. The ball was placed into the cylinder after printing, and then the top part was installed on the valve inlet. The check valve must be coated with a thin layer of parylene in order to make it biocompatible and safe for use in the human body [23]. Figure 3 shows a photo of the fabricated check valve.

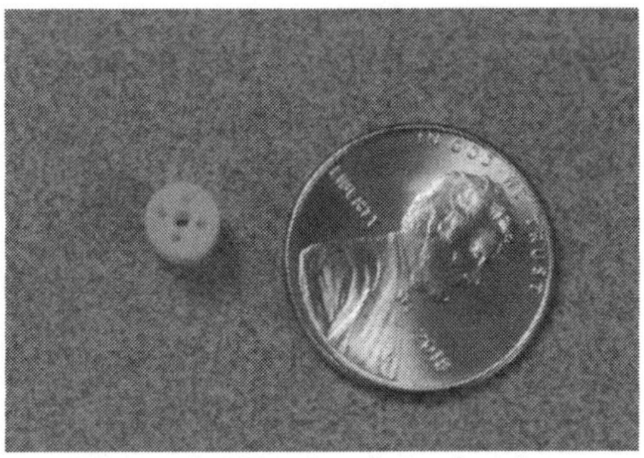

Figure 3. 3D printed check valve.

4. Testing

Several tests were conducted on the check valve to physically prove that the valve would prevent any backward flow and allow for a fluid to freely pass through it. The valve was fabricated in two different sizes, once at actual aforementioned size, and once at three times larger size. The larger valve was

IOP Conf. Series: Materials Science and Engineering **417** (2018) 012041 doi:10.1088/1757-899X/417/1/012041

used to visually represent what was occurring in the actual-sized valve, whereas the actual-sized valve was used to gather test results and data. Both valves were subject to the same experiments during the entire test process.

The first test conducted was simply to see if liquid would flow freely through the valve with very little pressure. The valve was secured to a wall and the inlet was placed upwards, whereas the outlet was placed downwards. Colored fluid was then placed in the valve by a syringe and it was observed that fluid was able to pass through the valve with no issues. This test proved that the valve would allow forward flow. Next, the valve was flipped so that the outlet was facing upwards. Fluid was then added in the same fashion, and it was observed that no fluid was able to pass through. This proved that, when not subjected to any additional pressure, the valve has a solid seal so that no fluid can flow backwards. The test results were conclusive that the valve design was functional enough to further the testing procedures.

The next test was to subject the valve to a certain pressure. These tests were conducted to prove that the valve would not fail and that the backward flow seal would not falter under certain pressures. The pressure that the valve had been subjected to is much greater than the pressure it would experience in the field, but the reasoning for this is to ensure that the valve would not fail under a normal operating pressure. Indeed, the valve would only be subject to pressures slightly greater than the atmospheric pressure, simply because the I.V. transfusion and similar applications that the valve will be used for is not a very high pressure application. The design on the valve allows for it to open when pressure at the inlet is greater than the outlet, which means that even the slightest difference in pressure will allow the valve to open or close, depending on which side of the valve has a higher pressure. Figure 4 shows the experimental setup and the aforementioned tests in forward and backward flow directions.

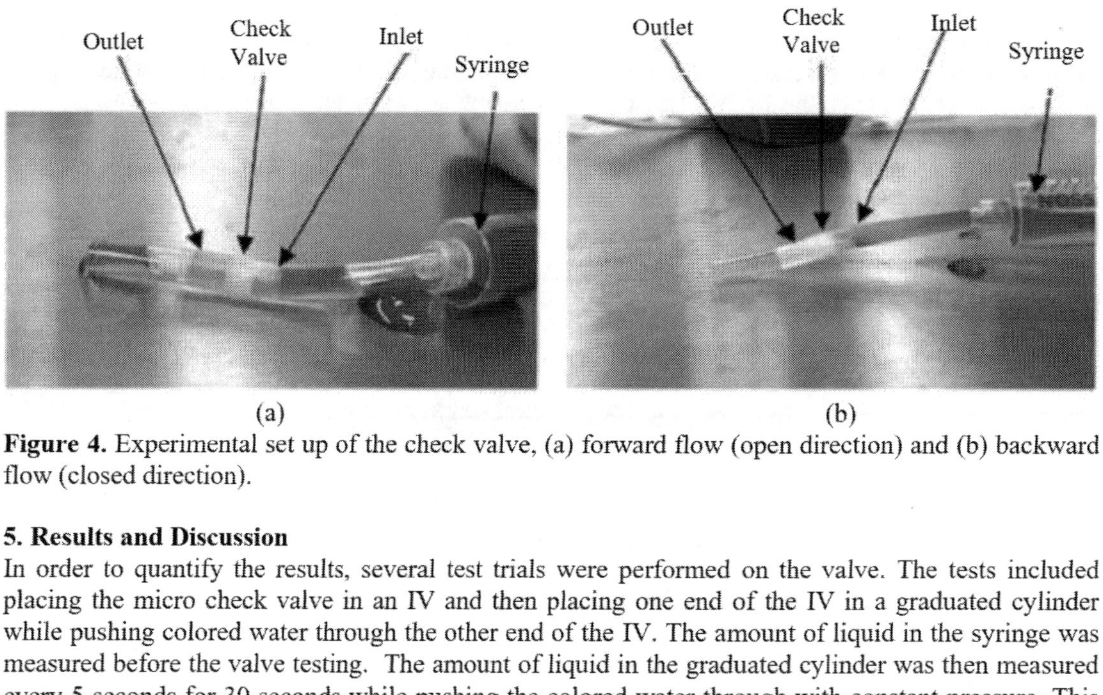

Figure 4. Experimental set up of the check valve, (a) forward flow (open direction) and (b) backward flow (closed direction).

5. Results and Discussion

In order to quantify the results, several test trials were performed on the valve. The tests included placing the micro check valve in an IV and then placing one end of the IV in a graduated cylinder while pushing colored water through the other end of the IV. The amount of liquid in the syringe was measured before the valve testing. The amount of liquid in the graduated cylinder was then measured every 5 seconds for 30 seconds while pushing the colored water through with constant pressure. This test was repeated to test for both forward and backward flow through the valve. Figure 5 shows the volume of the fluid passing through the valve as a function of time in forward and backward flow directions. This test not only showed whether or not there could be backward flow but it also showed the rate at which liquids could flow through the valve. As shown in Figure 6, the check valve design allowed for zero backward flow while also allowing flow through the device in the proper direction at a rate of 98.6 μl/sec.

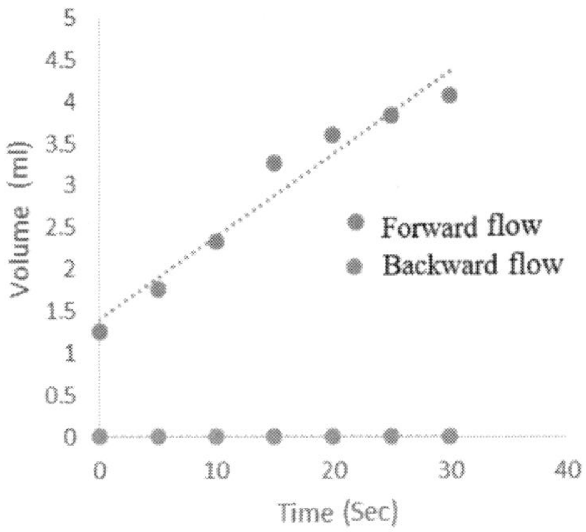

Figure 5. Volume of colored water passing through the micro check valve over time in forward flow and backward flow directions.

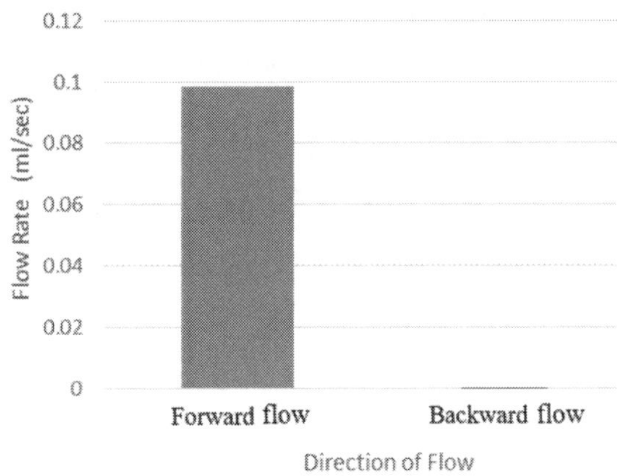

Figure 6. Flow rate in forward flow and backward flow directions of the micro check valve.

6. Conclusions

This work demonstrated the development of a novel micro check valve for biomedical applications, particularly I.V. transfusion. The design, fabrication and characterization of the check valve was discussed in this paper. In the check valve design, a movable ball was used to allow fluid flow in one direction while prevent any backward flow in the other direction. 3D printing was used to fabricate the valve prototype. The qualitative and quantitative tests were performed to check the performance of the check valve. The tests showed a successful design with a zero backward flow while a 98.6 μl/sec flow rate in the open direction.

References

[1] Rebello K J 2004 Applications of MEMS in surgery *Proceedings of the IEEE* 92(1)43-55

[2] Grayson A C, Shawgo R S, Johnson A M, Flynn N T, Li Y, Cima M J and Langer R 2004 A BioMEMS review: MEMS technology for physiologically integrated devices *Proceedings of the IEEE* 92(1)6-21

[3] Lavan D A, McGuire T and Langer R 2003 Small-scale systems for in vivo drug delivery *Nature biotechnology* 21(10)1184

[4] Staples M, Daniel K, Cima M J and Langer R 2006 Application of micro- and nano-electromechanical devices to drug delivery *Pharmaceutical Research* 23(5)847-863

[5] Bonhoeffer B, Kwade A and Juhnke M 2017 Impact of formulation properties and process parameters on the dispensing and depositioning of drug nanosuspensions using micro-valve technology *Journal of pharmaceutical sciences* 106(4)1102-10

[6] Yu Q, Bauer J M, Moore J S and Beebe D J 2001 Responsive biomimetic hydrogel valve for microfluidics *Applied Physics Letters* 78(17)2589-91

[7] Bassetti M J, Moore J S and Beebe D J 2002 Development of electrically triggered hydrogels for microfluidic applications *2nd Annual International IEEE-EMB Special Topic IEEE Conference on Microtechnologies in Medicine & Biology* 410-413

[8] Liu R H, Yu Q and Beebe D J 2002 Fabrication and characterization of hydrogel-based microvalves *Journal of microelectromechanical systems* 11(1)45-53

[9] Suzuki H and Yoneyama R 2003 Integrated microfluidic system with electrochemically actuated on-chip pumps and valves Sensors & Actuators: B Chemical 96(1-2)38-45

[10] Chen Y, Tian Y, Xu Z, Wang X, Yu S and Dong L 2016 Microfluidic droplet sorting using integrated bilayer micro-valves *Applied Physics Letters* 109(14)143510

[11] Gholizadeh A and Javanmard M 2016 Magnetically actuated microfluidic transistors: Miniaturized micro-valves using magnetorheological fluids integrated with elastomeric membranes *Journal of Microelectromechanical Systems* 25(5)922-8
[12] Singh R K, Kant R, Singh S, Suresh E, Gupta A and Bhattacharya S 2015 A novel helical micro-valve for embedded micro-fluidic applications *Microfluidics and Nanofluidics* 19(1)19-29
[13] Pourmand A, Shaegh S A, Ghavifekr H B, Najafiaghdam E, Dokmeci M R, Khademhosseini A and Zhang Y S 2017 Fabrication of whole-thermoplastic normally closed microvalve, micro check valve, and micropump *Sensors and Actuators B: Chemical* 2017 Dec 28
[14] Park C, Yang D, Cha J and Lee J 2016 Polymeric check valve with an elevated pedestal for precise cracking pressure in a glaucoma drainage device *Biomedical Microdevices* 18(1)20
[15] Siewert S, Sämann M, Schmidt W, Stiehm M, Falke K, Grabow N, Guthoff R and Schmitz K P 2015 Coupled analysis of fluid-structure interaction of a micro-mechanical valve for glaucoma drainage devices *Klinische Monatsblatter fur Augenheilkunde* 232(12)1374-80
[16] Min H, Hejun D, Shih-Fu Ling F and Bo L 2003 Development of an SOI-based micro check valve *International Journal of Computational Engineering Science* 4(3)691-694
[17] Nath T, Raut G, Kumar A, Khatri R and Palani I A 2015 Investigation on laser assisted actuation of shape memory alloy based micro-valve *IEEE International Conference on Robotics, Automation, Control and Embedded Systems (RACE)* 1-6
[18] Chen S, Lu S, Liu Y, Wang J, Tian X, Liu G and Yang Z 2016 A normally-closed piezoelectric micro-valve with flexible stopper *AIP Advances* 6(4)045112
[19] Bonhoeffer B, Boldrini M, Boiger G, Kwade A and Juhnke M 2017 Experimental characterization and simulation of a piezo-actuated micro dispensing valve *Journal of Fluids Engineering* 139(5)051105
[20] Gund V, Ruyack A, Ardanuc S and Lal A 2015 Graphene one-shot micro-valve: Towards vaporizable electronics *28th IEEE International Conference on Micro Electro Mechanical Systems (MEMS)* 1090-1093
[21] Mehner P J, Allerdißen M, Voigt A, Marschner U and Richter A 2015 Combined simulation of the closing behavior of a smart hydrogel micro-valve *ASME 2015 Conference on Smart Materials, Adaptive Structures and Intelligent Systems* V001T03A016-V001T03A016
[22] Truong T, Huang X and Nguyen N 2003 Polymeric stack-assembled micropump with SU-8 check valves International *Journal of Computational Engineering Science* 4(2)249-252
[23] Shin Y S, Cho K, Lim S H, Chung S, Park S J, Chung C, Han DC and Chang J K 2003 PDMS-based micro PCR chip with parylene coating *Journal of Micromechanics and Microengineering*, 13(5)768

IOP Conf. Series: Materials Science and Engineering **417** (2018) 012042 doi:10.1088/1757-899X/417/1/012042

Design, Modeling and Simulation of a Reconfigurable Wheelchair with Stand-sit-sleep Facilities

Sumit Desai, Shankar Mantha, Vikas Phalle, Sangram Patil

Department of Mechanical Engineering,
Veermata Jijabai Technological Institute,
Mumbai, India-400019

sumitdesai1234@gmail.com

Abstract. Standing feature of the wheelchair allows the disabled to be able to reach different heights and thus gives enhanced navigations in different conditions. Wheelchair with back seat adjustment helps in health preservation as well as daily living activities including trunk positioning, functional enhancement stability and gait balance. Thus, paper presents the design, modeling and simulation of a reconfigurable wheelchair with stand-sit-sleep facilities. Key areas focused on maximizing safety, optimizing the size with minimizing cost and weight of the newly designed wheelchair. The contribution of paper are (i) Kinematic concept of the reconfigurable wheelchair (ii) Synthesis of the proposed wheelchair mechanism (iii) CAD modeling using the Top down modeling approach (iv) Multi-body dynamics Simulation and stress simulation of model. Wheelchair design has been analyzed and verified by using MSC ADAMS and ANSYS simulation tools. The developed model will help in improving the functional capabilities of wheelchair users allowing enhanced independence and quality of life (QoL).

1. Introduction

Wheelchairs are one of the most commonly used assistive devices by many lower limb disabled, elderly subjects due to their incapability to stand and walk without help. People with disabilities in developing countries like India tackle different social, medicinal, educational and political difficulties [1]. Disability is a significant public health issue in India. Indian populace is equal to 17.86% of the entire world populace. National Sample Survey Organization (NSSO) reported the disabled populace in India as 18.5 million in 2001, which is 2.13% of the entire populace of the country. The disabled populace of India has increased by 22.4% from 2001 to 2011 [2]. A large section of disabled populace lives in rural areas with low family income. Furthermore, the World Health Organization (WHO) reported that any populace of world has a 10% disable people. Further, 10% of any disabled populace could facilitate from a wheelchair. As per WHO, there are around 10 million potential wheelchair users in India. These insights can be validated by latest nationwide census of India. There is an enormous requirement for assistive technologies (AT) for disabled people in India. However, assistive technologies are either imported (i.e. designed for developed countries) or prohibitively expensive (i.e. not suitable for low income profile people) in India. Accessibility of rehabilitation interventions and their delivery is exceptionally constrained in India. In many cases, use of such devices is restricted to the urban communities [2].

Comprehensive needs assessment study of extensive wheelchair users [2] reported several common problems such as (1) difficulties in transfer from a wheelchair to and from bed (2) difficulties in stand from seating position (3) difficulties using commode without caregiver support and (4) risk of

IOP Conf. Series: Materials Science and Engineering **417** (2018) 012042 doi:10.1088/1757-899X/417/1/012042

pressure sores due to extensive use of a wheelchair. A wheelchair with back seat adjustment can assist extensive wheelchair users in health preservation, balancing, trunk positioning, etc [3]. Standing feature of the wheelchair allows the disabled to be able to reach different heights and thus gives enhanced navigations in different conditions [4]. Although reconfigurable wheelchairs [5-8]were given a lot of attention, none of the contributions were intended on studying and building a wheelchair model integrated with Stand-sit-sleep capabilities.

The paper presents the design and development of a reconfigurable wheelchair with stand-sit-sleep configurations. Key areas focused on maximizing safety, optimizing the size with minimizing cost and weight of the wheelchair. Key areas focused on maximizing safety, optimizing the size with minimizing cost and weight of the wheelchair. The contribution of paper are (i) Kinematic concept of reconfigurable wheelchair (ii) Synthesis of new wheelchair mechanism (iii) CAD modelling using Top down modelling approach (iv) Analysis of model using multi-body dynamics and stress simulation. The wheelchair design has been analyzed and verified by applying MSC ADAMS and ANSYS simulation tools.

2. Design Specifications

QFD analysis [2] conducted in the early phase of the project, revealed that out of 11 design parameters recommended by cross-functional team based on needs assessment study of long-term wheelchair users (1) seat to stand transfer feature (13.90%), (2) the bedsore prevention feature (12.60%), (3) seat to sleep transfer feature (12.00%) and (4) adjustable backrest feature with multi-posture adjustment feature (11.20%) were received highest technical importance. Based on these findings, present research was focused on design and development of a reconfigurable wheelchair built-in with sit-stand-sleep capabilities. Major focus was on designing compact, safe and Self-use wheelchair for long-term wheelchair users to assist them to stand from a seated position, sit from stand position, sleep from seated position, and sit from sleep position. The design specifications (both functional and performance) that were considered in present study are as bellow:

1. The primary function of wheelchair is to help user to stand from a seated position, sit from stand position, sleep from seated position, and sit from sleep position.
2. Sufficient body supports to prevent the user from falling while using wheelchair.
3. Structure of wheelchair should be stable and light in weight with less complexity.
4. Ease of operate and transport
5. Ergonomic considerations L=900 mm, B=900 mm, H=1000 mm
6. Load sustainability: 100 kg (Minimum)
7. The device should be practical and affordable.

3. The kinematic concept

The kinematic mechanism concept (Fig. 1and Table 1) for a reconfigurable wheelchair consists of Watt's six-bar mechanism [9] (attached twice) and one slider crank mechanism (attached once).

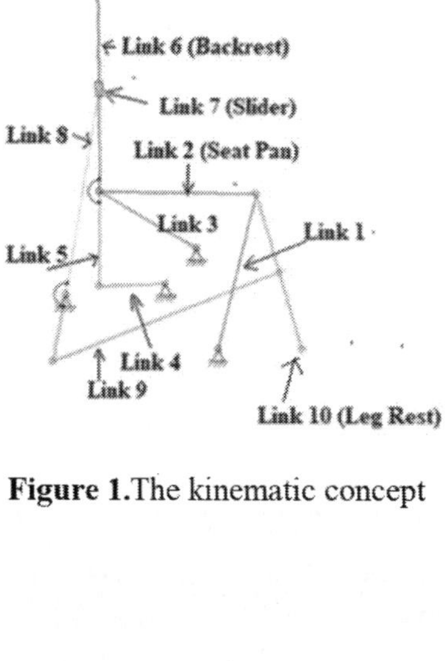

Figure 1. The kinematic concept

IOP Conf. Series: Materials Science and Engineering **417** (2018) 012042 doi:10.1088/1757-899X/417/1/012042

Table 1. Description of the kinematic concept

Link name	Function
Link 2	Seat Pan
Link 6	Backrest
Link 10	Largest
Link 1, seat (link 2), link 3 and frame	First four-bar linkage of Watt's six-bar mechanism
Link 3, link 4, link 5 and frame	The second four-bar of Watts six bar mechanism
Backrest (link6), Link 7, link 8 and frame	Slider mechanism

4. Mechanism Synthesis

The sit configuration is same in both the cases. The sitting configuration meets the standard wheelchairs dimensions. In the end configuration of sit to stand transformation, the backrest and seat pan are inclined in a way that it secures support for the users and provides safety to the user. The wheelchair must be fixed by wheelchair breaks during transformation to ensure safety. The sleeping configuration is intended to put the user in an almost horizontal pose. Fig. 2 and 3 illustrates the two-position graphical synthesis [9] of the four-bar mechanism used in kinematic concept of a wheelchair.

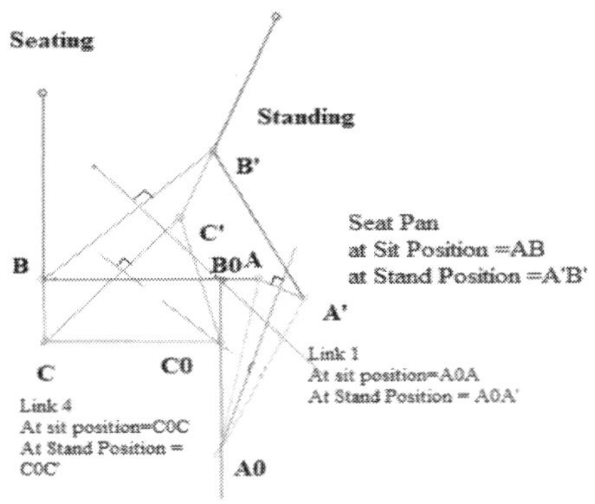

Figure 2. Synthesis of the four-bar mechanism used in Sit to stand transformation

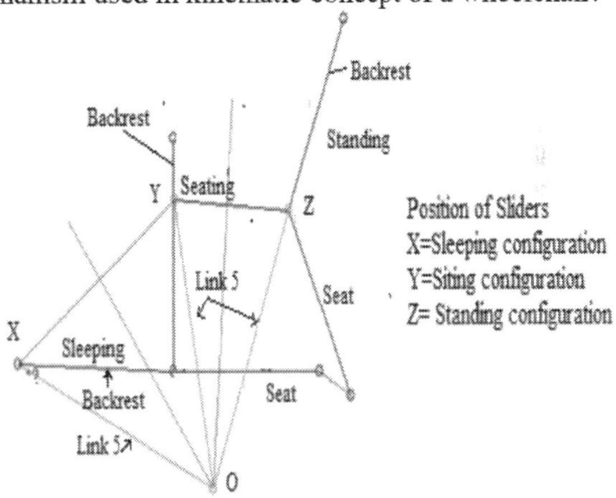

Figure 3. The Synthesis of the slider crank mechanism used in stand-sit-sleep transformation

5. Simulation study

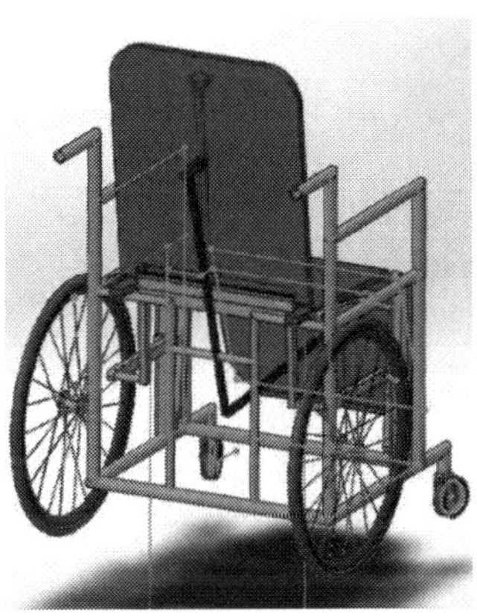

Figure 4. Linkage structure with frame

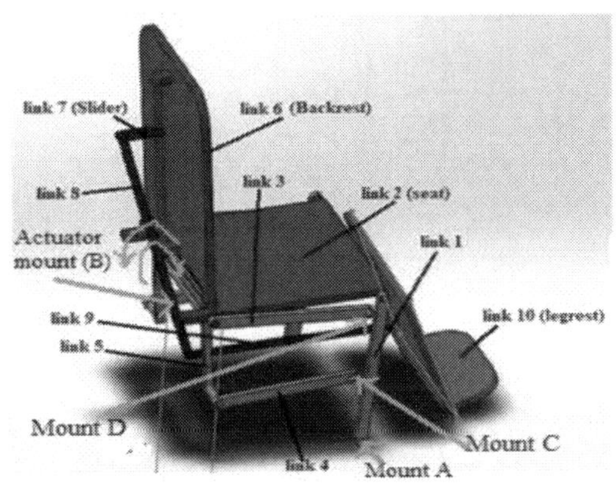

Figure 5. Mounting points of linkage

The main function of wheelchair is to support user to stand from a seated position, sit from stand position, sleep from seated position, and sit from sleep position. In addition, to minimize the overall wheelchair weight and the cost, it is vital to select a proper actuator with minimum energy utilization providing sufficient force to alter wheelchair configuration. Fig. 4 shows Linkage structure with frame for seat position. Fig. 5 shows the mounting points, where actuator rotates link 8 by 70^0 (Sit-Sleep) while in clockwise direction and by 22^0 in anticlockwise direction (Sit-Stand). The mechanism is designed such that the two identical six bar mechanism are mounted on the two sides of a frame and slider crank mechanism at middle. A harness is provided to support the back and is attached to the backrest.

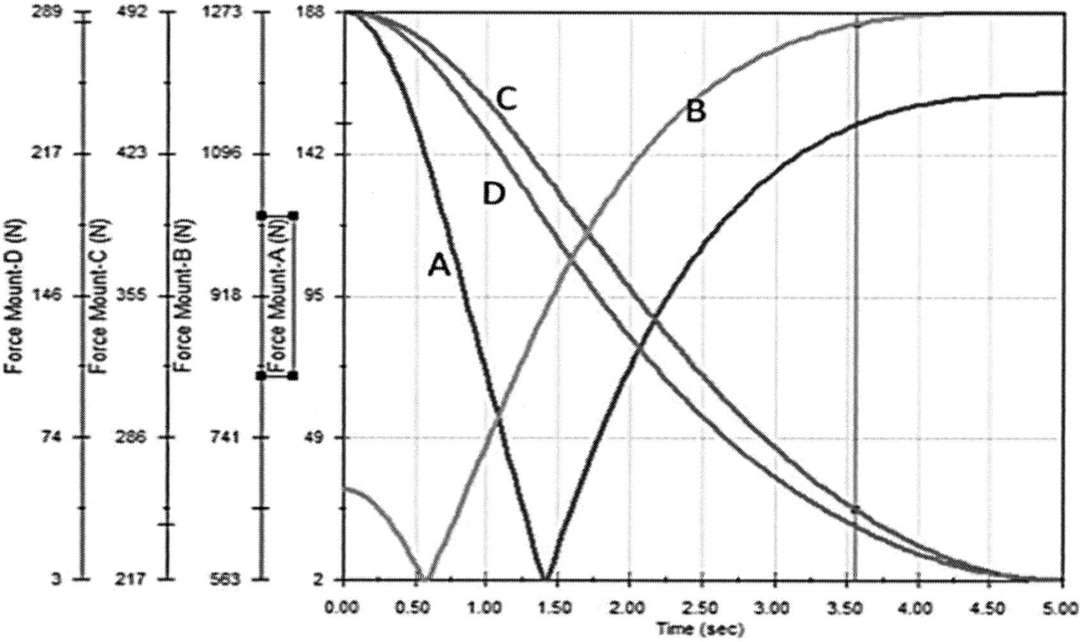

Figure 6. Forces exerted at various mount points during Sit to stand

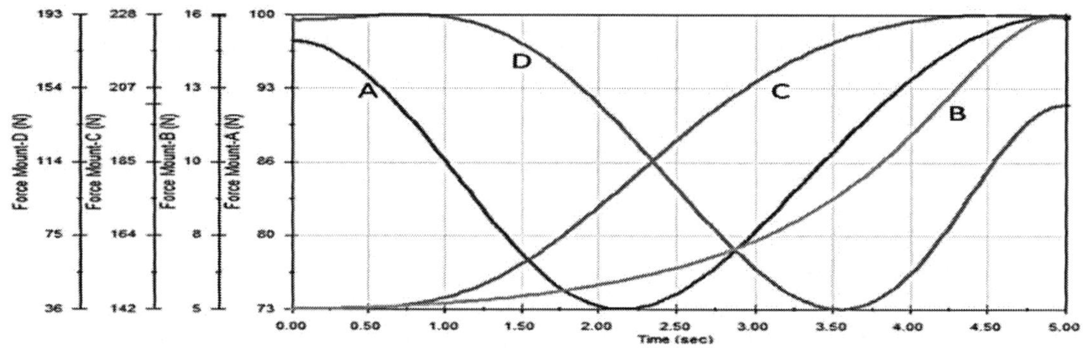

Figure 7. Forces exerted at various mount points during Sit to sleep

To calculate the required actuator torque, MBD simulations were done in ADAMS simulation software. A maximum actuator torque of 515 Nm is found during the entire cycle of operation. Considering a factor of safety, a suitable actuator with minimum 800 Nm torque motor is required for the final wheelchair design. AISI 1018 MS was chosen for the wheelchair parts. The newly modeled frame verified for the structural stability using ANSYS. To find the external load to be applied on the mechanism mounting points (points A, B, C and D (Fig. 5)) on the wheelchair frame, a dynamic simulation is conducted over the working cycle. Considering the load of 80 Kg of wheelchair user and an actuator torque, forces exerted at various mount points (points A, B, C and D) on the frame are plotted in Fig.6, Fig.7.Stress analysis using ANSYS was conducted on the frame employing on the maximum forces exerted on the mounting points (A, B, C, D) during the entire cycle of operation. The maximum Von Mises stress of 171.31 MPa (below the yield stress (i.e. 365 MPa)) is generated at the intersection of the link 8 and the bar supporting the actuator as seen in Fig. 8.

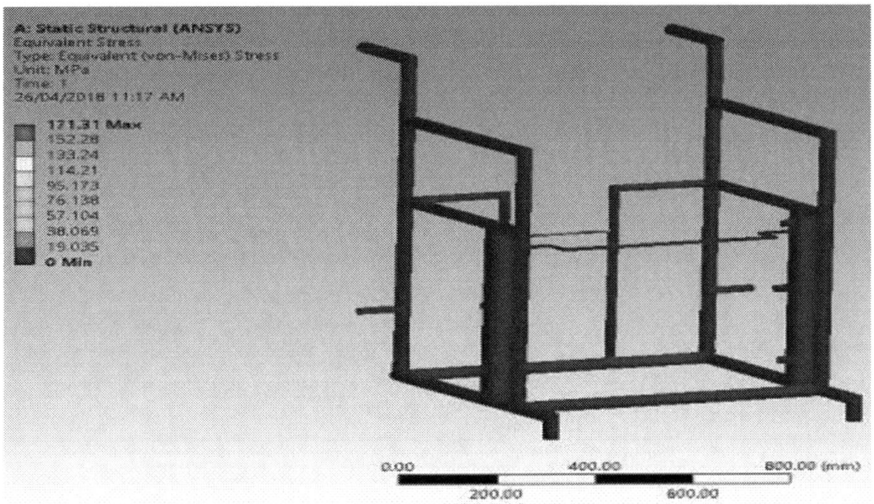

Figure 8. Von Mises stress obtained after stress simulation

6. Conclusion

This paper presents an overall product design and development of a reconfigurable wheelchair with stand-sit-sleep configurations. Key areas focused on maximizing safety, optimizing the size with minimizing cost and weight of the wheelchair. The contribution of paper are (i) Kinematic concept of reconfigurable wheelchair (ii) Synthesis of new wheelchair mechanism (iii) CAD modeling using Top down modeling approach (iv) CAE Analysis of model. The process is supported by CAE tools like MSC ADAMS and ANSYS. The multi-body dynamics and the Finite Element Analysis the wheelchair model were used in actuator selection and frame design. The inventive solution to the problem of long term wheelchair has a promising potential for improving healthcare support to disabled in developing nations.

References

[1] E.R.Vieira, and L. Miller2008 Facing the challenge of patient transfers: using ceiling lifts in healthcare facilities *HERD: Health Environments Research & Design Journal*, **2**(1) 6-16
[2] S. Desai, S. Mantha, and V. Phalle 2017 Comprehensive Needs Assessment Study and Deployment of QFD Targeted at New Wheelchair Design *Iranian Rehabilitation Journal***15**(4) 377-388
[3] B.E. Dicianno, J. Arva, J.M. Lieberman, M.R. Schmeler, A. Souza, K. Phillips M. Lange, R. Cooper, K. Davis, and K.L. Betz 2009 RESNA position on the application of tilt, recline, and elevating legrests for wheelchairs*Assistive Technology*, **21**(1), 13-22
[4] J. Arva, , G. Paleg, M. Lange, J. Lieberman, M. Schmeler, B. Dicianno, M. Babinec, and L. Rosen 2009 RESNA position on the application of wheelchair standing devices *Assistive Technology*, 21(3), 161-168
[5] C.C. Auel 1991 All-purpose rocking, swiveling, reclining, and lifting chair*U.S. Patent 5,024,486*
[6] E.D. Henderson 1992 Elevator chair apparatus*U.S. Patent5,165,753*
[7] Komura, S., Kaneda, T., Yamashita, K., Adachi, K. and Mizuseki, I., Komura Corp 2004 Elevation chair*U.S. Patent 6,783,179*
[8] Goher, K. M. 2016 A reconfigurable wheelchair for mobility and rehabilitation Design and development *Cogent Engineering* **3**(1), 1-23
[9] McCarthy, J. M., and Soh, G. S. 2010 Geometric Design of Linkages *Springer*, New York.